Lecture Notes in Physics

385

A. Treves G.C. Perola L. Stella (Eds.)

Iron Line Diagnostics in X-ray Sources

Proceedings of a Workshop
Held in Varenna, Como, Italy
9–12 October 1990

Springer-Verlag
Berlin Heidelberg GmbH

Editors

Aldo Treves
Scuola Internazionale Superiore di Studi Avanzati
Trieste, Italy

Giulio Cesare Perola
Istituto di Astronomia, La Sapienza
Roma, Italy

Luigi Stella
Osservatorio Astronomico di Brera
Milano, Italy

ISBN 978-3-662-13817-5 ISBN 978-3-540-47583-5 (eBook)
DOI 10.1007/978-3-540-47583-5

Originally published by Springer-Verlag Berlin Heidelberg New York in 1991
Softcover reprint of the hardcover 1st edition 1991

2153/3140-543210 – Printed on acid-free paper

Preface

The 1976-1977 discovery of an iron K-line at about 7 keV in the spectra of clusters of galaxies had profound implications for the understanding of the emission mechanism in these sources and the origin of the intracluster gas. This is an example of the great diagnostic potential of iron features in the spectra of X-ray sources, which stems from their unique properties. In particular, the strength and cleanliness of the iron K-shell features allow us to draw important conclusions even from experiments with modest energy resolution. As these features fall within the typical range of proportional counters, they have been discovered by now in almost every class of both extended and compact X-ray sources. For these reasons the detailed study of K-shell and L-shell iron features represents a major drive for many future experiments, with a rewardingly ample scope on all scales of energy resolution.

Powerful diagnostics in astrophysics are usually useful in rather different contexts, and often provide inspiring lines of research across different subjects. This workshop had two specific aims: firstly, to assess the state of the art of the theoretical and observational knowledge of iron X-ray lines over the widest range of topics and phenomena; secondly, to put the use of this diagnostic tool into the perspective offered by future missions.

The workshop was initiated from within the Working Group set up to define the core observation program of the Italian X-ray mission SAX. It was hoped that it could also be the first occasion for the presentation of highlights of the BBXRT experiment. Even though this did not happen, the successful results obtained some months later by BBXRT in the scientific area covered by the workshop are a demonstration of its timely occurrence.

Milan
March 1991

Aldo Treves
G. Cesare Perola
Luigi Stella

Scientific Organizing Committee

Acknowledgements

It is a pleasure to acknowledge the collaborative efforts of the other members of the Scientific Committee, the invaluable help provided by Prof. E. Sindoni, director of the international School of Plasma Physics "Piero Caldirola", and by Ms. E. Ficara of the Department of Physics of Milan University, the contribution of Ms. A. van Vloten of the Italian Physical Society for the preparation of the Conference poster, and finally the hospitality provided by Villa Monastero.

Financial support came from the Italian Space Agency (ASI), the National Research Council (CNR) and the Physics Department of the University of Milan.

CONTENTS

PARTICIPANTS

M. Balucinska-Church
P. Barrett
T. Belloni
J. Bleeker
A.C. Brinkman
W. Brinkmann
A. Celotti
L. Chiappetti
M. Colpi
B. Czerny
A. Fabian
J. Felsteiner
F. Fiore
I. M. George
G. Ghisellini
E. Girardi
M. Gottwald
F. Haardt
F. Haberl
G. Hasinger
J. Hughes
R. Jager
L. R. Jones
J.S. Kaastra
P. Kahabka
S. Kahn
B. J. Kellett
K. Koyama
M. Kowalski
H. Kunieda
A. Laor
D. Leahy
G.M. Madejski
G. Manzo
L. Maraschi
G. Matt
R. Mewe
S. Molendi
D. Molteni
F. Nagase
P. Nandra
H. Norgaard-Nielsen
A. Norton
F. Paerels
C. Paizis
R. Pallavicini
A. Parmar
S. Pellegrini
C. Perola
L. Piro
T. Ponman
B. Rosner
M. Salvati
R. Sambruna
C.L. Sarazin
N.I. Shakura
G. Shaviv
H. Schnopper
D. Schwartz
W. Snyder
L. Stella
G. Stewart
Y. Tanaka
L. Titarchuk
A. Treves
T.J. Turner
O. Vilhu
R. Walter
S. Warwick
N. Westergaard
N. White
M. Wise
K. Wood

I. Thin Plasma: SN Remnants, Stellar Coronae and Clusters

IRON L-SHELL LINE FORMATION IN DIVERSE ASTROPHYSICAL ENVIRONMENTS

Steven M. Kahn and Duane A. Liedahl
Departments of Physics and Astronomy, and
Space Sciences Laboratory
University of California
Berkeley, California 94720

ABSTRACT: Iron L-shell transitions, which lie in the energy range 0.7 - 1.5 keV, can provide a wealth of important information about physical conditions in diverse astronomical sources. Although few data sets of adequate statistical quality are currently available for analysis, the next series of X-ray satellite experiments will provide order-of-magnitude improvements in both resolution and sensitivity for this kind of work. We review the important physical processes that underlie the diagnostic potential of these features. We specifically consider coronal plasmas, X-ray photoionized nebulae, and the ionizing plasmas believed to be present in young supernova remnants.

1. INTRODUCTION.

To most X-ray astronomers, the title of this conference, *Iron Line Diagnostics in X-Ray Sources*, connotes a particular energy range within the X-ray band of the spectrum, 6 - 7 keV. Certainly, the iron K lines which fall in that range have been detected for a large number of cosmic sources and can contribute much to our understanding of X-ray emission in astrophysical settings. However, in terms of the potential for truly diagnosing physical conditions in cosmic plasmas, the iron L-shell lines (transitions to $n = 2$) may eventually prove far more important. Iron L lines provide the dominant radiative luminosity for at least eight ionization stages, Fe XVII - XXIV, covering a very wide range in temperature. Moreover, since these lower ionization species are multiple electron systems, the iron L spectra are considerably more "rich" than those of the corresponding K-shell series. A virtual forest of iron L lines is expected in the spectra of most cosmic X-ray sources.

As yet, relatively little has come from detailed analyses of iron L features in existing astronomical X-ray spectra. This is a consequence of instrumental limitations; iron L lines are located in the spectral region between 0.7 and 1.5 keV, a difficult band for the non-dispersive instruments that have been primarily utilized to date. Even the best solid-state detectors offer a resolving power, $E/\Delta E$, of only 10 - 20 in this band, inadequate for clean separation of the most important iron L features.

However, this situation will change dramatically in the not-too-distant future when the Advanced X-ray Astrophysics Facility (AXAF) and the X-Ray Multi-Mirror Mission (XMM) are launched. AXAF will carry two transmission grating experiments with resolving powers of several hundred in this band (Weisskopf 1988, Schwartz 1990), more than sufficient for resolving nearly all the iron L features. XMM will carry a reflection grating spectrometer with similar resolution and very high sensitivity (Brinkman *et al.* 1989, Bleeker 1990). Collectively, these mission wsill provide high resolution, high statistics iron L spectra for nearly all classes of cosmic X-ray sources.

The diagnostic potential of the iron L features and the new observational capabilities offered by AXAF and XMM have motivated us to initiate a program, collaborative with William Goldstein and Albert Osterheld of the Lawrence Livermore National Laboratory, to recalculate iron L-shell spectra for conditions expected in diverse astronomical sources. While a number of previous calculations of this type have been performed for collisional plasmas like the solar corona and tokamaks, very few exist for parameters appropriate to other kinds of sources where severe departures from collisional equilibrium are expected. We have exploited an atomic physics code originally developed by M. Klapisch and collaborators at Hebrew University (Klapisch 1971; Klapisch *et al.* 1977). The atomic structure is calculated using the relativistic, multiconfiguration, parametric potential method. The model is fully collisional-radiative. Radiative transitions between all levels are calculated explicitly, down to magnetic quadrupole terms. Collisional rates are calculated using the quasi-relativistic distorted wave method. The code is very efficient because of a clever factorization theorem which allows the collisional matrix element to be factorized into radial and angular parts, even after inclusion of exchange effects (Bar-Shalom, Klapisch, and Oreg 1988). The time consuming radial integral then only needs to be calculated once for a large subset of transitions.

Below we discuss some of the key iron L-shell diagnostic features which have emerged from this work. We begin with the well-studied coronal case and then move on to consider X-ray photoionized plasmas and ionizing plasmas in young supernova remnants.

2. CORONAL PLASMAS.

In the coronal approximation, the plasma is said to be in collisional equilibrium. The heat input is coupled directly to the ions and free electrons, the line and continuum emission are optically thin, and the density is low enough that most ions are predominantly in the ground state. Examples of plasmas in which these approximations are valid include stellar coronae, intermediate-age and older supernova remnants, and the hot gas in galaxies and clusters. In these situations, the ionization balance is determined almost entirely by electron collisional ionization and radiative and dielectronic recombination. Line excitations are dominated by electron impact excitation and de-excitations are dominated by spontaneous radiative decay. Discrete X-ray spectral models for coronal plasmas have been calculated by a number of authors (Raymond and Smith 1977; Mewe, Gronenschild, and van den Oord 1985, Mewe, Lemen, and van den Oord 1986) and are now widely applied in the analysis of recent astrophysical X-ray observations.

It is generally well-known that the iron L-shell lines provide temperature diagnostics under coronal conditions. The temperature sensitivity is primarily associated with the ionization structure, which varies dramatically with temperature throughout the range 10^5 - 10^8 K (Mewe 1988). Since the "central wavelength" of the L-shell emission complex decreases monotonically with increasing ionization, the discrete line spectrum provides a thermometer for the source emission region, even when measured at relatively low spectral resolution.

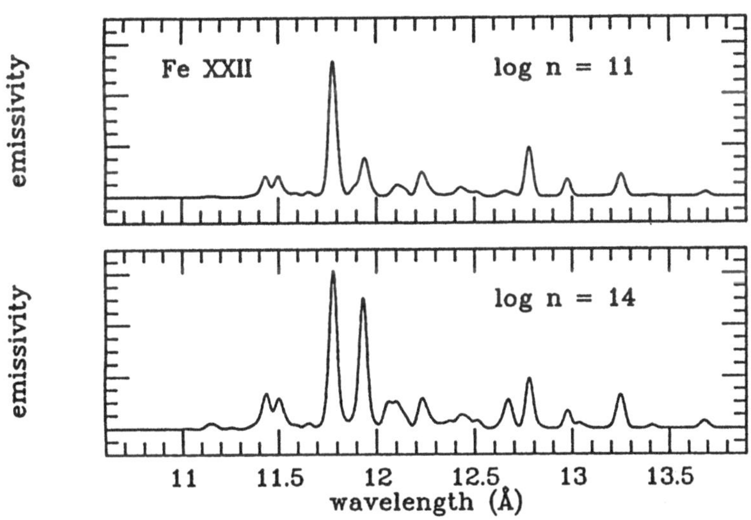

Figure 1: Density sensitivity of B-like Fe XXII with $kT =$ 850 eV. Atomic model includes the 125 levels through $n = 3$. *Top:* $n_e = 10^{11}\ cm^{-3}$, representative of the low-density limit. *Bottom:* Increasing metastable state populations in ground configuration act as sources for $n = 3$ excited states through electron impact. This process pumps the line at 11.93 Å, shown here at $n_e = 10^{14}\ cm^{-3}$.

It is not so widely appreciated that the iron L complexes also provide good density diagnostics in collisional equilibrium. The mechanism behind this is associated with the fact that the ground state configurations of the fluorine-like through boron-like species, Fe XVIII - XXII, are quasi-degenerate; they are split by fine structure interactions leading to the presence of a number of low-lying metastable states. As the electron density increases, electron impact excitation from the ground state begins to feed these metastable levels until the populations approach their thermal equilibrium values at densities $\sim 10^{12}$ - 10^{14} cm^{-3}. The metastable states then act as sources for excited states, again through electron impact, leading to the appearance of new lines which are not easily excited from the absolute ground state. This effect was first studied by Mason *et al.* (1979; see also Mason and Storey 1980, Mason *et al.* 1984) and has been verified by our more recent calculations.

An example is shown in Figure 1 for the boron-like Fe XXII atom. The L-shell discrete spectra for this ion at densities of 10^{11} and 10^{14} cm^{-3} are plotted as a function of wavelength in the upper and lower panels respectively. The resolution assumed for the spectrum is $\Delta\lambda = 0.05$ Å, characteristic of the reflection grating experiment on XMM (Brinkman *et al.* 1989). There is an obvious density dependence in many of the lines. These are all transitions which involve allowed decays to metastable states. Unlike some more familiar density diagnostics which rely on collisional depopulation, the relevant lines *appear* at higher density rather than disappear. The critical densities for most of the iron L diagnostics are $\sim 10^{12}$ cm^{-3}. There is, however, one weaker feature present in the Fe XX spectrum which is density-sensitive down to 10^{9} cm^{-3}, perhaps a more interesting number for cosmic sources. Note as well that since density diagnostics of this kind are present in boron-like all the way up to fluorine-like species, they can be invoked over a wide range in temperature.

3. X-RAY PHOTOIONIZED NEBULAE.

Spectral diagnostics for coronal plasmas are certainly useful for astrophysical X-ray spectroscopy. However, for many of the more exotic sources, the assumptions of collisional equilibrium are unlikely to be valid. Frequently, the departures from equilibrium are of interest in and of themselves, since they can convey crucial information about the source environment and/or history. An especially important case involves X-ray photoionized plasmas, which are likely to be present in accretion-powered systems like X-ray binaries, cataclysmic variables, and possibly active galactic nuclei. These sources possess intense X-ray continua, produced down near the compact object, which irradiate the surrounding medium. The transfer of the X- radiation field outward through the cooler, less dense accreting gas, produces a number of discrete observable spectral features, including the iron L-shell lines, in the soft X-ray band (cf. Kahn, Seward, and Chlebowski 1984).

The iron L-shell line spectra of X-ray photoionized plasmas are expected to differ significantly from those of coronal plasmas primarily because of the relationship between the electron temperature and the mean level of ionization. In a coronal plasma, the sole source of ionization and excitation is electron collisions. The lines which appear brightest are those with large collisional coupling to the ground state. For the photoionized case, on the other hand, the presence of the intense X- ray continuum makes the gas severely overionized relative to the local electron temperature. For example, a substantial ion fraction in Fe XVII would require a temperature of greater than 2×10^6 K in a collisional plasma, however, in a photoionized plasma, it can exist at temperatures $\sim 10^5$ K (Kallman and McCray 1982). Since the electrons are cool, they are incapable of collisionally exciting X-ray transitions. Instead, the excited levels are populated predominantly by radiative recombination or by radiative cascade following recombination.

As shown by Liedahl *et al.* (1990), this yields a characteristic spectral signature of X-ray photoionization in the iron L-shell line structure. In particular, collisional spectra are dominated by contributions from $3d$ - $2p$ transitions, whereas X-ray nebular spectra are made up almost entirely of $3s$ - $2p$ transitions. The effect can be understood in the following way: For the collisional plasma, the relative line intensities are governed largely by the collisional excitation rates from the ground state to the appropriate levels. Because of the greater "overlap" with the $2p$ orbitals, the collisional excitation of the $3d$ states is considerably larger than that of the $3s$ states, which are populated predominantly through cascade following excitation of higher-lying states. In contrast, for photoionized plasmas, collisional excitation is negligible for *both* the $3d$ and the $3s$ states. Instead, cascade following recombination is the dominant mechanism. Since direct radiative recombination does not favor $3d$ states over $3s$ states and since nine of the twelve levels in the $3d$ manifold (j = 0, 2, 3, or 4) cannot make allowed transitions to the ground state, most of the recombination to the $3d$ levels leads to subsequent decays to $3p$ and eventually to $3s$ anyway. Put simply, the large $3s/3d$ line ratios are expected since the $3s$ states are lower in energy, hence lower in the cascade chain, than the $3d$ states. Essentially all of the flux into the $n = 3$ shell, whether from radiative recombination or cascades from higher n-levels, ends up in the $3s$ state, resulting in intense emission in the $3s$ - $2p$ lines.

Iron L-shell lines also provide good density diagnostics for photoionized plasmas. It is easy to see that the low-lying metastable states, which are responsible for density sensitivity in the coronal case, can have a similar effect here. However, given the predominance of radiative recombination in the iron L line formation for X- ray photoionized plasmas, it is the quasi-degeneracy of the ground configuration of the one times more ionized species (Z+1) that determines the density diagnostics for any given ion stage Z. At low density, the Z+1 ion is always in the absolute ground state. Since radiative recombination tends to preserve the core configuration of the recombining ion, the only states of the Z ion that can get populated are those that have an inner shell configuration identical to that of this Z+1 ground state. Transitions from these levels are therefore the only ones that can be observed. As the density is increased, the Z+1 low lying metastable levels get populated by electron collisions. Recombination can then occur to other states possessing these excited core configurations. The subsequent radiative decay for the Z ion therefore exhibits many new lines.

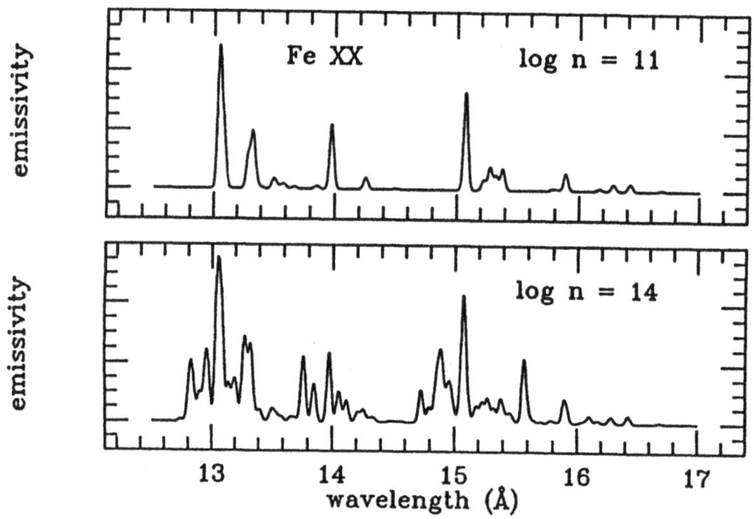

Figure 2: Density sensitivity of N-like Fe XX for photoionization conditions. *Top:* The Lα spectrum with $kT = 10$ eV and $n_e = 10^{11}$ cm^{-3}. Lines originate from ground state of Fe XXI through recombination. *Bottom:* Recombination spectrum at $n_e = 10^{14}$ cm^{-3}. The Fe XXI low-lying metastable state populations are enhanced at high densities, providing virtually independent sources for Fe XX excited states.

This is illustrated in Figure 2 which shows our calculated emission spectrum for a photoionized nitrogen-like iron plasma (Fe XX) at densities of 10^{11} and 10^{14} cm^{-3} respectively. The carbon-like iron ion (Fe XXI) possesses a multiplicity of low lying metastable levels which is especially rich. Hence, the recombination-dominated spectrum of the nitrogen-like species shows tremendous detail at higher densities. Similar effects are found for many of the other iron L-shell species. The critical densities for these diagnostics again tend to be $\sim 10^{13}$ - 10^{14} cm^{-3}. However those values may not be out of range, at least for galactic accretion-powered sources where X-ray photoionized nebulae are most likely to be found.

At first sight, it might seem unlikely that the iron L-shell spectra can provide good temperature diagnostics for photoionized plasmas as well. In particular, the detailed line structures produced by cascade following recombination are relatively insensitive to temperature in this regime. In addition, the ionization balance is determined by the radiation field, so ratios of lines from different ions are not very sensitive to temperature either. However, we have found that tight constraints on the electron temperature can be derived from measurements of the recombination continua in this setting. Because the electrons are cool, the recombination continua are narrow and have the appearance of discrete features. The emissivities of the continua scale as $T^{-3/2}E^2\ \sigma(E)\ \exp[-(E-E_0)/kT]$ (photon units) where E is the photon energy, E_0 is the energy of the recombination edge, σ is the photoionization cross-section, and T is the electron temperature (Osterbrock 1989). Hence, measurement of the intrinsic width or peak emission associated with these features yields an estimate of the temperature. Of course, the strengths of these features in absolute units also scale like the volume emission measure. To break the ambiguity, one needs to compare the recombination continuum feature to the discrete lines from the same ion. This is illustrated in Figure 3, where we show the full spectrum of Fe XXI at electron temperatures of 10 eV and 30 eV. As can be seen, the detailed structure of the discrete line spectrum is insensitive to temperature, however, the ratio of the peak intensity of the L-continuum to that of any or all of the lines is a strong function of temperature.

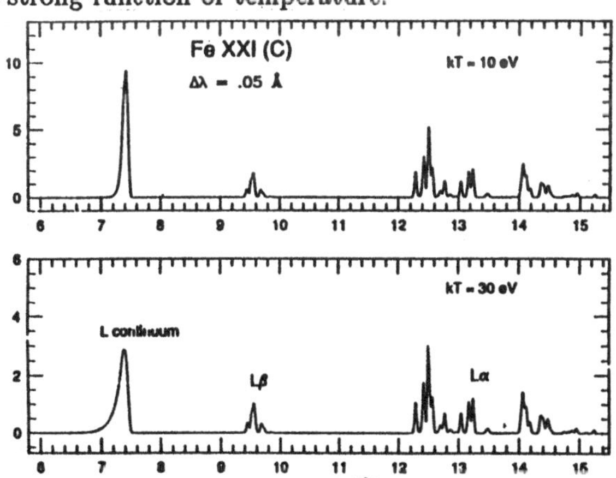

Figure 3: Temperature dependence of C-like Fe XXI recombination spectrum showing the narrow recombination spectrum near 7.4 Å as well as the discrete line spectrum between 12 and 14 Å. *Top:* The L-shell spectrum with $kT = 10$ eV. *Bottom:* $kT = 30$ eV.

4. IONIZING PLASMAS IN YOUNG SUPERNOVA REMNANTS.

A second nonequilibrium case that is of some interest involves the X-ray emitting plasmas in young supernova remnants. It was recognized a number of years ago (Itoh 1977), that the shocked gas in young remnants is likely to exhibit severe departures from ionization equilibrium, largely because the timescales for collisional ionization at the extremely low densities characteristic of these environments can greatly exceed the age of the remnant. Evidence for this effect has been deduced from detailed spectral fits to low resolution data on a number of remnants (cf. Charles *et al.* 1981), as well as from a few high resolution spectra acquired with the crystal spectrometer flown on the *Einstein* Observatory (cf. Winkler *et al.* 1981).

In one sense, this case is the opposite of that encountered for the X-ray photoionized nebulae; the gas is *underionized* relative to the local electron temperature, rather than *overionized*. For example, in a remnant ~ a few hundred years old, the electron temperature in a region where Fe XVII is the prevalent ion can be several times higher than that encountered for a plasma in collisional equilibrium. As in the case of X-ray photoionized plasmas, this can change the balance between the various processes contributing to a particular line excitation. Processes favored at higher electron energies dominate and may enhance a select set of discrete spectral features.

An especially interesting case is provided when the time-dependent ionization favors the presence of sodium-like Fe XVI. In coronal plasmas, Fe XVI is never an important source of line emission, since its ionization potential is much below that of its very stable neon-like neighbor, Fe XVII. However, in an evolving supernova remnant, the shocked gas must spend some non-negligible amount of time in an Fe XVI state as it slowly makes its way toward collisional equilibrium. When both Fe XVI and Fe XVII are present in appreciable fractions, a number of competing processes can contribute to the line excitation. First, the hot electrons can collisionally excite Fe XVII to allowed levels, which then decay radiatively, yielding the bright $3d$ - $2p$ line at 15 Å. Fe XVI can also be collisionally excited, however since the electron temperature exceeds the ionization potential, this occurs preferentially through autoionizing states of the form $2s^2 2p^5 3snd$, where $n \geq 3$. Many of these stabilize via electron emission. However some have large radiative branching ratios, leading to $3d$ - $2p$, $4d$ - $2p$, and $5d$ - $2p$ emission lines near 16 Å, 13 Å, and 12 Å, respectively. Finally, the hot electrons can also collisionally ionize Fe XVI out of the $2s$ or $2p$ orbitals, leading to an excited Fe XVII ion which decays via a $3s$ - $2p$ line near 17 Å.

A comparison of a simulated supernova remnant spectrum with that of a coronal plasma shows vast differences as can be seen in Figure 4. Many of these are associated with a different ion mix. For example the presence of the Fe XVI lines and the absence of the Fe XVIII lines. However, notice that the Fe XVII emission structure alone is very different, particularly the $3s/3d$ line ratio (17 Å/15 Å). This cannot be mimicked by any multi-temperature situation and serves as a clear indication of nonequilibrium ionization.

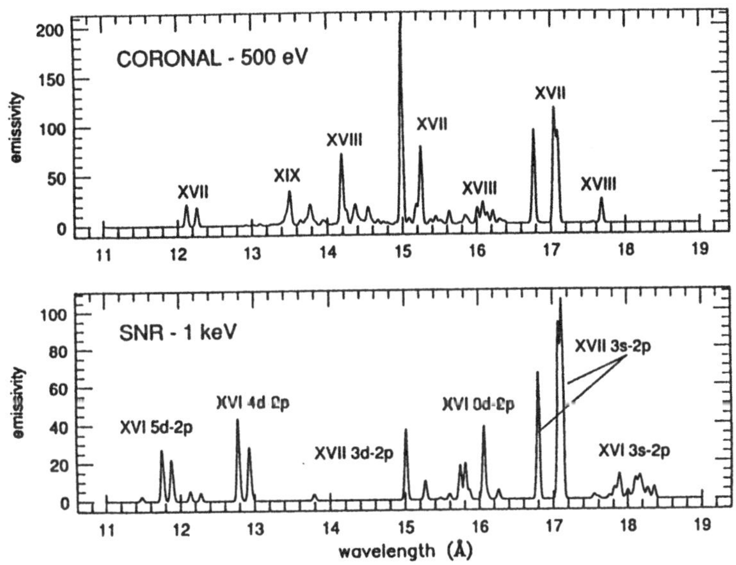

Figure 4: Comparison of Fe L-shell spectrum in coronal equilibrium with that of an ionizing Fe XVI/XVII plasma. *Top:* Fe XVI-XIX in coronal equilibrium with $kT = 500$ eV. *Bottom:* Conditions representative of transient ionizing phase, $n(\text{XVI})/n(\text{XVII}) = 10$, $kT = 1$ keV. Note contribution of Fe XVI transitions and inversion of Fe XVII (15 Å)/(17 Å) line ratios.

These features can be utilized to place tight constraints on conditions in the source emission region. Because collisional ionization and collisional excitation have different temperature dependences over most of the temperature range of interest, the $3s/3d$ ratio is a strong function of both the electron temperature, and the Fe XVI/Fe XVII ionization ratio. The Fe XVI $3d$/Fe XVII $3d$ line ratio (16 Å/15 Å) also depends on these two quantities, so measurement of all three lines yields explicit determination of both the electron temperature and the iron ionization structure. These can then be used to deduce the thermal history of the diffuse gas.

It might seem that this is a rather "specialized" situation. However, it may apply to one of the most interesting remnants in our own galaxy, SN1006. The shocked ejecta in that remnant are believed to be passing through the Fe XVI phase with $T_e \sim 10^7$ K (Hamilton, Sarazin, and Szymkowiak 1986, Hamilton 1990). When high resolution Fe L-shell spectra become available, we will be able to very accurately test the prevailing models for this source.

This work was supported in part by grants from the NASA Astrophysics Data Program, the NASA Innovative Research Fund, and the Institute for Geophysics and Planetary Physics at the Lawrence Livermore National Laboratory. We wish to thank W. Goldstein and A. Osterheld for extensive help with the atomic calculations, and A. Hamilton for discussions concerning ionizing plasmas in young supernova remnants.

REFERENCES

Bar-Shalom, A., Klapisch, M., and Oreg, J., 1988, *Phys. Rev. A*, **38**, 1773.

Bleeker, J., 1990, this volume.

Brinkman, A.C., *et al.*, 1989, in *EUV, X-Ray, and Gamma-Ray Instrumentation for Astronomy and Atomic Physics*, ed. C.J. Hailey and O.H.W. Siegmund, *Proc. S.P.I.E.*, **1159**, 495.

Charles, P.A., Kahn, S.M., Mason, K.O., and Tuohy, I.R., 1981, *Ap. J. (Letters)*, **246**, L121.

Hamilton, A.J.S., 1990, private communication.

Hamilton, A.J.S., Sarazin, C.L., and Szymkowiak, A.E., 1986, *Ap. J.*, **300**, 698.

Itoh, H., 1977, *Pub. Astr. Soc. Japan*, **29**, 813.

Kahn, S.M., Seward, F.D., and Chlebowski, T., 1984, *Ap. J.*, **283**, 286.

Kallman, T.R., and McCray, R., 1982, *Ap. J. Suppl.*, **50**, 263.

Klapisch, M., 1971, *Computer Phys. Comm.*, **2**, 239.

Klapisch, M., Schwab, J.L., Fraenkel, J.S., and Oreg, J., 1977, *J. Opt. Soc. Am.*, **61**, 148.

Liedahl, D.A., Kahn, S.M., Osterheld, A.L., and Goldstein, W.H., 1990, *Ap. J. (Letters)*, **350**, L37.

Mason, H.E., Bhatia, A.K., Kastner, S.O., Neupert, W.M., and Swartz, M., 1984, *Solar Phys.*, **92**, 199.

Mason, H.E., Doschek, G.A., Feldman, U., and Bhatia, A.K., 1979, *Astr. Ap.*, **73**, 74.

Mason, H.E., and Storey, P.J., 1980, *M.N.R.A.S.*, **191**, 631.

Mewe, R., 1988, in *Astrophysical and Laboratory Spectroscopy*, eds. R. Brown and J. Lang, (Scottish Univ. Summer School in Phys. Publ.), p. 129.

Mewe, R., Gronenschild, E.H.B.M., and van den Oord, G.H.J., 1985, *Astr. Ap. Suppl. Ser.*, **62**, 197.

Mewe, R., Lemen, J.R., and van den Oord, G.H.J., 1986, *Astr. Ap. Suppl. Ser.*, **65**, 511.

Osterbrock, D.E., 1989, *Astrophysics of Gaseous Nebulae and Active Galactic Nuclei*, (Mill Valley, CA: University Science Books), p. 18.

Raymond, J.C., and Smith, B.W., 1977, *Ap. J. Suppl.*, **35**, 419.

Schwartz, D.A., 1990, this volume.

Weisskopf, M.C., 1988, *Space Sci. Rev.*, **47**, 47.

Winkler, P.F., Canizares, C.R., Clark, G.W., Markert, T.H., Kalata, K., and Schnopper, H.W., 1981, *Ap. J. (Letters)*, **246**, L27.

X-ray Lines in Stellar Coronae

R. Mewe
SRON-Laboratory for Space Research, Utrecht
Sorbonnelaan 2, 3584 CA Utrecht, The Netherlands

ABSTRACT

Because the emission line spectra and continua from optically thin plasmas are fairly well known, high-resolution X-ray spectroscopy has its most obvious application in the measurement of optically thin sources such as the coronae of stars. In particular X-ray observations with the *EINSTEIN* observatory have demonstrated that soft X-ray emitting coronae are a common feature among stars on the cool side of the Hertzsprung-Russell diagram, with the probable exception of single very cool giant and supergiant stars and A-type dwarfs. Observations with the spectrometers aboard *EINSTEIN* and *EXOSAT* have shown that data of even modest spectral resolution ($\lambda/\Delta\lambda$= 10-100) permit the identification of coronal material at different temperatures whose existence may relate to a range of possible magnetic loop structures in the outer atmospheres of these stars. The improved spectral resolution of the next generation of spectrometers aboard *XMM* and *AXAF* is needed to fully resolve the temperature structure of stellar coronae and to enable density and velocity diagnostics. In this paper spectral results from *EINSTEIN* and *EXOSAT* are discussed. A few simulations of high-resolution Fe L, K, and $2s$–$2p$ spectra with *AXAF*, *XMM*, and several detectors such as calorimeter and Nb-junction are shown to demonstrate the capabilities for plasma diagnostics.

1. INTRODUCTION

X-ray observations with the *EINSTEIN* Observatory have demonstrated that stars of almost all spectral types are X-ray emitters, with the probable exception of single very cool giant and supergiant stars and A-type dwarfs (e.g. Vaiana *et al.* 1981). There is a dichotomy between early-type stars (O and B) and late-type stars (F to M) with regard to the dependence of coronal emission on basic stellar parameters. The X-ray luminosity of early-type stars appears to be strongly correlated with bolometric luminosity and virtually independent of rotation rate, whereas for late-type stars, the emission strongly depends on rotation (e.g. Pallavicini *et al.* 1981). Thus the possible mechanisms of coronal formation and heating are probably fundamentally different for these two classes of stars (for recent reviews see e.g. Rosner *et al.* 1985, Linsky 1985, Schmitt 1988, Pallavicini 1989).

While the existence of X-ray emitting coronae throughout the Hertzsprung-Russell diagram has been well established, only poor spectral information is available about these X-ray sources because the majority of the stellar observations was done with only limited or no spectral resolution and in many cases with modest signal-to-noise ratio. Early space instruments such as proportional counters allowed to determine merely the overall shape of the X-ray spectra of stellar coronae and it was not until the 1980s that the introduction of solid state, Bragg crystal, and transmission grating spectrometers on the *EINSTEIN* and *EXOSAT* satellites with modest spectral resolution ($\lambda/\Delta\lambda \simeq$ 10–100) provided for the first time a taste of more detailed spectroscopy of stellar coronae and motivated the use of more sophisticated emission models, which include the line emission expected from hot optically thin plasmas in collisional ionization equilibrium (e.g. Raymond and Smith 1977, Mewe *et al.* 1985, 1986a). However, since most of the spectral lines are not spectrally resolved, the spectral constraints derived are still highly model-dependent.

In particular the next generation of spectroscopic space missions near the end of the century, viz. NASA's Advanced X-ray Astrophysics Facility (*AXAF*) (Weisskopf 1987) and ESA's X-ray Multi-Mirror Mission (*XMM*) (Barr *et al.* 1988) will give astronomers the opportunity to measure many emission lines with a resolution of $\lesssim$ 0.05 Å in the wavelength range $\sim$ 1–140 Å, allowing them to much better resolve the temperature and density structure of stellar coronae.

In this paper I briefly discuss the spectral model of an optically thin plasma and compare this with the results of broad-band and medium-resolution spectral X-ray observations with *EINSTEIN* and

EXOSAT. I present a few simulations of high-resolution Fe L, K, and $2s$–$2p$ spectra with *AXAF*, *XMM*, and various types of detectors to demonstrate the capabilities for plasma diagnostics.

2. OPTICALLY THIN THERMAL PLASMAS

2.1. Coronal model, ionization and line excitation

The X-ray emission from a stellar corona is typically optically thin thermal plasma emission from a very hot, tenuous, and highly ionized gas in which the atoms have all, or almost all, of their electrons stripped due to collisions with energetic electrons. Temperatures are typically above one million degrees Kelvin. The plasma is sufficiently transparent that the transfer of radiation can be neglected. This is the standard *coronal model* that was first applied by Elwert (1952) to the solar corona (cf. also S. Kahn, this conference). It assumes a tenuous plasma in a steady state in which electron collisions control the ionization state and emissivity of the gas. The gas is in a state of statistical equilibrium both for the bound atomic states and for the ionization balance and the plasma electrons and the ions are relaxed to Maxwellian energy distributions with a common temperature, T, a free parameter controlled by external processes (for a discussion of deviations from the coronal model due to effects of photoionization, optical depth, density, non-Maxwellian electron distributions, and transient ionization, see Raymond (1988) and Mewe (1990a)). The line emission - that dominates the soft X-ray spectrum and the cooling of coronae at temperatures up to $\sim$ 10 MK - comes mainly from the excitation by the impact of a free electron of an electronic transition between bound levels within a highly ionized atom, followed by the spontaneous radiative decay from the upper line level. Continuum emission is produced by the interaction of a free electron with an ion either by a free-free transition (between two continuum states of the ion), a free-bound transition (capture of the electron), or a two-photon process (excitation of a metastable level in a hydrogen- or helium-like ion followed by the simultaneous emission of two photons).

The overall appearance of the X-ray spectra will be dominated by the ionization structure which in the coronal model is determined by a balance between electron impact ionization (including sometimes a contribution from autoionization) and radiative plus dielectronic recombination. The ionization structure varies dramatically with temperature throughout the range 0.1–100 MK (e.g. Mewe 1990a).

In extremely hot (> 100 MK) plasmas all abundant elements are nearly fully ionized and the X-ray emission is dominated by the free-free continuum from hydrogen and helium. With decreasing temperature the heavier trace elements are only partly ionized, beginning with iron, and spectral emission lines excited by electron collisions show up and will begin to dominate the X-ray spectrum. This is especially true for the Fe L-shell lines around 10 Å, the Fe K-shell lines at $\sim$ 2 Å and the Fe $2s$–$2p$ lines around 100 Å. In a wide temperature range (0.01–10 MK) the X-ray spectra of optically thin sources are rich in emission lines from many ions. The wavelength band 1–140 Å contains a multitude of prominent lines from nearly all ionization stages of cosmically abundant elements, including the K-shell transitions of carbon through iron, and the L-shell transitions from silicon through iron (for a discussion of the ionization and recombination processes, the excitation mechanisms, and the formation of optically thin spectra with examples, see e.g. Raymond (1988, 1990) and Mewe (1988, 1990a,b)). The spectral lines are broadened mainly by Doppler broadening. If suitably resolved these lines are powerful diagnostics of plasma parameters because the line strengths are generally very sensitive to the electron temperature, the elemental abundances, and in some cases to the electron density. In certain circumstances (e.g. transient plasmas) the line intensities are dependent on deviations from the ionization equilibrium or to deviations from Maxwellian electron energy distributions (e.g. Mewe 1990a). The emission line spectra and continua from optically thin plasmas are fairly well known from calculations (for reviews, e.g. Raymond 1988, Mewe 1990a).

2.2. Diagnostics of plasma parameters

I briefly review the diagnostics of coronal temperature, ionization balance, emission measure (distribution), abundances, densities, and velocities.

The soft X-ray spectrum sensitively depends on the ionization structure throughout the temperature range 0.1–100 MK so that the mere detection of many lines provides an accurate "thermometer" for the source emission region. The electron temperature T can be diagnosed from the intensity ratio of two collisonally excited lines from different ions of the same element assuming coronal stationary ionization equilibrium, but this technique is limited by the assumption on the ionization equilibrium. To get rid of this, one can use the ratio of two lines depending on the same ionization stage, but differently dependent on temperature, e.g. two collisionally excited lines with different excitation energies such as the Lyman α and β lines. Alternatively, one can measure the ratio of lines at nearly the same wavelength for which the excitation functions have a different dependence on T, e.g. in the helium-like $2 \rightarrow 1$ triplet. This consists of a resonance (r) $2^1P \rightarrow 1^1S$ line, a forbidden (f) $2^3S \rightarrow 1^1S$ line and an intercombination (i) $2^3P \rightarrow 1^1S$ line. The intensity ratio $r/(i+f)$ varies with T. Another possibility is to use the intensity ratio of the collisionally excited resonance line and a nearby dielectronic recombination satellite line (Gabriel 1972).

In many sources, however, the X-ray spectrum will not be a unique function of one single temperature, but instead will be determined by a distribution in temperature across the emission region. In those cases, multi-temperature fits to the entire spectrum can yield the differential emission measure (DEM) distribution appropriate to the source (e.g. Mewe (1990b)). Such an analysis has been performed on the *EXOSAT* transmission grating spectra of late-type stars to constrain the basic properties of stellar magnetic loops (Lemen *et al.* 1989, Schrijver *et al.* 1989). This method can be applied equally well to investigate the temperature structures of other kinds of optically thin plasmas such as supernova remnants and cooling flows in clusters of galaxies so that in many cases the DEM analysis will be the primary handle to test models for the physics of the source emitting region.

In both coronal equilibrium and nonequilibrium situations, ions of different elements which have similar ionization potentials usually coexist in close proximity. Thus the ratios of the prominent emission lines from ions of different elements yield in a relatively model-independent way elemental abundances. By using a suitable set of lines from different ionization stages from one given element, e.g. iron, we can determine the temperature distribution by fitting the line intensities. Then by selecting suitable sets of lines from different elements, relative elemental abundances can be determined and by determining intensity ratios of lines to neighbouring, unblended continuum bands, the absolute abundances can also be derived.

From spectral fits to the line and/or continuum emission one can determine the emission measure $\varepsilon = \int n_e^2 dV$ but this alone is not sufficient to model a stellar corona. We need the electron density, n_e, to determine the emitting volume V, hence the product Lf of loop length and filling factor, and to constrain coronal models. Expected coronal densities (and temperatures) probably range from $n_e \sim 10^9$–10^{10} cm^{-3} ($T \sim$ 1–3 MK) under quiescent conditions up to $\sim 10^{13}$ cm^{-3} ($\lesssim$ 50 MK) under flaring conditions. Electron densities can be measured using density-sensitive spectral lines originating from metastable levels or using innershell excitation satellites to resonance lines (e.g. Mewe 1988). In the first case the helium-like $2 \rightarrow 1$ triplet system lines are particularly important (Gabriel and Jordan 1969, 1972, Pradhan and Shull 1981, Pradhan 1982, 1985, Mewe *et al.* 1985). The intensity ratio f/i varies with electron density due to the collisional coupling between the metastable 2^3S upper level of the forbidden line (f) and the 2^3P upper level of the intercombination line (i). The f/i line intensity ratio of helium-like ions from carbon through magnesium in the wavelength region 9–42 Å can be used to diagnose coronal plasmas in the density range $n_e = 10^8$–10^{13} cm^{-3} and corresponding temperature range $T \sim$ 1–6 MK. However, for density diagnostics in active late-type stars one must observe lines formed at temperatures around $T \sim$ 10 MK. For this purpose Mewe *et al.* (1985, 1991) have considered many density-sensitive lines in the Fe L-shell complex, i.e. $2\ell \!\!-\!\!\geq 3\ell'$ transitions from ions Fe XVII–XXIV (and some corresponding nickel lines from Ni XXI–XXIV) covering the wavelength region 7–13 Å, lines between 90–140 Å from 2ℓ–$2\ell'$ transitions in Fe XVIII–XXIII, and lines between 170–275 Å from 3ℓ–$3\ell'$ transitions in Fe IX–XIV ions. The density dependence is because the upper line level can be excited from various sublevels within the ground state which become collisionally coupled at increasing densities (the same holds for the satellite lines). The Fe and Ni lines can be used as tools for diagnosing plasmas in the density range $\sim 10^{10}$–10^{15} cm^{-3} and temperature range $\sim$ 0.5–15 MK.

In particular high-resolution ($\Delta\lambda \simeq 0.05$ Å) spectroscopy at longer wavelengths ($\gtrsim$ 100 Å) with the Low Energy Transmission Grating Spectrometer (LETGS) on *AXAF* (Brinkman *et al.* 1987) will permit for the first time detailed studies of spectral line profiles in the X-ray region. This may provide direct measurements of the wind expansion velocities of stellar coronae, the flow velocities along magnetic loops in stellar active regions, the orbital velocities in X-ray binaries (allowing to determine the activity level of each of the binary components), and the dynamics (upflow velocities) of stellar flares, etc. (Linsky 1990, Mewe *et al.* 1991).

3. STELLAR CORONAL X-RAY OBSERVATIONS

3.1. Early-type stars

It is now widely accepted that the winds of early-type stars are driven by radiation pressure (e.g. Cassinelli 1979), but the origin of the X-ray emission of early-type stars is at present still unclear. Some models rely on the existence of a hot (few millions Kelvin) corona underlying the stellar wind. For example, the detection of emission features near 2 keV in the spectrum of ζ Ori A taken with the Solid State Spectrometer (SSS) on *EINSTEIN* has been interpreted by Cassinelli and Swank (1983) as line emission of Si XIII and S XV from a very hot ($\gtrsim$ 10 MK) source component at the base of the wind. However, the substantial mass loss does imply difficulties in transferring the X-ray radiation through the wind. Moreover the lack of convection near the stellar surface prevents a mechanism similar to the solar dynamo to be operative. Alternatively, it seems more probable that the heating (and the associated X-ray emission) is located in the hot post-shock regions of atmospheric gas blobs moving in the stellar wind or trains of shocks which are accelerated by the intense radiation field in the atmosphere and wind (e.g. Lucy 1982). These gas blobs develop from small acoustic disturbances via radiation-driven instabilities (Castor 1987, Rybicki 1987). Consequently the heating in early-type stars depends on the bolometric luminosity and not on rotation. For a discussion of models see Schmitt (1988). The available experimental constraints from *EINSTEIN* SSS spectra (Cassinelli and Swank 1983) and from the measured short-term variability of the soft X-ray emission (Collura *et al.* 1984) seem to favor the latter model, but better spectral resolution is needed to distinguish between different models. More specifically, the spectral resolution and the high throughput of an instrument such as the Reflection Grating Spectrometer (RGS) on *XMM* (Brinkman *et al.* 1989) will allow to search for absorption edges produced by the overlying wind and in case of the base coronal model, to search for emission lines of a possible very hot component, or to search for strong Doppler broadenings of emission lines by turbulent motions at speeds up to $\sim$ 3000 km/s if shocks are indeed formed high in the stellar wind.

3.2. Late-type stars

On the contrary, for the late-type stars with convective envelopes the solar analogon appears to be applicable, because the relationship between X-ray luminosity and stellar rotation suggests that surface magnetic fields - generated by a dynamo mechanism - are responsible for the heating of a coronal plasma with temperatures ranging from one to several tens of millions Kelvin. The majority of late-type coronae is thought to be solar-like in the sense that the same physical processes as in the Sun are likely to be relevant for the formation and heating of coronae, though in many cases the level of X-ray activity can be orders of magnitude higher than on the Sun (e.g. Linsky 1985). By analogy with the Sun, the tenuous, hot coronal plasmas are presumed to be heated by magnetohydrodynamic processes, with much of the plasma confined in loop-like structures connecting opposite magnetic polarities observed in the photosphere.

X-rays have been detected from many cool stars, with surface flux densities ranging from below 10^3 to 10^8 erg cm^{-2} s^{-1}, i.e. from about five times weaker than a solar coronal hole up to ten times brighter than solar active regions (e.g. Schrijver 1983). These coronal fluxes are tightly related to the chromospheric and transition-region radiative losses, apparently independent of the stellar spectral type or mass (e.g. Ayres *et al.* 1981, Schrijver 1983, 1985, Schrijver *et al.* 1984). This suggests the existence of a single activity parameter that uniquely determines the surface flux levels at distinct temperature regimes in the stellar outer atmosphere and probably plays a decisive role in the heating of the hot coronal gas.

It is generally considered (e.g. reviews by Serio (1985) and Rosner *et al.* (1985)) that, in analogy to the picture of the solar corona, the heating of coronae is related to the presence of magnetic fields, amplified and brought to the surface by a dynamo similar to the one at work in the interior of the Sun. The dynamo action is probably caused by the interaction of magnetic fields with convective fluid motions, thus coupling stellar rotation and convective motions. Observed X-ray fluxes increase with the rotation rate. The role of rotation in determining the activity level of the outer stellar atmosphere is stressed by the bright X-ray luminosities of synchronized close binary systems, which can maintain a high rotation rate despite strong magnetic braking because the system can feed on its energy stored in the orbital motion.

The few observations with the spectrometers aboard the *EINSTEIN* Observatory (Objective Grating Spectrometer (OGS), Solid State Spectrometer (SSS), Focal Plane Crystal Spectrometer (FPCS))

(Giacconi *et al.* 1979) and aboard *EXOSAT* (Transmission Grating Spectrometer (TGS)) (Brinkman *et al.* 1980) have shown that spectra of even modest spectral resolution ($\lambda/\Delta\lambda$ = 10–100) allow the identification of coronal material at different temperatures whose existence may relate to a range of possible magnetic loop structures in the outer atmospheres of these stars (for recent review on solar and stellar loop structures, cf. Mewe 1991). These data already showed a glimpse of what can be expected if full advantage is taken of the wealth of information contained in the X-ray spectral lines.

FPCS spectra with a resolving power $\lambda/\Delta\lambda \gtrsim 100$ have been taken from Capella (Vedder and Canizares 1983) and σ^2 CrB (Agrawal *et al.* 1985) in a few small band passes ($\Delta\lambda/\lambda \sim 3$ %), centered on some interesting lines such as Fe XX (13 Å), Fe XVII, XIX (15 Å), and O VIII (19 Å) (the latter only for Capella). In both cases the results are consistent with a temperature $T \simeq 6.5$ MK. OGS transmission grating spectra with a resolution $\Delta\lambda \sim 1$ Å between 5–30 Å were obtained by Mewe *et al.* (1982) for Capella. A two-component fit yielded temperatures of 5 and 10 MK. With the SSS spectrometer, sensitive in the energy range 0.5–4.5 keV with a resolution of $\Delta E \simeq 160$ eV about twenty coronal sources have been observed: e.g. 7 RS CVn stars such as Capella and σ^2 CrB, Algol, two M dwarfs (e.g. AD Leo), an active G star (π^1 Uma) and several O stars (Holt *et al.* 1979, Swank *et al.* 1981, Swank 1984, Cassinelli and Swank 1983). The spectral fits to the RS CVn-type spectra yielded two temperature components, at ~ 5 MK and at ~ 30–50 MK. Generally these spectra show helium-like features from magnesium, silicon, and sulphur.

The *EXOSAT* TGS spectra with a resolution $\Delta\lambda \sim 3$ Å have been obtained between 5–200 Å for three late-type stars Capella, σ^2 CrB, and Procyon (e.g. Brinkman *et al.* 1985, Schrijver 1985, Mewe *et al.* 1986b, Lemen *et al.* 1989, Schrijver *et al.* 1989) and between 50–400 Å for three hot white dwarfs HZ 43, Sirius B, and Feige 24 (Paerels *et al.* 1986 a,b). These spectra cover the wavelength range between 90–140 Å which has great diagnostic potential for optically thin plasmas with temperatures around $T \sim 10$ MK (Mason *et al.* 1984, Mason 1990) (i.e. just the temperature at which most of the emission is concentrated for active late-type stars (Schmitt 1990)), but which has been sadly neglected. Only few solar flare spectra from *OSO-5* are reported (Kastner *et al.* 1974) and analyzed by Mason *et al.* (1984). Schmitt (1990) and Mewe *et al.* (1985, 1991) have discussed the possibilities for temperature ($T \sim 10$ MK) and density ($n_e = 10^{10}$–10^{15} cm^{-3}) diagnostics in this region, where the lines from Fe XVIII–XXIII are relatively strong, the spectral density is relatively low, hence a spectral resolution of ~ 0.1 Å is sufficient to avoid ambiguities due to line blending. In this wavelength region stellar coronal sources at distances d up to about 200–300 pc can be observed (Schmitt 1990). Figure 1 shows the TGS results for three late-type stars (Mewe 1990b). The active stars Capella and σ^2 CrB show a prominent Fe XXIII line at 133 Å which is typical for hot plasmas with $T \sim 10$–20 MK. This line is absent in the spectrum of the much cooler corona ($T \simeq 0.6$ MK) of the star Procyon, which contains instead a strong blend of lines around 170 Å, that can be identified as a group of spectral lines from Fe IX–XI ions (formed at $T \sim 0.6$–2 MK), and which is not present in the spectra of Capella and σ^2 CrB. This implies that only relatively little plasma cooler than 2 MK can be present in the coronae of these stars. The intensity ratio of the 15 Å Fe XVII-XVIII line complex and the Fe XVIII–XXIII line complex between 90–140 Å demonstrates that the spectra cannot be emitted by an isothermal plasma, but instead requires at least a two-temperature interpretation. On the basis of a detailed differential emission measure (DEM) analysis Lemen *et al.* (1989) confirm that the emission of the corona of each of these stars is dominated by plasma in two relatively narrow temperature intervals: 5 MK and 25 MK for Capella and σ^2 CrB, and 0.6 MK and ~ 2 MK for Procyon. By analogy with the Sun, the outer atmospheres of cool stars with a convection zone are generally assumed to consist of magnetic loop structures filled with hot plasma, which connect opposite polarities in the photosphere. The DEM distributions as provided by such loop models suggest model coronae comprising two distinct ensembles of quasi-static magnetic loops with maximum temperatures around the dominant temperatures given above and with the cool loops probably expanding with increasing height over the photosphere (Schrijver *et al.* 1989).

Proportional counters (e.g. the Imaging Proportional Counter (IPC) on *EINSTEIN* and the Medium Energy (ME) detector on *EXOSAT*) provide only a crude spectral resolution ($\lambda/\Delta\lambda \lesssim 5$ below ~ 7 keV) which is not sufficient to resolve lines. These low-resolution spectra of stellar coronae are essentially featureless and a coronal temperature can only be approximately derived from modeling with theoretical spectral models (e.g. Schmitt 1988, 1990, Schrijver *et al.* 1984). Only in the case of a very strong line complex such as the iron K-shell blend at 6.7 keV (1.9 Å) the detection of a line feature is sometimes possible (e.g. ME observations of Algol (White *et al.* 1986, Van den Oord and Mewe 1989) and UX Ari (Pasquini *et al.*

1988) and *GINGA* LAC observations of UX Ari (Tsuru *et al.* 1989) (cf. also R. Pallavicini and G. Tagliaferri, this conference).

4. SIMULATIONS OF HIGH-RESOLUTION SPECTRA

4.1. Fe L-shell and Fe $2s$–$2p$ spectra with *AXAF* and *XMM*

At the resolutions of the *EINSTEIN* and *EXOSAT* transmission gratings and of the SSS the main emission line complexes from helium- and hydrogenlike ions of cosmically abundant elements and from the iron L-shell (Fe XVII–XXII) ions become distinct, while at much higher resolution ($\gtrsim$ 1000) as can be reached in future experiments with *AXAF* and *XMM* the helium-like triplets and possibly also the dielectronic satellites to resonance lines may be resolved thus providing us with powerful temperature and density diagnostics.

The low- and medium-resolution spectral X-ray results require at least a two-temperature interpretation, although differences in the derived temperatures may suggest a wide range of temperatures. The improved spectral resolution of the next generation of spectrometers aboard *XMM* and *AXAF* is needed to fully resolve the temperature structure of stellar coronae, as is shown in Figure 2 which compares between spectra of Capella measured with the TGS on *EXOSAT* and spectra simulated for the LETGS on *AXAF* and the RGS on *XMM*, using the source parameters derived from *EXOSAT* observations. The simulated spectra of the Fe L-shell blend between 10–18 Å and of the strong $2s$–$2p$ lines between 90–140 Å clearly demonstrate the diagnostic capabilities of the instruments on these future missions and show that spectrometers with a resolving power of $\gtrsim$ 500 up to $\sim$1000 or more resolve the L-blend in its many individual components, allowing more refined temperature, density, and velocity diagnostics.

An important diagnostic of stellar coronal activity is temporal variability of the coronal X-ray emission. Observations of time variations are essential in order to resolve the spatial structure of the coronae of eclipsing binaries (e.g. Walter *et al.* 1983), to infer locally distributed areas of activity on single, rapidly rotating stars, or to establish the size of loop structures confining the hot plasma in stellar flares (e.g. Van den Oord and Mewe 1989). To demonstrate what the future high-resolution Reflection Grating Spectrometer (RGS) on the high-throughput mission *XMM* will be able to accomplish I have chosen to simulate the spectrum of the eclipsing binary system AR Lacertae, a relatively bright RS CVn system consisting of a G2 IV and a K0 IV star and with an orbital period of 1.983 days. During primary eclipse (K star in front) the G star is completely occulted (total eclipse lasts 25000 s, but totality only 7600 s), but the secondary eclipse (G star in front) is annular. The physical properties of individual, not too complex, geometrical structures in the corona may be deduced by differencing the spectra before and after a structure is occulted. For the simulations we use the parameters as derived by Walter *et al.* (1983) from their analysis of *EINSTEIN* IPC data, which indicates local bright emitting regions close to the surface of both stars as well as an extended hot corona covering one hemisphere of the K star. I have calculated the spectra during eclipse (only K star) and during quadrature (G + K), integrated both over 6000 s and determined the difference spectrum (G star) (see Figure 3). Comparison of the results without noise (left panels) with those with noise (right panels) shows that the difference spectrum can be determined with a sufficient S/N ratio.

4.2. Diagnostics from Fe K-shell spectra

The high-resolution X-ray spectra from highly ionized iron and calcium provide superior diagnostics of very hot plasmas with temperatures in the range 10–100 MK. The availability of high-resolution spectra of the flaring Sun as obtained with the Bragg crystal spectrometers aboard the spacecrafts *P78-1* (instrument SOLFLEX (a)) (Doschek *et al.* 1979, Feldman *et al.* 1980), *SMM* (BCS (b)) (Culhane *et al.* 1981), and *Hinotori* (SOX1 (c), SOX 2 (d)) (Tanaka *et al.* 1982, Tanaka 1986) and of spectra obtained with hot laboratory sources (e.g. *Tokamak*, cf. Bitter *et al.* 1979) has stimulated the diagnostics to measure physical plasma parameters such as temperature, emission measure distribution, electron density, and departures from coronal ionization equilibrium. The wavelength intervals covered by these instruments are: 1.75–1.80 Å (b,c) (H-like Fe XXVI), 1.85–1.95 Å (a,b,c,d) (He-like Fe XXV), 3.00–3.05 Å (a,b) (H-like Ca XX), and 3.15–3.25 Å (a,b) (He-like Ca XIX).

The electron temperature can be diagnosed from the helium-like singlet/triplet ratio or for very hot plasmas from the intensity ratio of the Fe XXVI Ly α and Fe XXV w He-like resonance line. The technique of measuring the ratio of the resonance line and a nearby dielectronic recombination satellite line (see §2.2) has been widely used in these short wavelength regions to derive the solar flare temperature from very high-resolution X-ray spectra. The helium-like triplets with their satellites provide a very valuable diagnostics for electron temperature, emission measure, abundances, and ionization balance. However, an extremely high spectral resolution is required in order to avoid too much blending of the stronger line features that are important for the diagnostics of the plasma parameters by the hundreds of weaker unresolved satellite lines. In case of rather blending a more precise determination of the electron temperature requires a profile fitting of the entire spectrum (e.g. Antonucci *et al.* 1982, Tanaka 1986).

Consider the spectrum of the K-shell transition blend of helium-like iron around 1.85 Å. In order to demonstrate the effects of spectral resolution I have simulated spectra in the interval 1.84–1.92 Å for a series of resolving powers $R = \lambda/\Delta\lambda$ (cf. also Schmitt (1990)). A gaussian instrumental profile with a FWHM of $1.85/R$ Å is assumed and the lines are assumed to be broadened by thermal Doppler broadening. The electron temperature is chosen to be $T = 10$ MK. Figure 4 shows the results for the photon flux in 0.0002 Å bins integrated over 10^4 s for a temperature $T = 10$ MK and reduced emission measure $\varepsilon/d^2 = 10^{50}$ cm^{-3} pc^{-2}. I start with a very high instrumental resolving power of $R = 10^4$ (panel a) (typical for the SOLFLEX, BCS and SOX2 instruments). In this case the actual resolution is mainly determined by the Doppler line broadening ($R_D = 3300$). Line identifications are given following Gabriel's (1972) notation for the helium-like Fe XXV lines (w = resonance, x, y = intercombination, z = forbidden line) and lithium-like Fe XXIV lines. The dielectronic recombination (DR) Fe XXIV satellite/Fe XXV resonance line ratio j/w is particularly important, because it depends strongly on the electron temperature and is independent of the ionization balance. On the other hand, the innershell excitation (IE) Fe XXIV satellite/resonance line ratio q/w determines the ratio of the densities of the Fe XXIV and Fe XXV ions and can be used to measure departures from coronal ionization equilibrium. In the spectrum we also indicate the strongest pairs of IE and DR satellites from Fe XXIII–Fe XIX, which can also be used to determine T, ε, and (departures from) ionization equilibrium (notations from Doschek *et al.* 1981). The typical splitting of the lines is ~ 0.001 Å, so that we need a typical resolving power of $\gtrsim 2000$ to have full profit of the diagnostics. In the next step we degrade the resolution until $R = 3300$ (i.e. $\Delta E \simeq 2$ eV, typical for SOX1 and nearly corresponding to the theoretical limits to be attained for the microcalorimeter (e.g. Linsky 1990) and Nb-junction solid state detector) (see panel b). We see that for most lines little information is lost, though the q line becomes somewhat blended by nearby Fe XXIII satellites and the dielectronic recombination Fe XXIV k line. For $R = 1000$ (c) ($\Delta E = 7$ eV, e.g. possibly attainable for the calorimeter proposed for *AXAF*, cf. D. Schwartz, this conference) the important temperature diagnostic lines become blended and at a slightly lower resolution of 670 (d), they are only partially resolved, though we still can distinguish between ionization stages from Fe XX through Fe XXV. Simulations show that at still lower resolutions, e.g. that of the High Energy Grating (HEG) on *AXAF* (Canizares *et al.* 1987, Linsky 1990) ($R = 240$) most of the satellites are gone, and that at $R = 50$ ($\Delta E = 130$ eV), typically for charge coupled devices (CCDs) (e.g. European Photon Imaging Camera (EPIC) on *XMM* (Bignami *et al.* 1989)), no features are distinguishable. In the latter case only the centroid of the whole K-shell blend can be determined which gives a crude temperature information.

It should be remarked that though the CCD camera on *XMM* does not permit detailed spectroscopy within the Fe K blend, it is certainly capable to detect with good statistics for many relatively active late-type stellar coronae the iron line above the continuum background as I have tested with various simulations. In fact at these lower resolutions ($R \lesssim 500$) efficiency is more important than resolution. A few active late-type stars have shown the iron K-shell line complex in their spectra (see §3.2).

We consider the possibilities for electron density diagnostics in these spectra. The usual way is to derive n_e from the ratio of intensities of the forbidden and the intercombination line in the triplet of a helium-like ion (cf. §2.2). However, for high-Z ions the critical density n_{ec} above which the ratio becomes dependent on n_e is very high: $n_{ec} \sim 2\ 10^{16}$ cm^{-3} for Ca XIX and $7\ 10^{17}$ cm^{-3} for Fe XXV (Mewe *et al.* 1991), i.e. densities by far not reached in solar flares where typical densities are in the range 10^{11}–10^{13} cm^{-3}. Thus this density diagnostics is not applicable to stellar coronae in the short-wavelength K shell iron and calcium spectra. There is a limited possibility to diagnose stellar coronae for densities $\gtrsim 10^{11}$ cm^{-3} and temperatures $\lesssim 10$ MK from a number of Fe XIX–XXI satellite lines between 1.892–1.921 Å which vary with density in the range $n_e = 10^{11}$–10^{15} cm^{-3} (in particular the ratio of Fe XX lines at 1.9060 and 1.9086 Å) (Phillips *et al.* 1983, Lemen *et al.* 1986).

5. CONCLUSIONS

The helium-like triplets with their satellite lines in the K-shell spectra of highly ionized iron and calcium at short wavelengths ($\lesssim$ 3 Å) provide a very valuable diagnostics for electron temperature, emission measure, ionization balance, and (in limited cases also) electron density. However, to accomplish this goal, an extremely high spectral resolution ($\gtrsim$ 2000) will be required which with currently available technology can only be obtained with crystals and perhaps in the future with calorimeters and Nb-junctions if a resolution within a factor of two from the theoretically limit could be reached. But the wavelength region around the Fe L-shell blend (7–18 Å) and in particular the extreme ultraviolet regions at 90–140 Å (Fe $2s$–$2p$) and at 170–300 Å (Fe 3ℓ–$3\ell'$) are very promising for dispersive spectroscopy with a resolution of $\gtrsim$ 500–1000 which is attainable with present-day technology of transmission and reflection gratings and position-sensitive detectors. The application of such high-resolution X-ray spectroscopy as can be performed at the end of this century with the spectrometers onboard *AXAF* and *XMM* to observations of stellar coronae will be very promising in understanding more detailed physics of these sources. In particular line profile studies will be important in the studies of the activity of single and binary late-type stars and of the dynamics of stellar flares. Further, the possibility of detection of Doppler broadening and shifts of lines in the spectra of early-type stars can tell us whether these are indeed generated far out in the wind or alternatively in a hot corona at the base thus yielding a sound verification of the proposed models for X-ray emission.

This work has been supported by the Space Research Organization of the Netherlands (SRON).

REFERENCES

Antonucci, E., *et al.*: 1982, *Solar Phys.* **78**, 107

Ayres, T., Marstad, N., Linsky, J.: 1981, *Astrophys. J.* **247**, 545

Barr, P., *et al.*: 1988, *The High-Throughput X-Ray Spectroscopy Mission: The Mission Science Report*, ESA SP-1097

Agrawal, P.C., Markert, T.H., Riegler, G.R.: 1985, *Monthly Not. Roy. Astron. Soc.* **213**, 761

Bignami, F. *et al.*: 1989, *European (X-Ray) Photon Imaging Camera*, a Proposal submitted to ESA for the XMM Cornerstone Mission

Bitter *et al.*: 1979, *Phys. Rev. Letters* **47**, 921

Brinkman, A.C. *et al.*: 1980, *Appl. Optics* **19**, 1601

Brinkman, A.C. *et al.*: 1985, *Adv. Space Res.* **5**, Nr. 3, 65

Brinkman, A.C. *et al.*: 1987, *Astro. Lett. and Communications* **26**, 73

Brinkman *et al.*: 1989, in *SPIE EUV, X-Ray, and Gamma-Ray Instrumentation for Astronomy and Atomic Physics,* **1159**, p. 495

Canizares, C.R. *et al.*: 1987, *Astro. Lett. and Communications* **26**, 87

Cassinelli, J.P.: 1979, *Ann. Rev. Astron. Astrophys.* **17**, 275

Cassinelli, J.P., Swank, J.H.: 1983, *Astrophys. J.* **271**, 681

Castor, J.I.: 1987, in *Instabilities in Luminous Early Type Stars,* (ed. H.G.J.L.M. Lamers, C.W.H. de Loore), Reidel, Dordrecht, p. 159

Collura, A., Sciortino, S., Serio, S., Rosner, R.: 1984, in *X-Ray Astronomy '84,* (ed. M. Oda, R. Giacconi), Inst. of Space and Astronaut. Sci., Tokyo, p. 47

Culhane, J.L. *et al.*: 1981, *Astrophys. J.* **244**, L141

Doschek, G.A., Kreplin, R.W., Feldman, U.: 1979, *Astrophys. J.* **233**, L157

Doschek, G.A., Feldman, U., Cowan, R.D.: 1981, *Astrophys. J.* **245**, 315

Elwert, G.: 1952, *Z. Naturf.* **7a**, 432

Feldman, U., Doschek, G.A., Kreplin, R.W.: 1980, *Astrophys. J.* **238**, 365

Gabriel, A.H.: 1972, *Monthly Not. Roy. Astron. Soc.* **160**, 99

Gabriel, A.H., Jordan, C.: 1979, *Monthly Not. Roy. Astron. Soc.* **145**, 241

Gabriel, A.H., Jordan, C.: 1972, *Case Studies in Atomic and Collisional Physics* **2**, (ed. E.W. McDaniel, M.R.C. McDowell), N.-H. Publ. Co., Amsterdam, p. 209

Giacconi, R. *et al.*: 1979, *Astrophys. J.* **230**, 540

Holt, S.S. *et al.*: 1979, *Astrophys. J.* **234**, L65

Kastner, S.O., Neupert, W.M., Swartz, M.: 1974, *Astrophys. J.* **191**, 261

Lemen, J.R., Mewe, R., Schrijver, C.J., Fludra, A.: 1989, *Astrophys. J.* **341**, 474

Lemen *et al.*: 1986, *Astron. Astrophys.* **135**, 313

Linsky, J.L.: 1985, *Solar Phys.* **100**, 333

Linsky, J.L.: 1990, in *High Resolution X-ray Spectroscopy of Cosmic Plasmas* (eds. P. Gorenstein, M.V. Zombeck), Proc. IAU Coll. 115, Cambridge, U.S.A., Reidel Publ. Co., p. 94

Lucy, L.B.: 1982, *Astrophys. J.* **255**, 278; 286

Mason, H. *et al.*: 1984, *Solar Phys.* **92**, 199

Mason, H.: 1990, in *High Resolution X-ray Spectroscopy of Cosmic Plasmas* (ed. P. Gorenstein, M.V. Zombeck), Proc. IAU Coll. 115, Cambridge, U.S.A., Reidel Publ. Co., p. 11

Mewe, R.: 1988, in *Astrophysical and Laboratory Spectroscopy* (ed. R. Brown, J. Lang), Scottish Univ. Summer School in Phys. Publ., p. 129

Mewe, R.: 1990a, in *Physical Processes in Hot Cosmic Plasmas* (ed. W. Brinkmann, A.C. Fabian, F. Giovanelli), Kluwer Acad. Publ., p. 39

Mewe, R.: 1990b, in *Atomic Spectra and Oscillator Strengths for Astrophysics and Fusion,* (ed. J.E. Hansen), North-Holland, A'dam, p. 67

Mewe, R.: 1991, *Adv. Space Res.* , in press

Mewe, R. *et al.*: 1982, *Astrophys. J.* **260**, 233

Mewe, R., Gronenschild, E.H.B.M., van den Oord, G.H.J.: 1985, *Astron. Astrophys. Suppl. Ser.* **62**, 197

Mewe, R., Lemen, J.R., van den Oord, G.H.J.: 1986a, *Astron. Astrophys. Suppl. Ser.* **65**, 511

Mewe, R., Lemen, J.R., Schrijver, C.J.: 1991, *Astrophys. Space Sc.* , submitted

Mewe, R., Schrijver, C.J., Lemen, J.R., Bentley, R.D.: 1986b, *Adv. Space Res.* **6**, No. 8, 133

Paerels, F.B.S. *et al.*: 1986a,b *Astrophys. J.* **308**, 190; **309**, L33

Pallavicini, R.: 1989, *Astron. Astrophys. Rev.* **1**, 177

Pallavicini, R., Golub, L., Rosner, R. *et al.* :1981, *Astrophys. J.* **248**, 279

Pasquini, L., Schmitt, J.H.M.M., Pallavicini, R.: 1988, in *Activity in Cool Star Envelopes* , (ed. O. Havnes, B.R. Pettersen, J.H.M.M. Schmitt, J.E. Solheim), p.241

Phillips, K.J.H. *et al.*: 1983, *Astrophys. J.* **265**, 1120

Pradhan, A.K.: 1982, *Astrophys. J.* **263**, 477

Pradhan, A.K.: 1985, *Astrophys. J.* **288**, 824

Pradhan, A.K., Shull, J.M.: 1981, *Astrophys. J.* **249**, 821

Raymond, J.C.: 1988, in *Hot Thin Plasmas in Astrophysics* (ed. R. Pallavicini), Kluwer Acad. Publ., Dordrecht, p. 3

Raymond, J.C.: 1990, in *High Resolution X-ray Spectroscopy of Cosmic Plasmas* (ed. P. Gorenstein, M.V. Zombeck), Proc. IAU Coll. 115, Cambridge, U.S.A., Reidel Publ. Co., p. 1

Raymond, J.C., Smith, B.W.: 1977, *Astrophys. J. Suppl.* **35**, 419

Rosner, R., Golub, L., Vaiana, G.S.: 1985, *Ann. Rev. Astron. Astrophys.* **23**, 413

Rybicki, G.B.: 1987 in *Instabilities in Luminous Early Type Stars,* (ed. H.G.J.L.M. Lamers, C.W.H. de Loore), Reidel, Dordrecht, p. 175

Schmitt, J.H.M.M.: 1988 in *Hot Thin Plasmas in Astrophysics* (ed. R. Pallavicini), Kluwer Acad. Publ., Dordrecht, p. 109

Schmitt, J.H.M.M.: 1990, in *High Resolution X-ray Spectroscopy of Cosmic Plasmas* (ed. P. Gorenstein, M.V. Zombeck), Proc. IAU Coll. 115, Cambridge, U.S.A., Reidel Publ. Co., p. 110

Schrijver, C.J.: 1983, *Astron. Astrophys.* **127**, 289

Schrijver, C.J.: 1985, *Space Sci. Rev.* **40**, 3

Schrijver, C.J., Lemen, J.R., Mewe, R.: 1989, *Astrophys. J.* **341**, 484

Schrijver, C.J., Mewe, R., Walter, F.M.: 1984, *Astron. Astrophys.* **138**, 258

Serio, S.: 1985, in *Proc. ESA Workshop: Cosmic X-Ray Spectroscopy Mission,* Lyngby, Denmark, ESA SP-239, p. 59

Swank, J.H.: 1984, in *The Origin of Nonradiative Heating/Momentum in Hot Stars* , (ed. A.B. Underhill, A.G. Michalitsianos), NASA Conf. Publ. 2358

Swank, J.H., White, N.E., Holt, S.S., Becker, R.H.: 1981, *Astrophys. J.* **246**, 208

Tanaka, K.: 1986, *Publ. Astron. Soc. Japan* **38**, 225
Tanaka, K., Watanabe, T., Nishi, K., Akita, K.: 1982, *Astrophys. J.* **254**, L59
Tsuru, T. *et al.*: 1989, *Publ. Astron. Soc. Japan* **41**, 679
Vaiana, G.S. *et al.*: 1981, *Astrophys. J.* **245**, 163
Van den Oord, G.H.J., Mewe, R.: 1989, *Astron. Astrophys.* **213**, 245
Vedder, P.W., Canizares, C.R.: 1983, *Astrophys. J.* **245**, 163
Walter, F.M., Gibson, D., Basri, G.: 1983, *Astrophys. J.* **267**, 665
White, N.E. *et al.*:1986, *Astrophys. J.* **301**, 262
Weisskopf, M.C.: 1987, *Astro. Lett. and Communications* **26**, 1

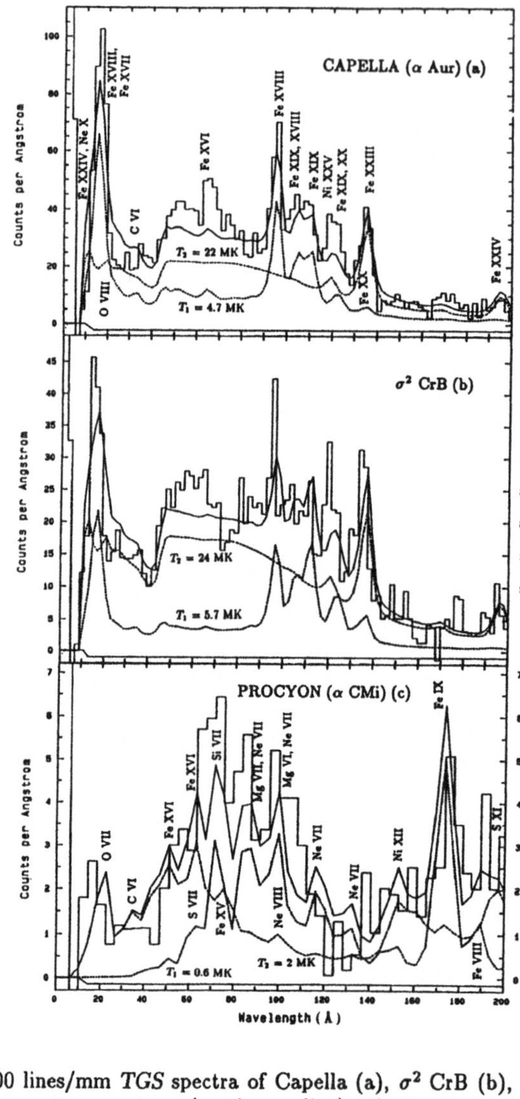

Figure 1. *EXOSAT* 500 lines/mm *TGS* spectra of Capella (a), σ^2 CrB (b), and Procyon (c), compared with the best-fit two-temperature spectrum (continuous line). The two temperature components 1 and 2 are shown separately by a dotted and a dash-dotted line, respectively. The most prominent lines are labelled with the corresponding ion species.

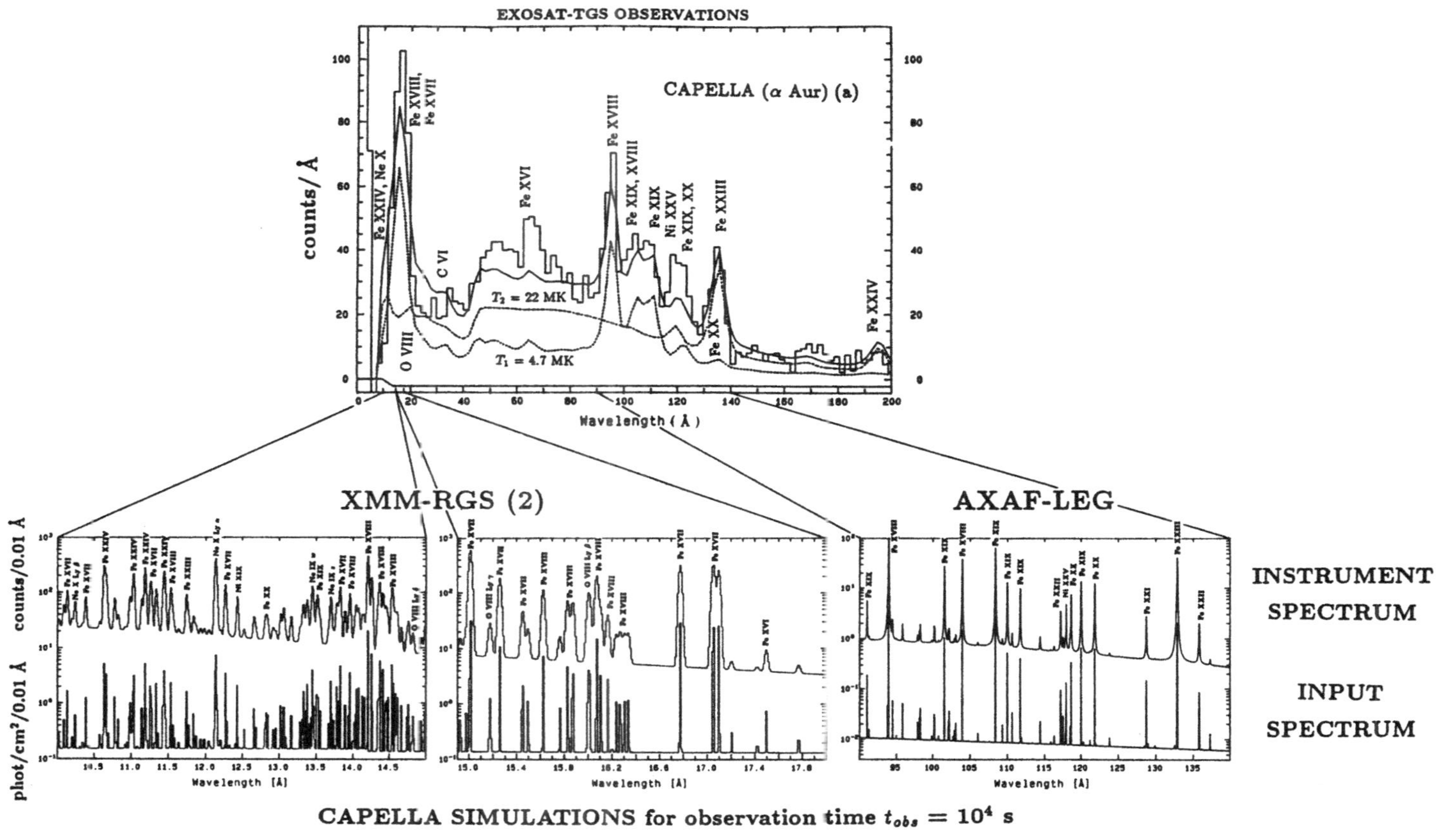

Figure 2. Comparison between *EXOSAT–TGS* observations of Capella and simulations in a blow up in the wavelength regions 10–18 Å (*XMM–RGS* (second order)) and 90–140 Å (*AXAF–LETGS*) which shows that the Fe L-shell blend (2ℓ–$\geq 3\ell'$ lines around 15 Å) and the iron $2s$–$2p$ lines around 100 Å are clearly resolved by the *XMM* and the *AXAF* instruments. The exposure time for the simulations is t_{obs}=10^4 s and the source parameters as derived from *EXOSAT–TGS* observations are for a two-temperature model: temperatures T_1=5 MK, T_2=25 MK, reduced emission measures ε_1/d^2=3 10^{50} cm^{-3} pc^{-2}, ε_2/d^2=9 10^{50} cm^{-3} pc^{-2} (d is source distance); line-of-sight hydrogen column density N_H=5 10^{18} cm^{-2}.

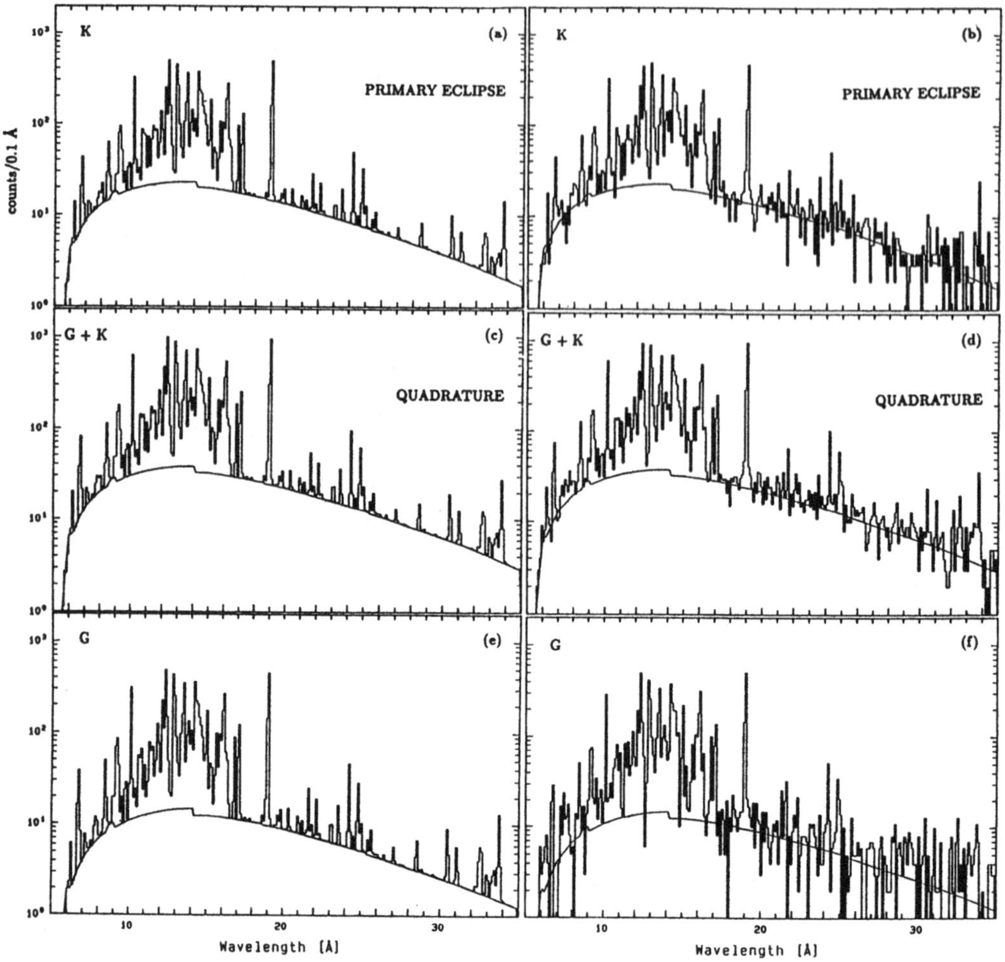

Figure 3. Simulated *XMM-RGS* first-order spectra of the eclipsing binary system AR Lacertae consisting of a G2 IV and a K0 IV star for an exposure time of 6000 s each. Left panels (a,c,e) show the spectra during totality of the primary eclipse when the K star is only observed, during quadrature (43000 s later) when both stars are observed, and the derived difference spectrum (of the G star). The source parameters are: T_{K1}=8 MK, T_{K2}=30 MK, ε_{K1}/d^2=3 10^{50} cm^{-3} pc^{-2}, ε_{K2}/d^2=1.2 10^{50} cm^{-3} pc^{-2}, T_G=8 MK, ε_G/d^2=3 10^{50} cm^{-3} pc^{-2}; N_H=10^{19} cm^{-2}). Right panels (b,d,f) show the corresponding spectra with random noise added.

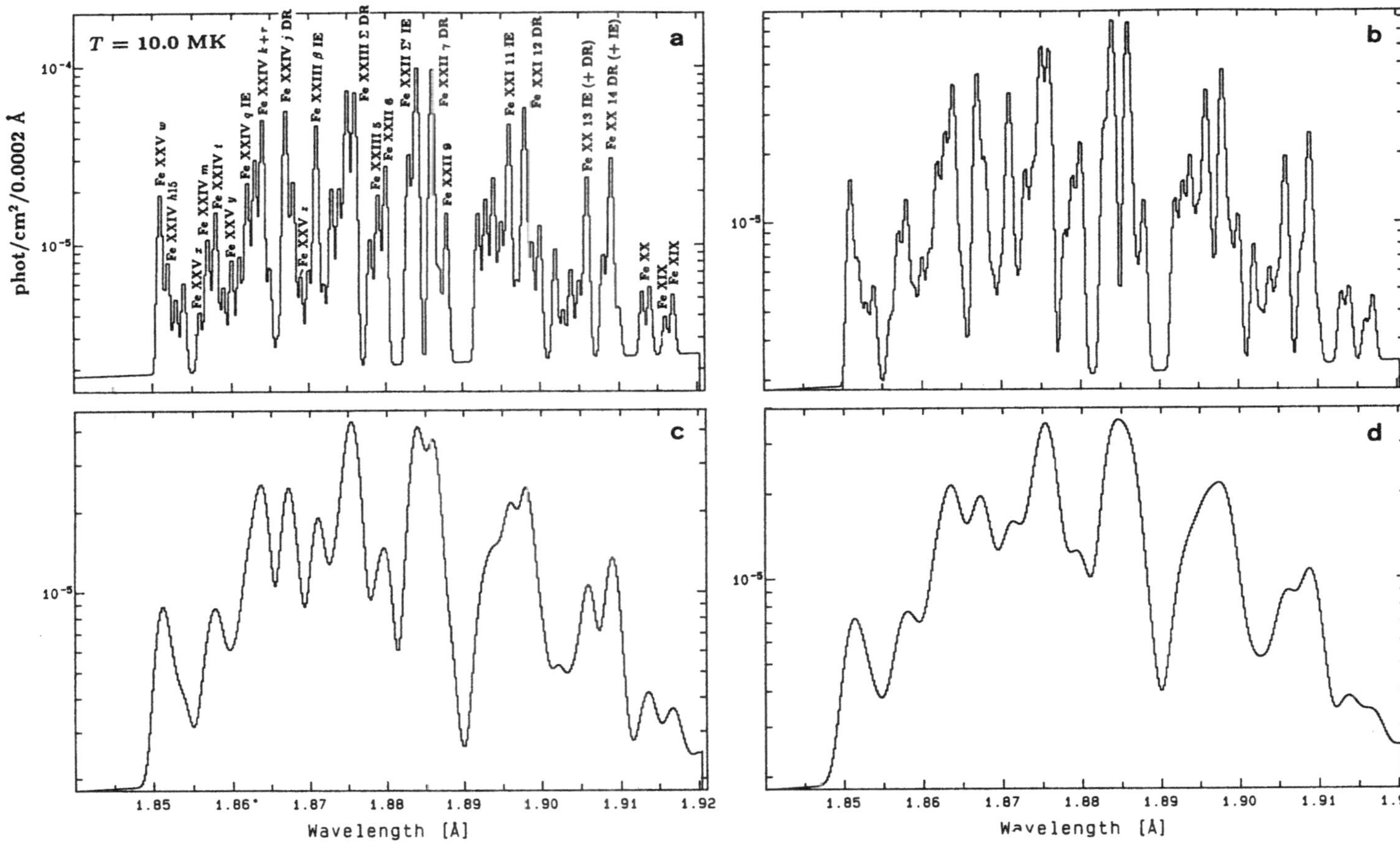

Figure 4. Simulated Fe K-shell spectra from an optically thin coronal plasma at a temperature $T = 10$ MK ($\epsilon/d^2 = 10^{50}$ cm^{-3} pc^{-2}, $t_{obs} = 10^4$ s) for several resolving powers R ($\equiv \lambda/\Delta\lambda$): 10000 (panel a); 3300 (b); 1000 (c); 670 (d), corresponding approximately to those obtained in: *P78*-SOLFLEX, *SMM*-BCS, *Hinotori*-SOX2 (a); SOX1 and the theoretical limit for calorimeter and Nb-junction (b); possible performance of *AXAF* calorimeter (c); current performance of calorimeters (d). For notations of line identifications see text.

LOW-RESOLUTION SPECTRA OF STELLAR CORONAE AND FLARES WITH EXOSAT: THE IRON K LINE

R. Pallavicini[1] and G. Tagliaferri[2]

[1]Osservatorio Astrofisico di Arcetri, Florence, Italy
[2]EXOSAT Observatory, ESTEC, Noordwijk, The Netherlands

1. Introduction

The Medium Energy (ME) experiment on EXOSAT obtained spectra of stellar coronal sources over the range $\approx$ 1-10 *keV* with a resolution of 21% at 6 *keV*. This resolution does not allow individual lines to be resolved; however, by fitting the observed spectra with appropriate theoretical models, it is possible to derive physical parameters such as electron temperature and emission measure. In a few cases, the sensitivity was high enough to allow detection of the Fe K-line complex at 6.7 *keV*.

The EXOSAT ME sample of stellar coronal sources comprises:
a) spectra of the quiescent emission of RS CVn binaries and Algol;
b) time-resolved spectra of flares on M dwarfs and RS CVn binaries;
c) spectra of quiescent and/or flaring emission from a few other types of coronal sources.

At present we are reanalysing these spectral data in a consistent manner. Here, we report briefly on the results of spectral fits by focussing on the detection of the Fe K-line complex and on the comparison with recent results from *Ginga*.

2. Results of spectral fits

2.1. Quiescent emission from RS CVn binaries

ME spectra of RS CVn binaries during "quiescence" are well fitted by single temperature line+continuum thermal models such as those computed by Raymond and Smith (1977) and Mewe et al. (1985). The derived temperatures are in the range $\approx$ 20 - 30 $\times 10^6 K$ and the emission measures are EM $\approx 10^{53} - 10^{54} cm^{-3}$.

If (0.05-2) *keV* Low Energy fluxes are used in conjunction with the ME data, a second low-temperature component (at T $\approx$ 4-8 $\times 10^6 K$) is often (but not always) required. This low-temperature component is poorly constrained since fitting of LE data depends critically on the usually unknown value of the interstellar absorption N_H. A two-temperature model for RS CVn systems is strongly supported by *Einstein* SSS spectra (Swank et al. 1981), by EXOSAT higher resolution TGS spectra (Lemen et al. 1989, Mewe 1990), and by EXOSAT observations of eclipsing systems (White et al. 1990, Culhane et al. 1990).

2.2. Flares on M dwarfs and RS CVn binaries

Time-resolved ME spectra were obtained for several flares on dMe stars, on RS CVn binaries and on Algol. The derived peak temperatures are typically $\approx 30 - 40 \times 10^6 K$ reaching values as high as $\approx 60 \times 10^6 K$ in a flare on Algol (White et al. 1986). The temperature decreases during the flare decay indicating cooling of the plasma by radiation and/or conduction. Strong flares show the presence of the Fe K-line complex, as we will discuss more extensively in the next section.

3. The Iron K-line

3.1. Ginga observations

Before discussing the EXOSAT data, it is appropriate to mention some results obtained recently by the LAC experiment on the Japanese satellite *Ginga*. The LAC has detected the Fe K-line complex in a few stellar coronal sources, including two flares on UX Ari (Tsuru et al. 1989) and a large flare on Algol (Stern et al. 1990).

An interesting aspect of the *Ginga* observations is that the equivalent width (EW) of the Fe K-line has been found to be significantly smaller (by a factor 2 to 10) than predicted from an optically-thin solar-abundance thermal plasma in ionization equilibrium. Moreover, large variations of the Fe-line EW were observed from flare to flare and even in the course of the same event. Previous observations of HR 1099 by *Temna* also indicated a Fe line weaker than expected (< 0.5 times solar, Koyama 1985 quoted in Tsuru et al. 1989). Doyle et al. (1990) report a Fe K-line 30% weaker than solar in a flare on II Peg observed with *Ginga*.

The reasons for the anomalously low Fe-line EW observed by *Ginga* are not understood. Clearly, this cannot be simply an abundance effect. Tsuru et al. (1989) suggested resonance scattering as a possible explanation. This mechanism may be relevant for sufficiently long and/or dense structures which can reach optical depth ≥ 1 at line center (Acton 1978).

3.2. EXOSAT observations

A quick inspection of the EXOSAT data suggested the presence of the Fe-line K complex in a number of spectra of stellar coronal sources (EQ Peg, Algol, HR 1099 and possibly UX Ari). In order to check whether the detection was significant, and to determine the equivalent width of the line, we fitted the ME spectra with both a simple bremsstrahlung spectrum and with bremsstrahlung plus a gaussian line centered at 6.7 keV. For comparison we also fitted the spectra with the line+continuum thermal model of Mewe et al. (1985). Since all sources are at distances less than 50 pc, the interstellar absorption N_H was fixed at zero in all cases.

The Fe K-line was detected significantly in flares observed from the dMe flare star EQ Peg (on Aug 6, 1985), from the RS CVn star HR 1099 (on Oct 17, 1983) and from Algol (on Aug 19, 1983). In all these cases, a bremsstrahlung + Fe line spectrum gave a significantly better fit than bremsstrahlung alone (the reduced χ^2 typically changed from ≥ 2 to ≤ 1). Acceptable fits were also obtained with the line+continuum thermal model of Mewe et al. (1985) using solar abundances.

In the case of quiescent emission from UX Ari (observed on 25-26 Sept 1985), we found no significant difference when fitting the spectrum with a simple bremsstrahlung model and with the line+continuum model of Mewe et al. (the χ^2 changed from 19 to 17 for 25 d.o.f. in the two cases, respectively). This indicates that the detection of the Fe-line complex was not significant in this case, in contrast to what had previously been reported (Pasquini et al. 1988). Observations of HR 1099 on 2-4 February 1986 indicate the presence of the Fe-line also during quiescence, but the statistics is too low to allow a reliable estimate of the EW.

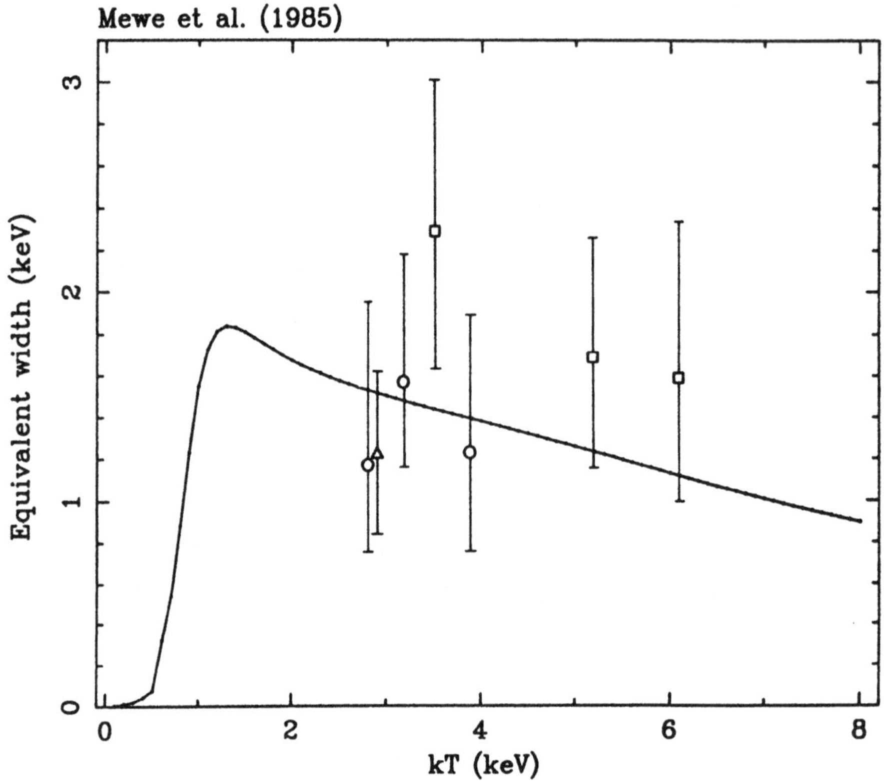

Figure 1. Observed and predicted Fe-line equivalent widths for stellar sources observed by EXOSAT. Circles are for EQ Peg, squares for Algol, and the triangle for HR 1099.

4. Discussion

Our results are summarised in Figure 1 where we plot the derived Fe-line EW's *vs.* the best-fit continuum temperature. The errors bars are 90% confidence levels. For comparison we also plot the expected Fe-line EW for a thermal plasma in ionization equilibrium and solar abundances, computed according to Mewe et al. (1985).

As shown in Figure 1, the Fe-line equivalent widths derived from the EXOSAT data are consistent within the errors with the predictions for a solar abundance thermal plasma. This is in agreement with the acceptable fits we have obtained by fitting the EXOSAT data with the Mewe et al. (1985) model and solar abundances.

The results obtained from EXOSAT contrast with those obtained by *Ginga*. For the latter, EW's smaller than expected are the rule, rather than the exception; only in a few cases (e.g., in the late decay of the flare from Algol, cf. Stern et al. 1990), *Ginga* found EW's consistent with expectations. The conflicting results obtained by EXOSAT and *Ginga* support the conclusion that the anomalously low Fe-line EW's occasionally observed for stellar coronal sources are not due to lower-than-solar Fe abundance. More likely, the low EW's are due to line opacity effects within the source itself and/or to inaccurate positionining of the continuum, for instance because of the presence of a non-thermal component.

If line opacity is produced by resonance scattering (as suggested by Tsuru et al. 1989), an optical depth $\tau \geq 1$ in the Fe XXV resonance line is reached for $nL \geq 3 \times 10^{21} cm^{-2}$. For densities $n \approx 3 \times 10^{10} cm^{-3}$ (as typically inferred from flare decay times), $\tau \geq 1$ occurs for a path length $L \geq 10^{11} cm$. This is comparable to a stellar radius in a typical RS CVn star. If so, the Fe K-line, when observed with high sensitivity and resolution, could become a useful diagnostic of the geometry and size of stellar flaring structures.

References

Acton, L.W.: 1978, *Astrophys. J.* **225**, 1069.

Culhane, J.L., White, N.E., Shafer, R.A., and Parmar, A.N.: 1990, *Montly Not. Roy. Astron. Soc.* **243**, 424.

Doyle, J.G., Kellett, B.J., Byrne, P.B., Avgoloupis, S., Mavridis, L.N., Seiradakis, J.H., Bromage, G.E., Tsuru, T., Makishima, K., and McHardy, I.M.: 1990, *Montly Not. Roy. Astron. Soc.*, in press.

Lemen, J.R., Mewe, R., Schrijver, C.J., and Fludra, A.: 1989, *Astrophys. J.* **342**, 474.

Mewe, R.: 1990, these proceedings.

Mewe, R., Gronenschild, E.H.B.M., and van den Oord, G.H.J.: 1985, *Astron. Astrophys. Suppl.* **62**, 197.

Pasquini, L., Schmitt, J.H.M.M., and Pallavicini, R.: 1988, in *Activity in Cool Star Envelopes* (O. Havnes et al. eds.), p. 241.

Raymond, J.C., and Smith, B.: 1977, *Astrophys. J. Suppl.* **35**, 419.

Stern, R.A., Haisch, B.M., Nagase, F., Uchida, Y., and Tsuneta, S.: 1990, in *Cool Stars, Stellar Systems and the Sun* (G. Wallerstein ed.), p. 224.

Swank, J.H., White, N.E., Holt, S.S., and Becker, R.H.: 1981, *Astrophys. J.* **246**, 208.

Tsuru, T., Makishima, K., Ohashi, T., Inoue, H., Koyama, K., Turner, M.J.L., Barstow, M.A., McHardy, I.M., Pye, J.P., Tsunemi, H., Kitamoto, S., Taylor, A.R., and Nelson, R.F.: 1989, *Publ. Astron. Soc. Japan* **41**, 679.

White, N.E., Culhane, J.L., Parmar, A.N., Kellett, B.J., Kahn, S., van den Oord, G.H.J., and Kuijpers, A.N.: 1986, *Astrophys. J.* **301**, 262.

White, N.E., Shafer, R.A., Horne, K., and Culhane, J.L.: 1990, *Astrophys. J.* **350**, 776.

GINGA Observations of AB Doradus

O. Vilhu [1], T. Tsuru [2], A. Collier Cameron [3]

[1]Observatory and Astrophysics Laboratory, University of Helsinki, Tahtitorninmaki, SF-00130 Helsinki, Finland

[2]Department of Physics, University of Tokyo, 7-3-1 Hongo, Bunkyo-ku, Tokyo 113, Japan

[3]Astronomy Centre, University of Sussex, Falmer, Brighton BN1 9QH, UK

Abstract: We report *GINGA* observations of the pre main sequence star AB Doradus (HD 36705), performed during 8 - 12 January, 1990. Some rotational modulation might be present. Four X-ray flares were detected. Three of these events were similar to the *EINSTEIN* HRI-flare (Vilhu and Linsky, 1987), with decay times around 25 min. The last flare had long rise and decay times (100 min), resembling the *EXOSAT* flares observed by Collier Cameron et.al. (1988). The mean flare spectrum can be fitted by a thermal bremstrahlung with temperature 5.0 keV, or by a power-law model with photon index 2.2. The 3σ upper limit of the Iron line equivalent width in the flare spectrum is 1 keV, weaker than predicted by thermal models. This Iron line anomaly was first discussed in the case of UX Ari by Tsuru et. al. (1989). However, normal equivalent widths can be derived from several *EXOSAT* spectra of active cool stars (Pallavicini and Tagliaferri, 1990). We discuss the possibility that the continuum from non-thermal electrons (producing also the microwave emission) could occasionally lower the apparent equivalent width. The mechanism works for reasonably low magnetic field strengths and electon power-law indexes. However, a large population of non-thermal electrons is needed (comparable to the thermal one). Stronger magnetic fields could explain the radio emission with less electrons, but then the non-thermal X-ray continuum remains small.

1 Introduction

AB Doradus (HD 36705) is a rapidly rotating young star in the Southern Hemisphere, showing a variety of interesting phenomena related to its magnetic activity (K1 IV, P = 0.514 day). In particular, strong X-ray flares were observed by the *EINSTEIN* HRI (see Vilhu and Linsky, 1987) and the *EXOSAT* LE and ME detectors (Collier Cameron et.al., 1988). The very long rise times (100 min) of the *EXOSAT* flares imply a long period of heating and flare loops that extend several radii above the star (see Linsky, 1990, including a good summary of all detected stellar X-ray flares). We planned coordinated *GINGA* and ground-based observations to study the flare statistics and a possible connection between flares and transient Hα clouds.

2 Rotational Modulation and the Flares

AB Doradus was observed with the *GINGA* LAC between 8 - 12 January, 1990. Here we report observations lasting 5 rotations of the star = 2.6 days (90/1/09 12:30 UT - 90/1/12 03:30 UT). The folded light curve (Fig.1) shows some marginal rotational modulation. However, when the two close-by supernova remnants in the LAC's field of view will be subtracted, the two maxima around phases 0.2 and 0.8 might become more evident. It is interesting to note that in simultaneous optical data minima were found at phases 0.1 and 0.8 (Banks et.al.,1990), and at 8.4 GHz maxima around somewhat earlier phases (Budding et.al.,1990).

Fig.2 gives the light curves, showing four strong flares. Using the conversion factor 3.2 10^{-11} erg cm^{-2} /count (channels 3 - 16, 1.73 - 9.78 keV), flare peak luminosities (2 - 3) 10^{30} erg/s can be obtained. Fig.3 shows the mean spectrum of the three first flares, after the quiescent spectrum was subtracted (to eliminate the contribution of the SN remnants in the field).

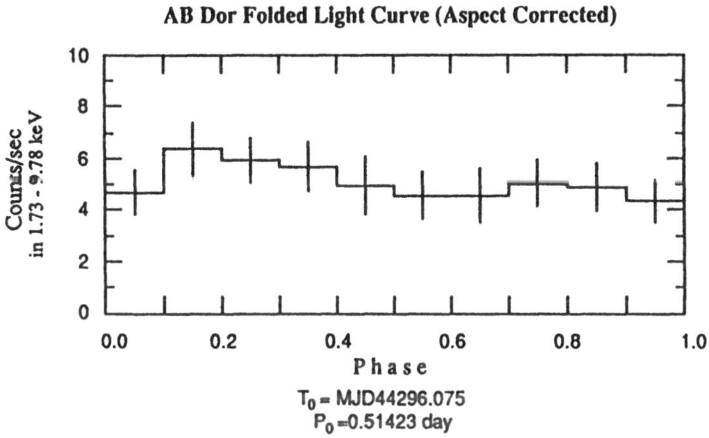

Fig. 1. The X-ray light curve folded by the spin period of 0.514 days (1.73 - 9.78 keV). The background has been subtracted, except the contribution from two supernova remnants (N49 and N49B) in the LAC's field of view (about 3 cts/sec).

3 The Possible Contribution of a Non-thermal Continuum to the Iron Line Anomaly

In solar flares and cool stars the microwave emission is ubiquitous. This is usually considered to be due to mildly relativistic electrons (with energies above 100 keV), spiralling around magnetic field lines (gyrosynchrotron radiation; see Dulk, 1985; Gary, 1985; Li and Emslie, 1990). We followed these studies and computed the non-thermal continuum (bremstrahlung) from these electrons, and compared it with the thermal component at the Iron line (6.7 keV). Both thin and thick target cases were considered with thermal electron density $n_e = 10^{11}$ cm^{-3}.

The stellar input parameters are the thermal plasma temperature T (keV), the emission measure EM_{54} (10^{54} cm^{-3}) and the microwave emission at 5 GHz (S, mJy). Data of three stars were considered: AB Dor (Collier Cameron et.al., 1988; Slee, 1990; this study), UX Ari (Tsuru et.al. 1989) and VW Cep (Vilhu, Caillault and Heise, 1988) with typical observed X-ray and micro-wave parameters for both the quiescent and flaring levels. Only for VW Cep the X-ray and micro-wave data were simultaneous.

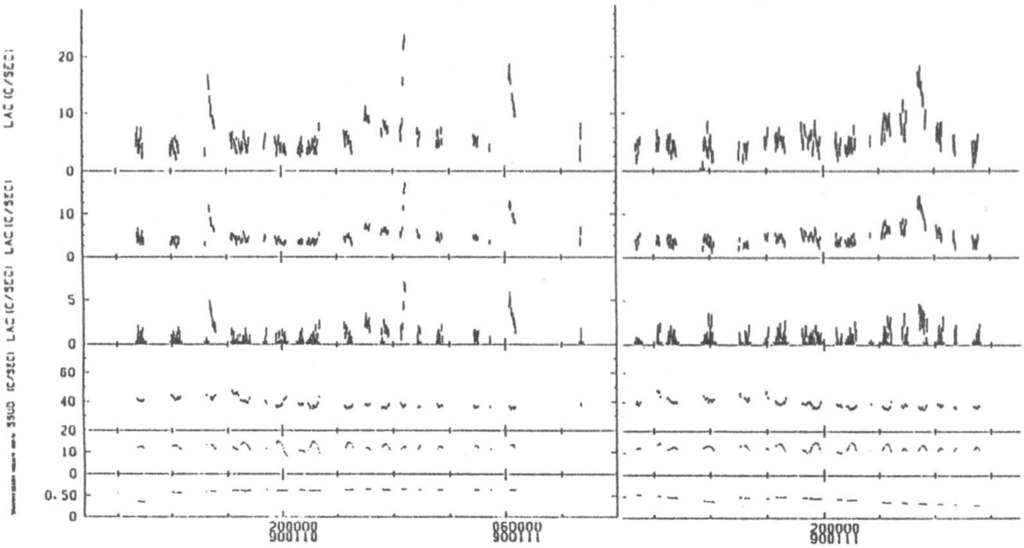

Fig. 2. The X-ray light curves of AB Doradus on 10 - 12 January, 1990. The four flares were observed at 90/01/10 16:42 UT, 90/01/11 01:32 UT, 90/01/11 06:11 UT and 90/01/12 00:00 UT (peaks). The panels from the top to the bottom are : a.) 1.73 - 9.78 keV , b.) 1.73 - 4.60 keV, c.) 4.60 - 9.78 keV , d.) and e.) background , f.) transmission.

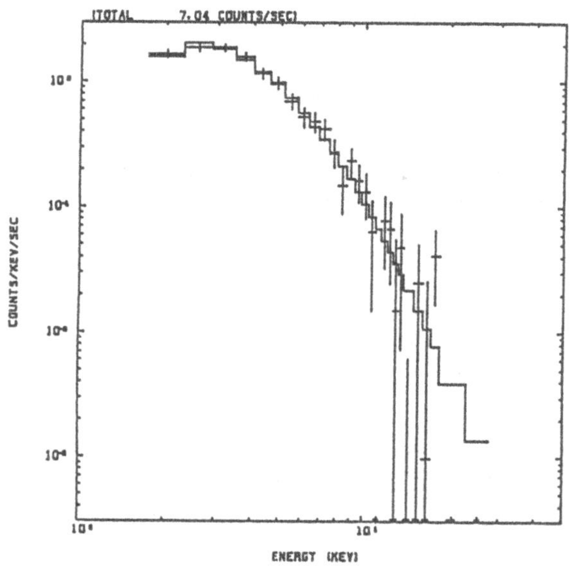

Fig. 3. The joint spectrum of the first three flares, subtracted by the average quiescent spectrum. A power-law fit, with the photon index $\gamma = 2.2$, is shown (reduced $\chi^2 = 1.12$). The 3σ upper limit of the Iron line equivalent width is 1 keV.

In Fig.4 the following data-sets of (T, EM_{54}, S) were used: AB Dor(quiesc) (1.6, 0.15, 5) ; AB Dor(flaring): (6.0, 0.15, 60) ; UX Ari(quiesc): (4.6, 1.9, 50) ; UX Ari(flaring) (6.5, 6.0, 200) ; VW Cep(quiesc): (1.7, 0.08, 0.3) ; VW Cep(flaring): (2.6, 0.3, 8). Distances 20 pc, 50 pc and 31 pc were used for AB Dor, UX Ari and VW Cep, respectively.

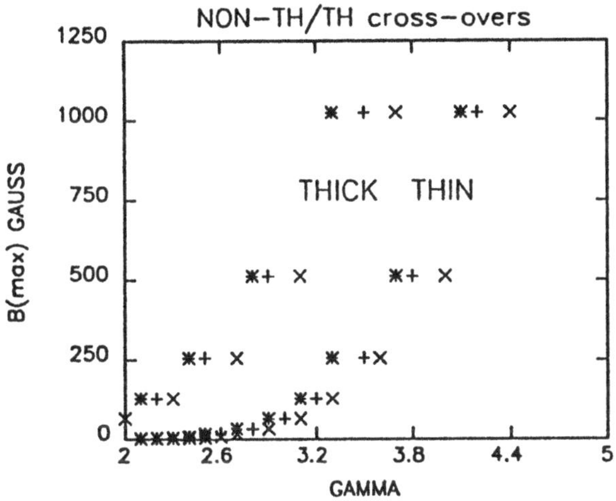

Fig. 4. The maximum magnetic field strength which guarantees that the non-thermal continuum (bremstrahlung from non-thermal electrons giving rise to the gyrosynchrotron radio emission) is equal or larger than the thermal one at 6.7 keV. The non-thermal photon index γ is related to the electron power-law index δ by $\delta = \gamma - 0.5$ (thin target) and $\delta = \gamma + 1$ (thick target). (*) = AB Dor. (+) = UX Ari. (x) = VW Cep.

4 Discussion

Although the resonance absorption (Tsuru et.al., 1989) or abundance effects (like in galaxy clusters; Rothenflug and Arnaud, 1985) might give a sufficient explanation for the Iron line anomaly, we considered the hypothetical effect of radio-electrons.

For fixed B and γ the non-thermal contribution at 6.7 keV is quite similar for all stars, during both the quiescent and flaring states (see Fig.4). However, this result is sensitive to the assumed magnetic field and micro-wave emission. These factors might also be very variable. Further, X-ray flares are not necessarily visible in micro-waves. For instance, Slee (1990) did not find any significant increase at 8.4 GHz during the flare No. 1 in Fig. 2.

Although the maximum magnetic field values in Fig.4 seem reasonable for stellar coronae (see Linsky, 1990), the number of accelerated electrons must be large. Along the cross-over values of Fig.4, the number of non-thermal electrons (above 100 keV) is 1 percent of the total number. If the power-law spectrum continues down to 10 keV, the total non-thermal population should be comparable to the thermal one. Then the coronal plasma can no longer be considered as a pure thermal one, but rather an equal mixture of thermal and non-thermal components.

References

Banks, T., Kilmartin, P.M., and Budding, E., 1990 (preprint).

Budding, E., Burgess, S., Chan, S., and Slee, O.B., 1990, preprint.

Collier Cameron, A., Bedford, D.K., Rucinski, S.M., Vilhu, O., and White, N.E., 1988, *M.N.R.A.S.*, **231**, 131.

Dulk, G.A., 1985, *Ann. Rev. Astr. Astrophys.*, **23**, 169.

Gary, D.E., 1985, *Astrophys. J.*, **297**, 799.

Li, P., and Emslie, G., 1990, *Solar. Phys.*, **129**, 113.

Linsky, J.L., 1990, 'X-rays from Stellar Flares', Colorado Astrophysics Preprint No.69, to appear in *Memoirs of the Italian Astronomical Society*.

Pallavicini, R., and Tagliaferri, G., 1990, these proceedings.

Rothenflug, R., and Arnaud, M., 1985, *Astron. Astrophys.*, **144**, 431.

Slee, B., 1990, private comm.

Tsuru, T., Makishima, K., Ohashi, T., Inoue, H., Koyama, K., Turner, M.J.L., Barstow, M.A., McHardy, I.M., Pye, J.P., Tsunemi, H., Kitamoto, S., Taylor, A.R., and Nelson, R.F., 1989, *Publ. Astron. Soc. Japan*, **41**, 679.

Vilhu, O., Caillault, J.P. and Heise, J., 1988, *Astrophys.J*, **330**, 922.

Vilhu, O. and Linsky, J.L., 1987, *Publ. Astr. Soc. Pac.*, **99**, 1071.

LINE SPECTROSCOPY IN SUPERNOVA REMNANTS

J.S. Kaastra
Laboratory for Space Research, P.O.Box 9504, 2300 RA Leiden, Netherlands
J.A.M. Bleeker
Laboratory for Space Research, Sorbonnelaan 2, 3584 CA Utrecht, Netherlands

1. Introduction

X-ray observations provide us with a wealth of information on the dynamics and evolution of supernova remnants and many related problems. First, we learn about the physical conditions at the shock site: the relation between thermal and nonthermal components, heating and acceleration processes and the role of the magnetic field, in particular when the X-ray observations are compared to radio data. Secondly, we gain insight into stellar evolution: about nuclear synthesis from analysis of chemical abundances in the stellar ejecta, about the explosion energy, progenitor mass and properties of the stellar wind by study of the temperature and mass distribution in the shocked regions. Moreover important information about the physics of shock propagation and heating is also contained in the temperature, density and ionisation distribution. Finally, the global structure of the interstellar medium can be studied, in particular its thermodynamics and (in)homogeneity.

Before the launch of ROSAT, some 145 supernova remnants were known in our galaxy, about 40 of which have been detected in X-rays. Another 38 are known in the Magellanic clouds. These remnants can be classified (except for a few peculiar objects like SS433 / W50) in 2 classes: Crab-like remnants and shell-like remnants. Crab-like remnants have a filled centre morphology, and usually contain a compact central energy source (e.g. a pulsar). The spectrum is a featureless synchrotron continuum. Since these objects are not of interest for spectral line studies we shall not discuss them here but focus on the shell-like remnants.

The X-ray images of these remnants reveal most of the emission close behind the shock (highest density), causing considerable limb-brightening. The X-ray emission process is optically thin thermal radiation from a plasma with typical temperatures of 10^6 – 10^8 K and densities of 0.1 – 10 cm^{-3}. The rather good correlation with radio images demonstrates that heating and acceleration processes in supernova remnants are closely related. The correlation with the optical filaments is in general much more complex.

2. The temperature

The passage of the supernova shockfront suddenly raises the ion temperature in the cool interstellar medium. Subsequently the electrons are heated due to Coulomb collisions. The equilibration time for this process is in general of the order of 10^4 year, implying that for most of the remnants $T_e \ll T_i$. There is some evidence against this, however. Processes like charge coupling, plasma turbulence, heat conduction and viscosity may lead to a much faster equilibration (see e.g. Mihalas and Mihalas 1984). This is evident by the X-ray observations showing the presence of a bremsstrahlung

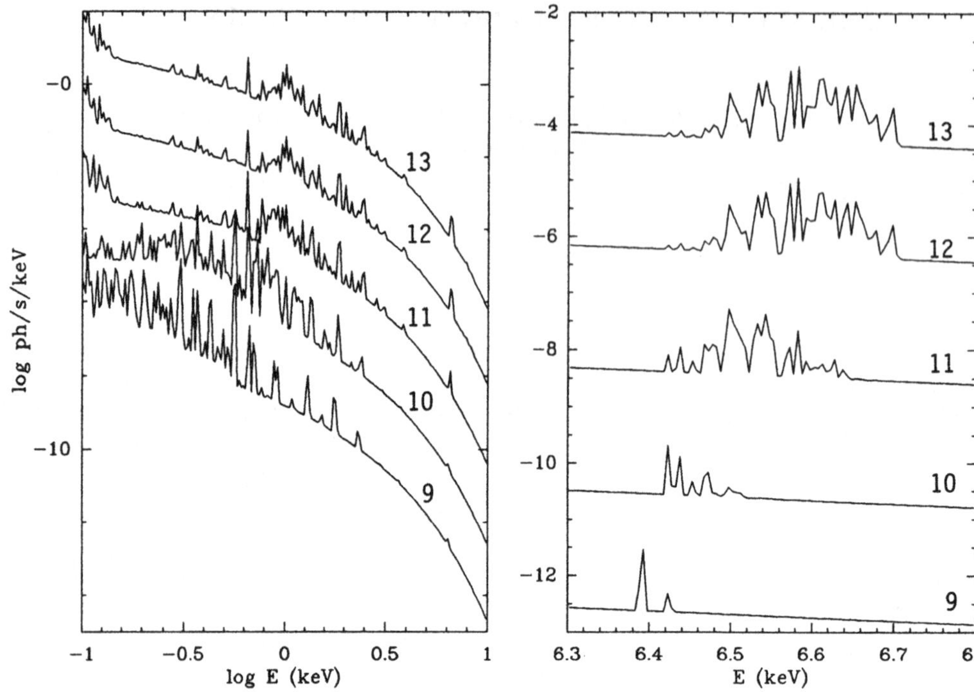

Fig. 1. The spectrum of a plasma with Hydrogen density 1 cm^{-3} which has been heated instantaneously to 10^7 K at $t = 0$, for values of t (s) (bottom to top) of 9, 10, 11, 12 and 13. The normalisation is arbitrary (steps of a factor 100). The spectrum is shown in a broad band from 0.1 to 10 keV (left panel) and around the Fe K-complex (right panel).

component with T_e of the order of a few keV, indicating that at least $T_e = O(T_i)$. In order to determine the real ratio T_e/T_i, one needs to measure both temperatures. T_e is determined from the bremsstrahlung continuum, and T_i can be measured by the thermal Doppler broadening of spectral lines, the typical thermal line width being:

$$\Delta E/E = \sqrt{T_i/(10^8\ \mathrm{K})}/300$$

which is detectable by future spectroscopic measurements on missions like AXAF or XMM, especially for the hotter remnants and spectral lines from the lighter elements (like O).

3. Non-equilibrium ionisation

The ionisation equilibrium time in many remnants is of the order of 10^4 year. This implies that most remnants observed in X-rays are not yet in equilibrium. Line emission constitutes the basic diagnostic mechanism to study the non-equilibrium ionisation (NEI). The main parameter used in general to characterise this non-equilibrium is the time integral of the electron density $\int n_e dt$, often denoted by $n_e t$ (ionisation parameter).

There are 2 main detectable effects when a plasma is ionising: first, as ionisation progresses, elements with a higher atomic number become ionised, resulting in a global shift of line complexes from low to high energy. The other effect is a shift in the effective

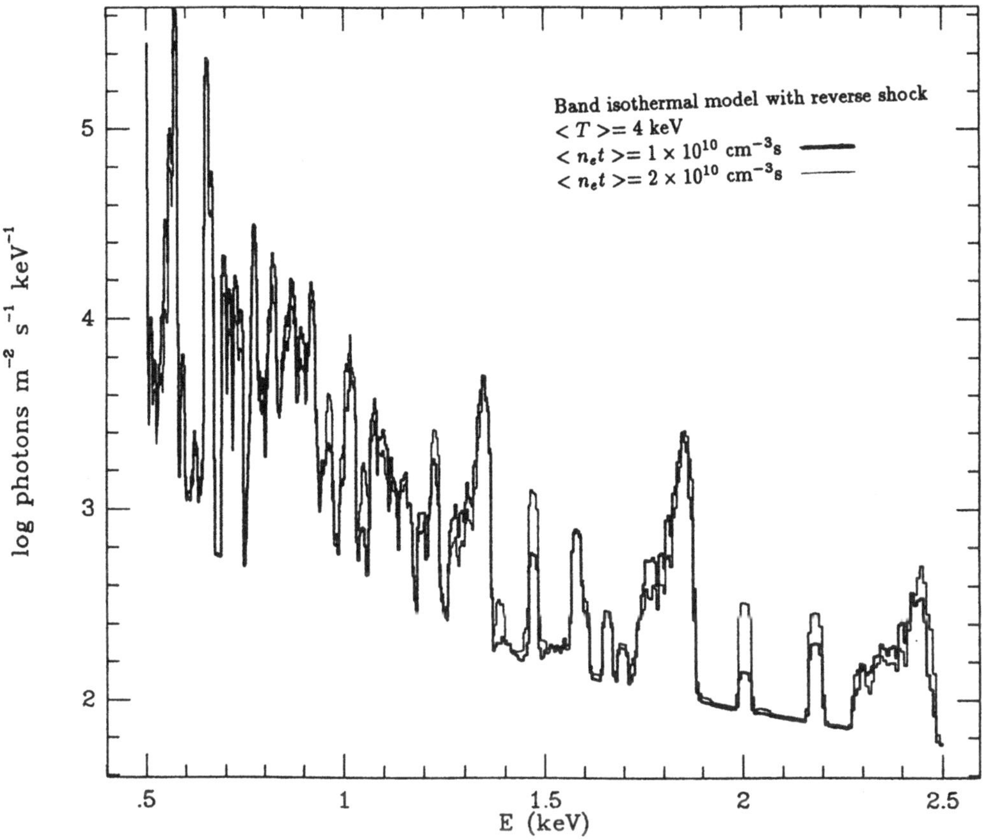

Fig. 2. The role of the ionisation parameter in the supernova remnant spectrum. The energy range (0.5–2.5 keV) and resolution ($E/\Delta E = 300$) are the typical values proposed for the XMM Reflection Grating Spectrometer.

line centroid for each element individually due to the increasing ionisation stage (e.g. from Fe I to Fe XXVI). Both are illustrated in figure 1: most of the line emission shifts from $\sim$ 0.5 keV to 7 keV when $n_e t$ increases from 10^9 to 10^{13} cm^{-3}s, and the line centroid and flux of e.g. the Fe-K complex change significantly.

Employing this energy shift makes it already possible to determine $n_e t$ with moderate resolution spectrometers, like the Gas Scintillation Spectrometers on TENMA and EXOSAT. For example Tsunemi et al. (1986) were able to derive $n_e t$ values for Cas A (10^{11} cm^{-3}s) and Tycho ($10^{9.5}$ cm^{-3}s) using the line centroids of the Si, S, Ar, Ca and Fe complexes. Typical energy differences of 0.1–0.2 keV compared to collisional equilibrium occur. With EXOSAT similar results were obtained by Jansen et al. (1988) for Cas A and Smith et al. (1988) for Tycho.

With higher spectral resolution it is possible to determine the ionisation parameter more accurately, and also to determine the elemental abundances with confidence. Figure 2 illustrates this clearly. Changing the ionisation parameter only by a factor of 2, the line strength of several strong spectral lines changes significantly. For each range of the ionisation parameter, there are always a few lines which are very sensitive to $n_e t$. For the case shown in figure 2 pronounced differences occur for Mg, Al, Si and S.

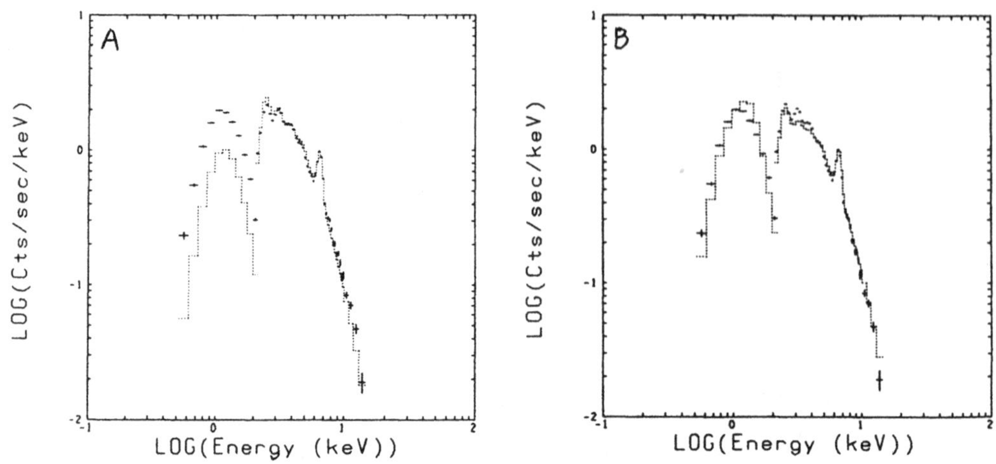

Fig. 3. The spectrum of Cas A as measured by the PSD and GSPC of EXOSAT. a: best fitting Sedov Spectrum (adiabatic, including NEI-effects without a reverse shock); b: Chevalier model (adiabatic, with a reverse shock; spectral fitting was not yet optimised). Calculations by Jansen (1988).

Observations with high resolution made with the Einstein Focal Plane Crystal of O VII and O VIII lines also showed the presence of strong NEI effects.

4. Reverse shock emission

Due to deceleration of ejected material by the ambient medium, a reverse shock propagating inward (with respect to the main shock front) is generated (McKee, 1974). Chevalier (1982) modelled this situation and predicted that the shocked ejecta have a considerably lower temperature than the shocked interstellar medium. This should show up as a dual temperature structure. However, NEI effects may mimic a low temperature component when the remnant is observed with low resolution spectrometers. It is therefore difficult to disentangle both effects. Jansen et al. (1988) clearly demonstrated the presence of a reverse shock in Cas-A using the EXOSAT imaging proportional counter (PSD) which possessed both moderate spectral and spatial resolution. This is also illustrated in fig. 3, where it is seen that the best fitting Sedov model at high energies completely fails to explain the observed low energy excess, while the model including a reverse shock explains all major features. Fig. 4 compares the contributions of the shocked ejecta and the shocked interstellar medium to the total X-ray spectrum, assuming solar abundances. It is evident from this figure, that especially the Fe K complex around 6.5 keV yields a good estimate of the interstellar Fe abundance, while the abundance in the ejecta may be estimated from the many lines in the Fe L complex around 1 keV. The line centroid of the Fe-K complex is higher in the shocked ejecta than in the shocked interstellar medium despite of the lower temperature of the ejecta. This is solely due to a higher ionisation parameter (see fig. 6, curve 3). Due to Doppler broadening the Fe-K complex appears rather featureless, despite the high spectral resolution.

5. Spatial inhomogeneities

Although spherically symmetric models give a zero-order approximation to a supernova remnant, they of course fail to explain the spatial inhomogeneities. As an example

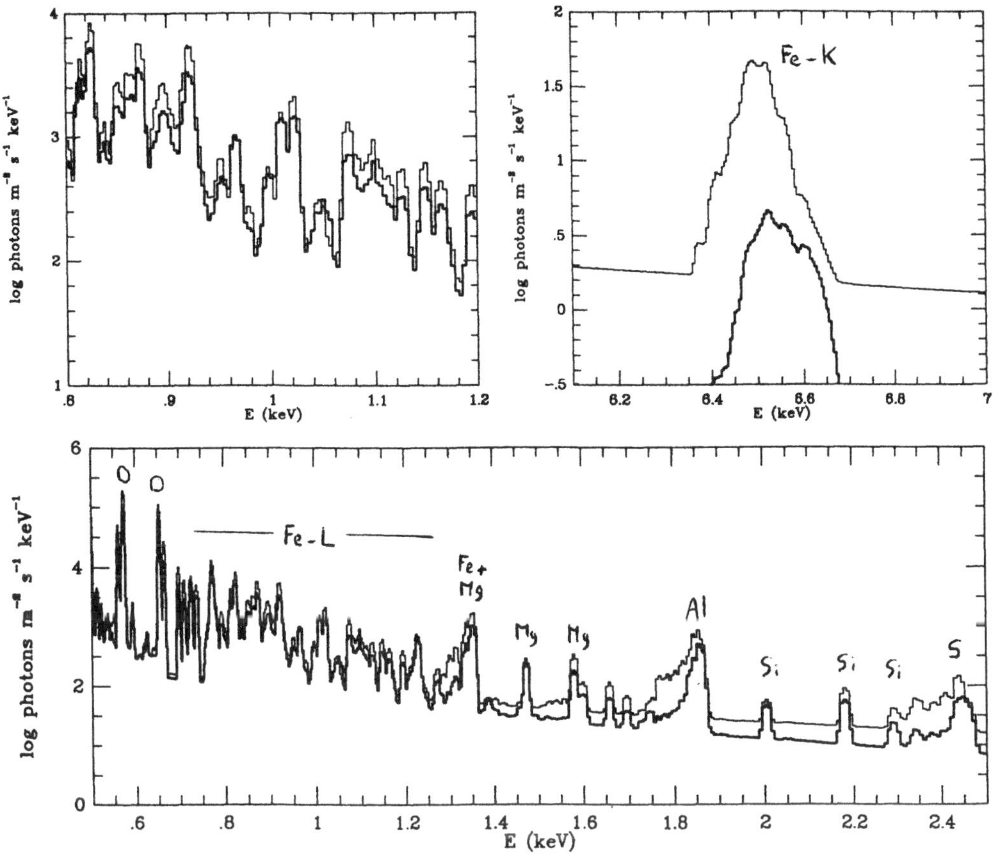

Fig. 4. Contribution of the shocked ejecta and the shocked interstellar matter to the X-ray spectrum. The spectrum is shown in 3 energy bands. An energy resolution $E/\Delta E = 300$ was chosen for the spectra between 0.5–2.5 keV, and a constant resolution $\Delta E = 0.005$ keV between 6.1–7 keV. A Chevalier model with constant interstellar density and stellar ejecta with $\rho \sim r^{-7}$ have been used with solar abundances. Thick line: shocked ejecta; thin line: shocked interstellar medium.

we consider here Puppis A, which by virtue of its large angular size ($45' \times 30'$) and large X-ray flux is well suited for such studies. Analysis of the EXOSAT CMA data (Jansen 1988) reveals substructure at scales of $2'$–$3'$ and sometimes even down to the resolution of $1'$. These structures sometimes appear to be filamentary. Data were obtained in two spectral bands. Analysing the data assuming collisional equilibrium yields a continuous temperature distribution between 10^6–10^7 K (figure 5a). However assuming a constant temperature and performing a NEI analyses, Jansen finds a double peaked distribution, with $\log n_e t$ (cm^{-3}s) peaking around 10.0 and 10.6 (figure 5b). This first component was also found by the Einstein FPCS, while the emission measure of the second component was probably too low for the FPCS.

It is evident from the above example that we need spatially resolved spectroscopy with at least moderate energy resolution, or a differential emission measure analysis using a high spectral resolution device in order to understand better these inhomogeneous structures.

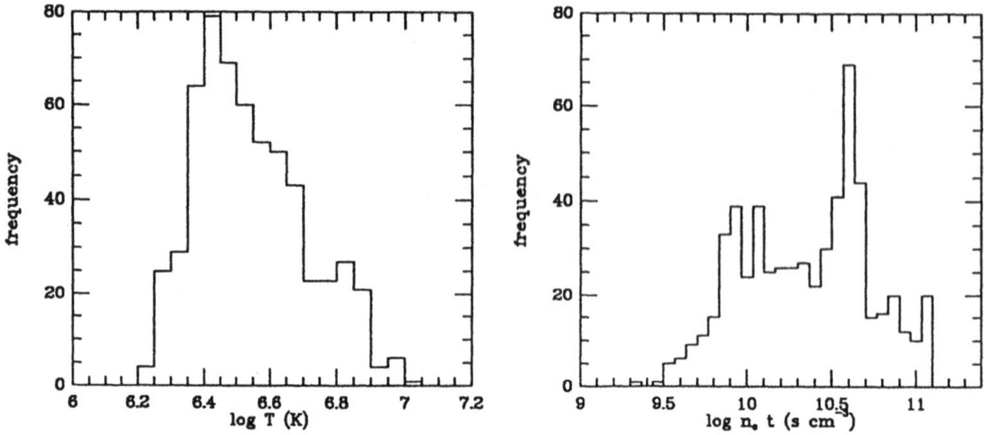

Fig. 5. a: temperature distribution of 599 spatial bins in Puppis A assuming collisional equilibrium, as derived from data obtained with the low energy detector of EXOSAT. b: distribution of $n_e t$ values assuming a constant temperature of 10^7 keV; the same data as in panel *a*. From Jansen (1988).

6. Physical processes and structure of a remnant

In the previous sections we have seen that observations have provided us with information about the average temperature and ionisation parameter in the supernova remnants. The density distribution in the shocked regions may be estimated from image data (Einstein IPC, HRI; EXOSAT CMA), although inhomogeneities and projection effects make the analysis not straightforward. It is much more difficult to obtain the temperature distribution and ionisation parameter distribution. These are needed in order to understand in detail the physical processes occurring in the remnant: how important is heat conduction, does plasma turbulence play a role, are the stellar ejecta mixed with the interstellar matter etc. Further, also the measurement of the radial expansion velocity using X-ray lines would be extremely usefull.

In order to illustrate this problem, we consider here a number of supernova models. Details will be given elsewhere (Kaastra and Jansen, 1991). The hydrodynamics is described by one of five self-similar models: two without a reverse shock for an adiabatic (Sedov 1959) and for an isothermal case (Solinger et al. 1975, Lerche and Vasyliunas 1976) and three with a reverse shock (adiabatic (Chevalier 1982), isothermal (Band 1988) and isentropic (Band 1988)). In all models, the interstellar density gradient and the density gradient of the stellar ejecta are taken as free parameters. The X-ray emission is calculated using an updated version of the plasma code of Mewe et al. (1985, 1986), taking NEI effects into account as described by Jansen (1988). Spectral line broadening due to thermal Doppler broadening and the radial expansion of the remnant are also taken into account.

We have chosen "typical", i.e. Cas A or Tycho-like parameters in the following calculations. An emission-measure averaged value for T_e of 4 keV and for $n_e t$ of 10^{10} cm^{-3}s were assumed, moreover solar abundances and zero interstellar density gradient were adopted. We have chosen a spectral resolution $E/\Delta E = 300$ for the 0.5–2.5 keV energy band. This is the typical resolution proposed for the reflection grating spectrometer of

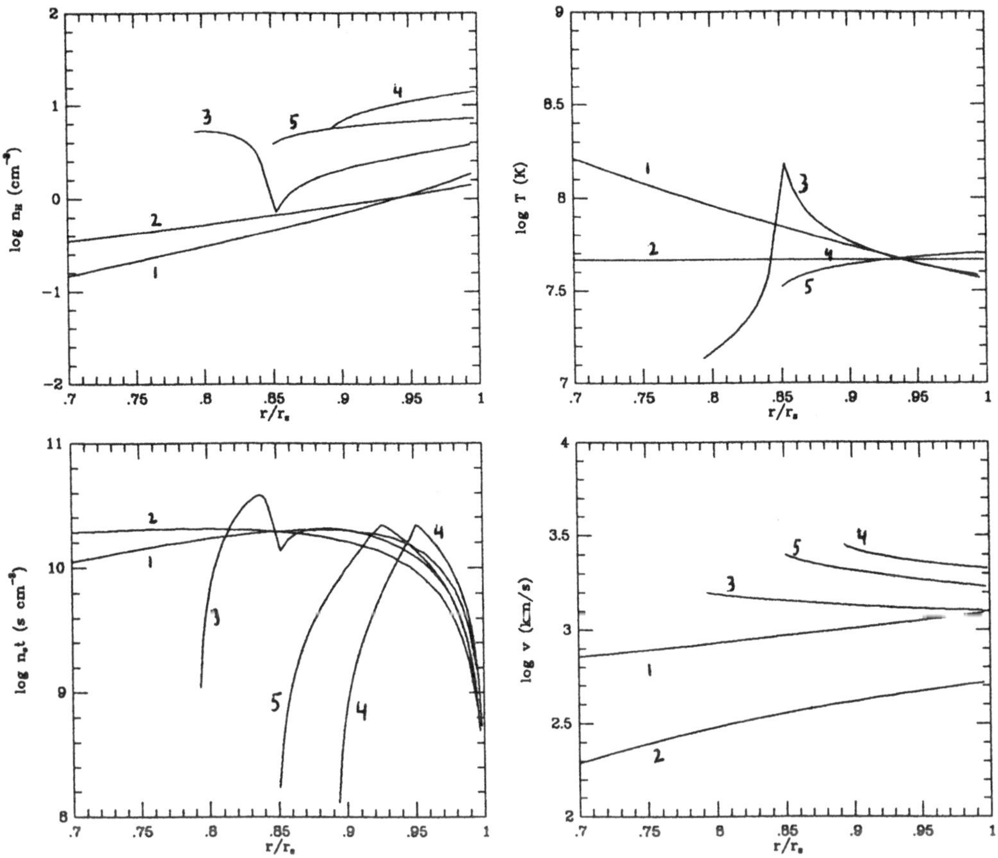

Fig. 6. Radial distribution of Hydrogen density, electron temperature, ionisation parameter and expansion velocity as a function of dimensionless radius (normalised to the main blast wave radius) for 5 different hydrodynamical models. An emission-measure averaged value for T_e of 4 keV and $n_e t$ of 10^{10} $cm^{-3}s$ was adopted. The 5 models are: 1: adiabatic without reverse shock (Sedov), 2: isothermal without reverse shock (Solinger et al.), 3: adiabatic with reverse shock (Chevalier), 4: isothermal with reverse shock (Band), 5: isentropic with reverse shock (Band). All models adopt solar abundances and have s=0, n=7 (see Chevalier, 1982).

XMM. Figure 6 illustrates the different radial profiles of density, temperature, ionisation parameter and expansion velocity for these 5 models. It is striking that the integrated spectra (figure 7) are so similar, despite the large differences in spatial properties.

Given the fact that a realistic supernova remnant will substantially deviate from an ideal, spherically symmetric system and often contains inhomogeneities, we conclude that even high spectral resolution observations of the integrated spectrum hardly allow discrimination between different global physical models, and only the temperature, ionisation parameter and abundances may be obtained this way. Spatially resolved spectroscopy with moderate spectral resolution offers a better approach as the following example shows.

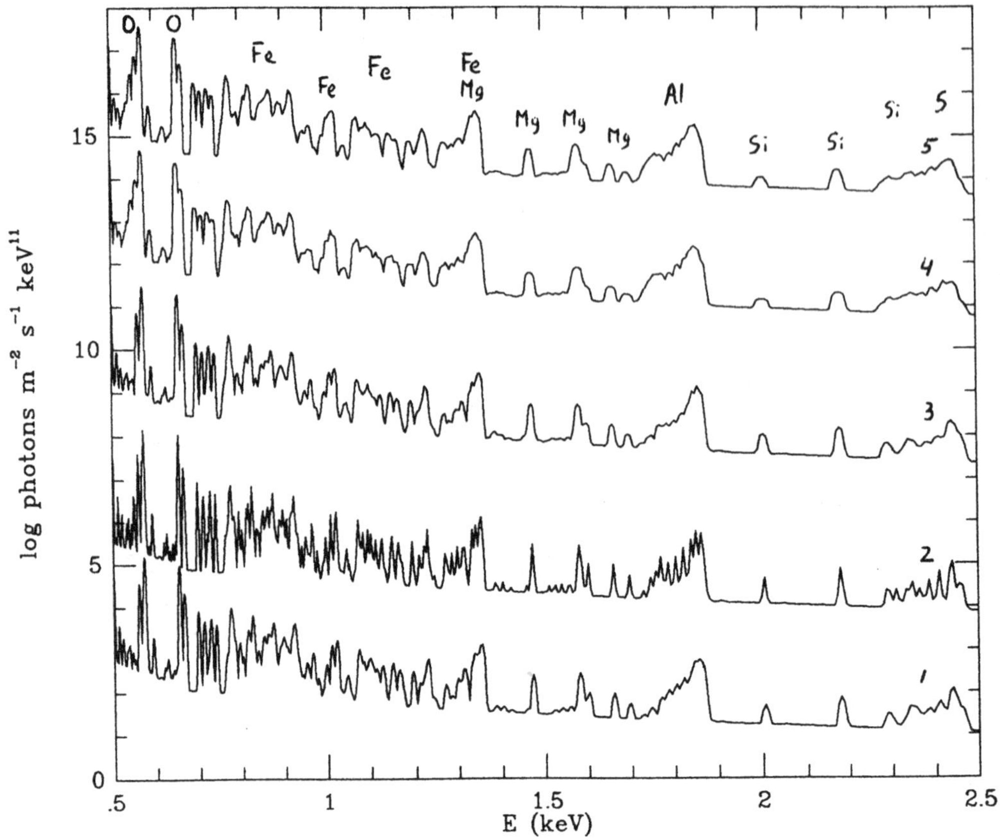

Fig. 7. X-ray spectra of the 5 hydrodynamical models of fig. 6. The normalisation is arbitrary. Energy resolution $E/\Delta E = 300$ as in figs. 2 and 4.

In figure 8 and 9 we show the spectrum for 2 of the hydrodynamical models discussed before integrated over 2 annuli on the sky. The Chevalier model and Band's isothermal model were chosen with the same parameters as in figure 6 and 7.

The first annulus (fig. 8) (0.80–0.85) contains for Chevalier's model substantial emission from shocked ejecta (cf. fig. 6), but because of projection effects also emission from shocked interstellar matter. The differences near the Fe L-complex are thus not very large. However the continuum flux level near the K-complex is quite different (due to the fact that in Chevalier's model the shocked ejecta are relatively cool), and for Chevalier's model the high energy wing of the K-complex near 6.6 keV has a larger flux (due to the large $n_e t$ value in the ejecta, cf. figure 6 for r/r_s between 0.80–0.85).

The second annulus (fig. 9) (shock radii 0.95–1.00) measures essentially the shocked interstellar matter. Apart from a global flux difference, the shape of the X-ray continuum for both models does not differ much (cf. the small difference in temperature for both models in this region, fig. 6). Again, the differences in the Fe L-complex near 1 keV are small, but there are significant changes for some other lines. The K-complex indicates the highest ionisation stage (highest effective line centroid) for Band's model (consistent with fig. 6). It is evident from figure 9 that the properties of the Fe K-complex are quite different in this zone, and can be distinguished already using

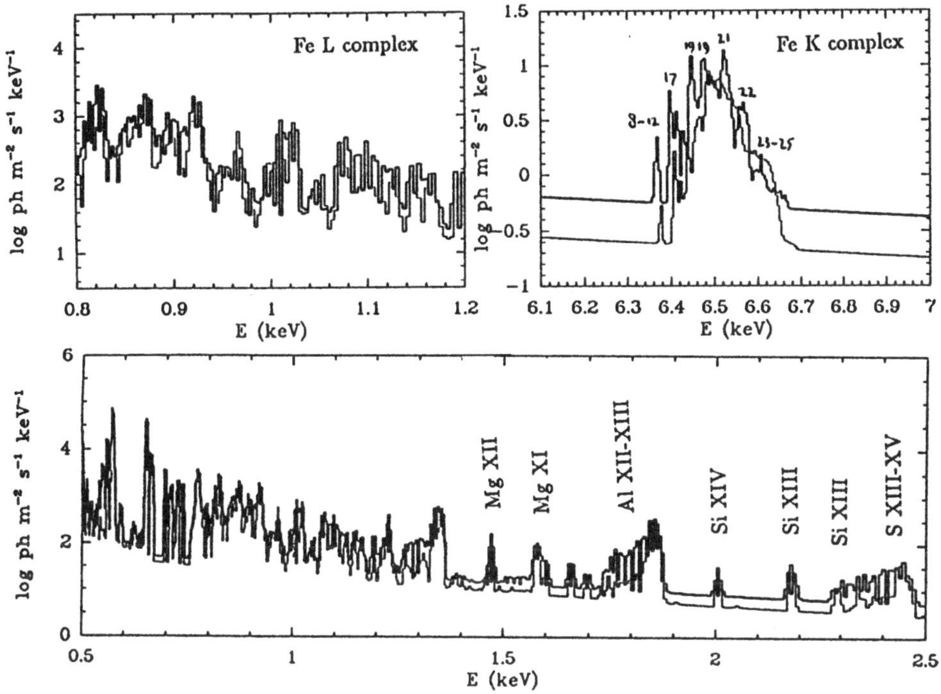

Fig. 8. Spectrum in 3 different energy bands, projected onto the sky and observed in the annulus between 0.80–0.85 shock radii. Thin line: adiabatic model (3). Thick line: isothermal model (4). Energy resolution as in fig. 4. The arabic numbers in the Fe-K frame are the ionisation stages of the most important lines.

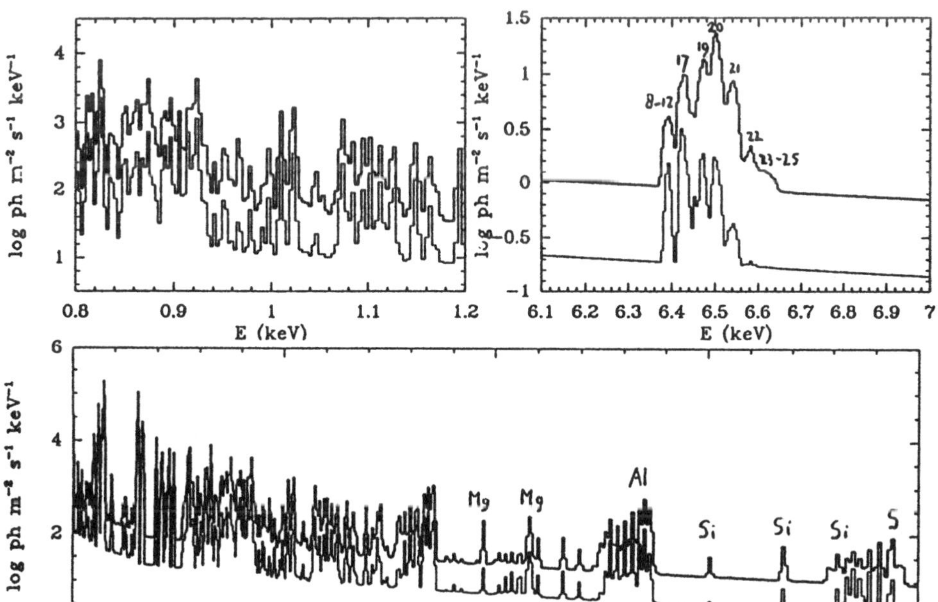

Fig. 9. As figure 8, but for the annulus between 0.95–1.00 shock radii.

moderate spectral resolution. Also the equivalent width of the Mg XI and Mg XII lines near 1.5 keV, the Si XIV lines near 2 keV and the Si XIII lines near 2.2 keV are very model dependent and therefore usefull diagnostic tools.

The Fe-K complex is better resolved here than in fig. 4 because we look at the outermost annuli where Doppler broadening is less important. Finally by comparing different annuli of the same hydrodynamical model (compare fig. 8 and 9) it is evident that the spectra differ considerably. We conclude that spatially resolved spectroscopy (angular resolution 5–10 % of the shock radius, spectral resolution a few %) is essential to obtain the temperature and ionisation parameter distribution within a supernova remnant.

7. Conclusion

For a better understanding of the structure and physics of supernova remnants, X-ray spectral line observations play a crucial role. High spectral resolution, as well as spatially resolved moderate spectral resolution observations are needed for questions like the T_e/T_i problem, the role of inhomogeneities and the physical state of the remnant. Future missions planned today help in resolving these problems. Iron lines are very important in this respect, but obviously line studies of other abundant elements like O, Mg, Si, S, Ar, Ca and Ni are required to properly diagnose the complicated temperature and density structures potentially present in many supernova remnants.

References

Band, D.L.: 1988, *Astrophys. J.* **332**, 842

Chevalier, R.A.: 1982, *Astrophys. J.* **258**, 790

Jansen, F.A.: 1988, Thesis, University of Leiden

Jansen, F.A., Smith, A., Bleeker, J.A.M., de Korte, P.A.J., Peacock, A., White, N.E.: 1988, *Astrophys. J.* **331**, 949

Kaastra, J.S., Jansen, F.A.: 1991, in preparation

Lerche, I., Vasyliunas, V.M.: 1976, *Astrophys. J.* **210**, 85

McKee, C.F.: 1974, *Astrophys. J.* **188**, 335

Mewe, R., Gronenschild, E.H.B.M., van den Oord, G.H.J.: 1985, *Astron. Astrophys. Suppl.* **62**, 197

Mewe, R., Lemen, J.R., van den Oord, G.H.J.: 1986, *Astron. Astrophys. Suppl.* **65**, 511

Mihalas, D., Mihalas, B.W.: 1985, *Foundations of Radiation Hydrodynamics*, Oxford University Press.

Sedov, L.: 1959, *Similarity and dimensional methods in mechanics*, Academic Press, New York

Smith, A., Davelaar, J., Peacock, A., Taylor, B.G., Morini, M., Robba, N.R.: 1988, *Astrophys. J.* **325**, 288

Solinger, A., Rappaport, S., Buff, J.: 1975, *Astrophys. J.* **201**, 381

Tsunemi, H., Yamashita, K., Masai, K., Hayakawa, S., Koyama, K.: 1986, *Astrophys. J.* **306**, 248

Thermal X-ray emission from centre-filled supernova remnants

L.R. Jones

Physics Department, University of Southampton, Southampton SO9 5NH

Summary

Two middle-aged supernova remnants, W44 (G34.6-0.5) and W28 (G6.4-0.1), have been studied using EINSTEIN IPC imaging data and EXOSAT ME spectra. Both remnants have a centre-filled X-ray morphology and a radio shell, but the X-ray spectra imply a thermal origin (in one case iron $K\alpha$ line emission is revealed). Deprojection analyses of the X-ray images have been performed under simplfying assumptions and radial density distributions derived. These are inconsistent with a simple Sedov model, but are more similar to predictions of models involving evaporation of cold ISM clouds in the remnant centre.

Images & spectra

EINSTEIN IPC images in the range 0.1-4.5 keV are shown in Figs. 1 & 2, overlaid on radio maps. These images were originally published by Smith *et al.* (1985) and Matsui & Long (1985). In both cases the X-ray emission is contained within the radio shell. This X-ray morphology is typical of SNRs containing a pulsar surrounded by a synchrotron nebula. However, in both SNRs the X-ray emission is found not be synchrotron in origin, but rather to be thermal, from a hot plasma in the SNR interior. The evidence comes from the poor fit of a power law to the spectra, and in particular, the presence of thermal emission lines (see Figs 3 & 4 and Szymkowiak 1980 for the EINSTEIN SSS spectrum of W44). An X-ray point source exists within W28 but contributes very little to the total diffuse flux; it is too faint to be resolved in this image.

Temperatures found from the EXOSAT ME spectra using Raymond & Smith (1977) model data are:

W44: $T = (1.1 \pm 0.1)\text{x}10^7 K$, $N_H = (8 \pm 2)\text{x}10^{21} cm^{-2}$ (90% confidence)
(Cosmic abundances, Reduced chi-sq = 0.93)

W28: $T = (2.5 \pm 0.5)\text{x}10^7 K$, $N_H < 6\text{x}10^{21} cm^{-2}$ (90% confidence)
(Iron abundance 2 x cosmic, Reduced chi-sq = 1.1)

Discussion

If the observed X-rays arise in the hot SNR interior, where are the X-rays from the limbs? The answer lies with the relatively high column densities to the SNRs ($\sim 6 - 8\text{x}10^{21} cm^{-2}$) and their age. With a radius of $\sim 12pc$ (7 times the radius of Cas A), expansion through the ISM of the galactic plane must have slowed the shock down and so the current post shock temperature must be relatively low. Typical values of 400 km/s and $2\text{x}10^6$ K in an ambient medium of density of 0.3 cm^{-3} would produce bright soft X-ray emission which would be almost completely absorbed along the line of sight to W44 and so invisible in the IPC image. The column density in the direction of W28 is less than that to W44 and here X-ray emission from the limb is visible at one point. The reason the interior of the SNRs is at a higher temperature then the limbs is that the shock passed through these regions at an earlier time, when it was travelling faster.

To investigate the structure of the hot gas further, the EINSTEIN images have been deprojected using the technique of Fabian *et al.* (1981), under the assumptions of constant pressure and ionization equilibrium throughout the remnant interiors. These assumptions are reasonable for SNRs of this age ($\sim$ 10000 years) over large regions of the interior. The value for the pressure is constrained by forcing consistency with the broad band EXOSAT spectra. Thermal bremmstrahlung and line emission from the major contributing elements (Fe,S,Si,O) are included. Interestingly, the pressure derived in both SNRs is P/k $\sim$ $6\mathrm{x}10^6 cm^{-3}K$. Plots of density versus radius are shown in Figs. 5 & 6. The general shape of these profiles (ie. density decreasing outwards) is insensitive to changes in the pressure, column density or abundances. The density near the shock cannot be calculated because of the lack of X-rays detected at the limbs, a factor which does however simplify the deprojection of the interior emission, since inhomogeneities at the limbs do not have to be taken in to account.

The density profiles are completely inconsistent with those predicted from the Sedov solution, in which the density increases by two orders of magnitude over the range of radii considered here. The most likely cause of the discrepancy is inhomogeneities in the ISM; in particular, the evaporation of cool clouds. These have been observed in several nearby SNRs but are not considered in the Sedov solution. In the models of Cowie, McKee & Ostriker (1981) (CMO) dense clouds are overtaken by the shock and remain evenly distributed in the remnant interior. Heating of this cool material reduces the temperature while increasing the density of the hot X-ray emitting gas. The interior temperature is predicted to be kept approximately constant by this mechanism since the "evaporation" rate of cool clouds is a strong function of the hot gas temperature, reducing it most where it is greatest.

The density profile for CMO's standard model (model 1, age =$1.1\mathrm{x}10^4$ yr) matches those observed resonably well (within a factor $\sim$ 2), compared with say the Sedov model. Given the number of model variables, for which CMO's standard values have been taken, (eg explosion energy, cloud mass and size distribution, efficiency of evaporation), the approximate agreement becomes more remarkable. The predicted pressure in the model is approximately constant with radius (within a factor $\sim$ 2), except in the deep interior where it is predicted to drop dramatically because most of the mass in the clouds was evaporated and swept up by the fast, hot shock as it passed through. Thus the comparison of data and model in this region $(R/R_s)<0.3$ is invalid.

The predicted mass averaged temperature of of the hot intercloud medium, expected to be radiating X-rays, is also in reasonable agreement with observation ($2\mathrm{x}10^7$K), as is the radius of both SNRs if the age above is assumed.

Conclusion

The diffuse X-rays from from the interior of W28 have been shown to be thermal in origin. W28 and W44 are part of a group of SNRs where the diffuse interior emission is emphasized because the softer limb emission is absorbed along the line of sight. Others in this group are W49B, 3C400.2, HB3, G33.6+0.1 and G327.4+0.4 (Kes 27).

In general the cloud evaporation model of CMO shows good agreement with the distribution and temperature of thermal X-ray emitting material within W44 and W28. Further work will allow a more detailed comparison to be made.

Acknowledgements

The IPC image of W28 was obtained from the EINSTEIN Database with the help of Fred Seward, Cfa.

References
Cowie, L.L., McKee, C.F. & Ostriker, J.P., 1981. Ap. J. **247**, 908.
Fabian, A.C. Willingale, R., Pye, J.P., Murray, S.S. & Fabbiano, G., 1981. Mon. Not. R. Ast. Soc., **193**, 175.
Matsui, Y. & Long, K.S., 1985. Proceedings of workshop on the Crab Nebula & related SNR, Fairfax, Virginia.
Raymond, J.C. & Smith, B.W. 1977. Ap. J. Suppl. **35**, 419.
Smith, A., Jones, L.R., Watson, M.G., Willingale, R., Wood, N., Seward, F. 1985. Mon. Not. R. Ast. Soc., **217**, 99.
Szymkowiak, A. 1980. Proceedings Texas workshop on Type I SN.

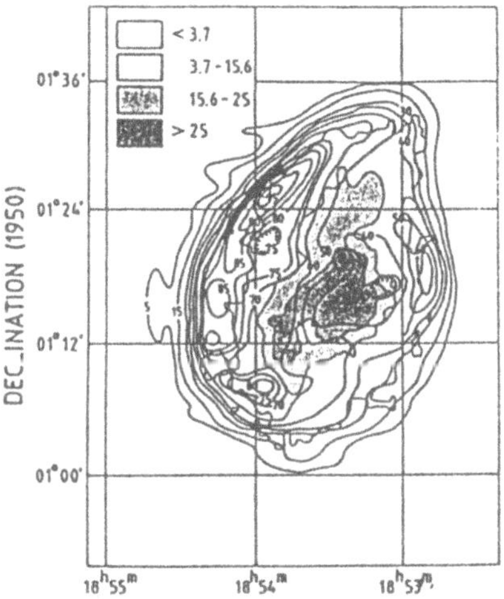

Fig. 1. EINSTEIN IPC image of W44 from Smith *et al.* (1985). Th shading corresponds to the IPC cts per arcmin^2. The image has been smoothed with a 30-arcsec Gaussian. The radio contours are from the 408 MHz map of Clark *et al.* (1975)

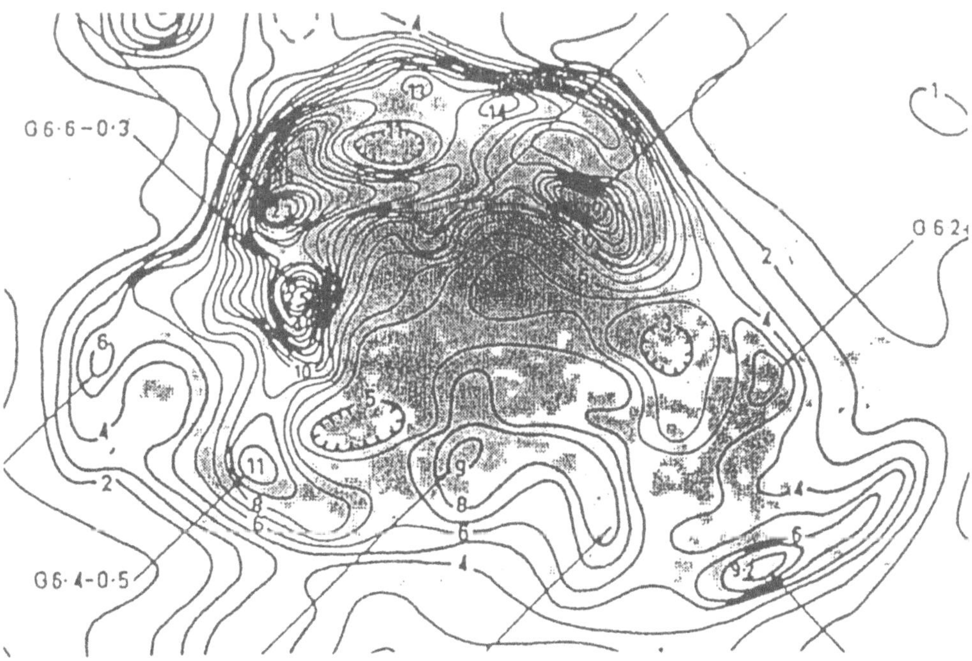

Fig. 2. EINSTEIN IPC image of W28. The image has been smoothed with a 24-arcsec Gaussian. The radio contours are from Shaver & Goss (1970).

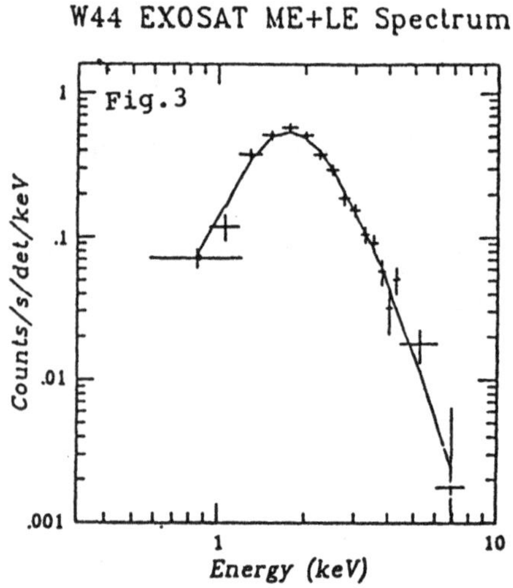

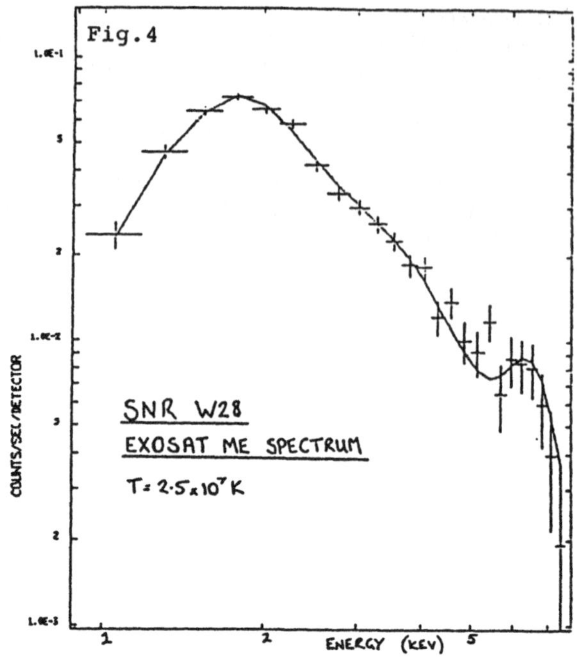

Fig. 3. EXOSAT ME + LE spectrum of W44. The best fit spectrum is an optically thin thermal (Raymond & Smith 1977) model with $T=1.1\times10^7$K and $N_H=8.4\times10^{21}$ cm^{-2}.

Fig. 4. EXOSAT ME spectrum of W28. The best fit spectrum is an optically thin thermal model with $T=2.5\times10^7$K.

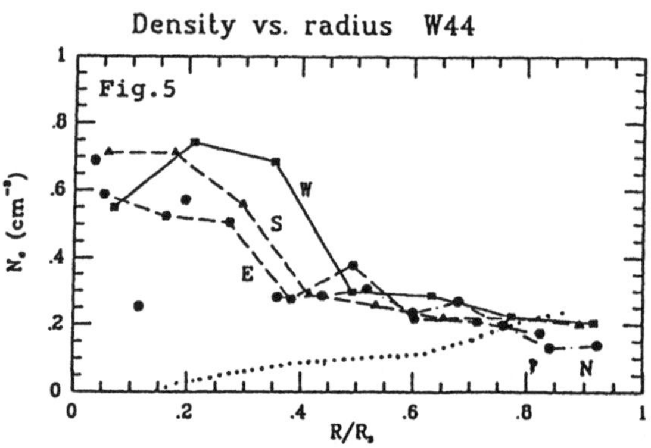

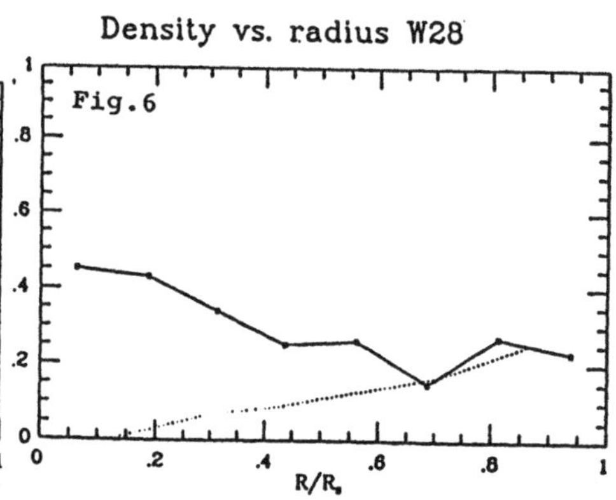

Fig. 5. Electron density as a function of radius relative to the shock radius for W44. This plot has been derived from a deprojection of the EINSTEIN IPC image in four sectors labelled from north to west. Densities predicted from the standard model of CMO (see text) are shown as a dotted line.

Fig. 6. As Fig 5, but for W28.

Non-Equilibrium, non-LTE Ionization in Supernova Remnants

W. Brinkmann
Max-Planck Institut für Physik und Astrophysik
Institut für extraterrestrische Physik
8046 Garching, W.-Germany

Introduction

Recent modellings of Tycho's supernova remnants by Brinkmann *et al.*, (1989) and Itoh *et al.*, (1988) find that this type I SNR can be described adequately in the framework of the currently favored deflagration model of a white dwarf (Nomoto *et al.*, 1984) with a total mass of the ejecta of $\lesssim$ 1.3 $M_\odot$. However, different ad hoc assumptions were made about the coupling of the ion temperature to the electron gas. Both approaches, assuming either that temperature equilibration between electrons and ions occurs instantaneously by non-linear plasma processes, or that the energy transfer from the shock heated ions occurs by Coulomb collisions only could explain the observations qualitatively, using different kinds of mixing of the ejecta. Both calculations suffered, however, from the inconsistency that the calculated centroids of the X-ray emission lines were found to be shifted systematically to higher energies than the observed values. The fact, that the corresponding high ionization stages of the relevant elements (Fe, S, Si) are obtained relatively early in the life of the SNR indicates some fundamental mismatch of the model with the actual physical conditions in the remnant.

The coincidence of nearly all occuring relevant time scales, from the macroscopic life time of the SNR to the microscopic ionization time scales and energy transfer time scales is well known (Shklovsky, 1961). Tis implies for the electron distribution function that elastic scattering and ionization processes must show strong interrelations especially at higher electron energies, where the electron–electron collision frequency decreases roughly as $\sigma_{ee} \sim E^{-3/2}$ whereas, according to the Lotz formula, the ionization rate is $\sigma_I \sim \ln(E/E_0)/E\ E_0$, with the ionization energy E_0.

We will address in this paper a previously widely neglected problem, the possible deviation of the local electron distribution function from the LTE, Maxwellian shape. We will discuss these effects in the astrophysical context of a plasma of pure iron - as SNR of type I are expected to consist mainly of Fe ($\approx 0.6 - 0.7$ $M_\odot$) and iron lines are an important diagnostic tool in X-ray observations.

Non - LTE ionization

Nearly all current ionization calculations assume implicitly that the electron distribution function is Maxwellian. Ionization and excitation rates are usually given as integrals of the relevant differential cross sections over the (Maxwellian) electron energy distribution function (for a review see Mewe, 1990). However, due to the above mentioned energy dependence of the cross sections inelastic collisions will reduce the energetic tail of the distribution function more than would be expected on the basis of the overall

collisionality. Nonthermal electron distribution functions will therefore result and the ionization structure as well as the radiation properties might change noticeably.
The physical state of the iron plasma is determined by two sets of equations. First, the set of equations determining the ionization structure of the plasma is given by

$$\frac{1}{n_e}\frac{\partial X_i}{\partial t} = S_{i-1}X_{i-1} - (S_i + \alpha_i)X_i + \alpha_{i+1}X_{i+1},$$

where the index i runs over all ionization stages of the iron ions, X_i is the fractional abundance of the i-th ionization stage. S_i is the total ionization rate of the i-th stage and α_i the recombination rate. Both rates are obtained by integrating the corresponding differential rates over the electron distribution function.
The temporal evolution of a spatially uniform isotropic electron distribution function is quite generally given by

$$\frac{\delta f}{\delta t} = \frac{\delta f}{\delta t}|_{el} + \frac{\delta f}{\delta t}|_{ion} + Q$$

Here, the first two terms on the rhs represents the evolution of the distribution function due to elastic collisions, i.e., in form of a Fokker-Planck collison operator (Rosenbluth *et al.*, 1957). The term Q describes the sources and sinks for particles. We will treat contributions from ionization, excitation and recombination processes only, neglecting bremsstrahlung losses of the electrons.
The differential cross sections for ionization and excitation obey scaling laws of the form $\sigma(E) \sim g(U)/U$ with $U = E/E_0$, and the gaunt factor $g(U)$ is approximated by the empirical interpolation formula (Mewe and Schrijver, 1978)

$$g(U) = A + BU^{-1} + CU^{-2} + 2DU^{-3} + Eln(U).$$

The coefficients for all relevant ionization processes were taken from the work of Arnaud and Rothenflug (1985), those for excitation from Mewe (private communication).
The changes of the phase space density of the electron distribution function n_e in the energy range from E to $E + dE$ due to collisional ionization of the ions can be described by the equation

$$\frac{\partial n_e(E)}{\partial t} = -\sum_i X_i\, n_e(E)\, \sigma_{ion}(E) + \sum_{(E')}\left[\beta_j \sum_j X_j\, n_e(E')\, \sigma_{ion}(E')\right]$$

The first term describes the loss of electrons from the distribution function at energy E, the second term is a source term for electrons coming from all other possible ionization reactions. The equations for the changes of the distribution function due to excitation and recombination are similar.
The recombination rates can be determined from the assumption of detailed balance. Using the relation for coronal ionization equilibrium

$$S_i X_i = \alpha_{i+1}X_{i+1},$$

the differential recombination coefficients $\alpha_i(E)$ are determined from an equilibrium ionization configuration, obtained by employing the known rates from a Maxwellian electron distribution function. Thus the calculated recombination coefficients include dielectronic, radiative and charge-exchange processes. Details and limitations of this procedure are discussed in Brinkmann (1991).

Results

We have calculated equilibrium configurations for iron plasmas between 1 keV and 10 keV. At temperatures below 10^7K there are hardly any sufficiently high energetic electrons in the distribution function to give rise to noticeable effects. Above 10^8K a relativistic treatment of the Fokker Planck equation should be used.

Starting from Maxwellian distribution functions we iterated the set of ionization and Fokker-Planck equations until the successive changes of the electron distribution functions remain negligibly small.

First, it must be noted that for equilibrium conditions the temperature of the ionic component is always considerably higher than that of the electron temperature such that the energy transfer from the hot ions to the cold electrons can balance the ionization and excitation losses. This means, that even under (quasi-) equilibrium conditions there must be a considerable electron - ion temperature difference in SNR.

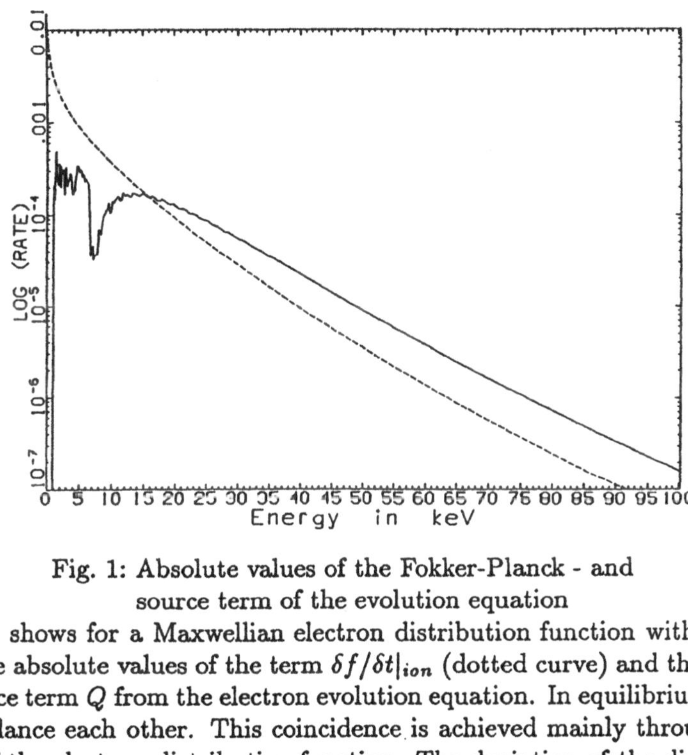

Fig. 1: Absolute values of the Fokker-Planck - and source term of the evolution equation

Figure 1 shows for a Maxwellian electron distribution function with a temperature of 10^8K the absolute values of the term $\delta f/\delta t|_{ion}$ (dotted curve) and the absolute value of the source term Q from the electron evolution equation. In equilibrium these two curves must balance each other. This coincidence is achieved mainly through changes of the shape of the electron distribution function. The deviation of the distribution function from a Maxwellian form, i.e., the quantity $|(f_{Maxwell} - f)/f_{Maxwell}|$ is shown in figure 2. At low energies, electrons are added while the high energetic tail is depleted. The graph shows, that at these high temperatures the changes are of the order of only a few percent. At lower temperatures, the absolute values of the modifications are much higher: at $T_e = 10^7\ K$ the phase space density of the tail changes by $\approx 15\%$ – however, absolutely there are initially less than 1 % of the electrons at energies above 8 keV.

This implies that in all cases only relatively few high energy electrons are redistributed to lower energies. The result is, that for the cases considered, the integrated inelastic interaction rates change only very little. Further, due to the temperature coupling between ionization- and recombination rates and due to the fact that in the ionization balance equation only neighbouring states are interacting, the ionization structure of the plasma remains nearly unchanged.

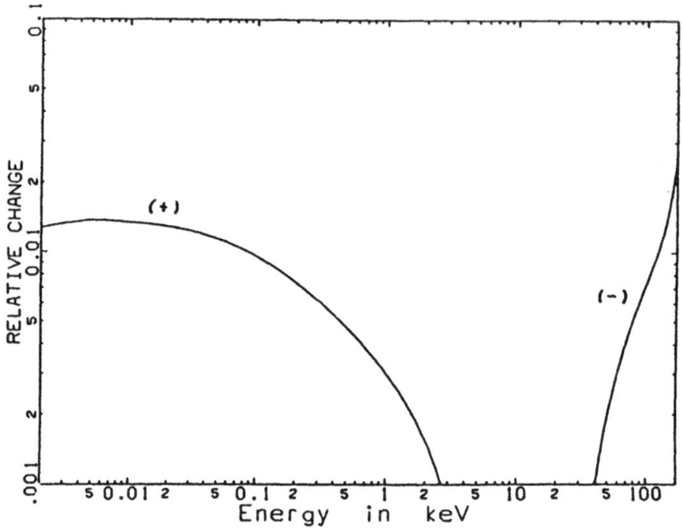

Fig. 2: Relative (absolute) changes of the electron distribution function; $T_e = 10^8\ K, T_i = 3.84\ 10^8\ K$

Another interesting question is that of the rate of energy transfer between the different species, as this controls the amount of heating of the electrons in the shock. As the absolute changes of the electron distribution function are only small no big differences to the classical (Maxwellian) energy exchange rate are expected. But in all cases the calculated, non-Maxwellian rates were higher than the corresponding Maxwellian rates, if only by a few percent. This means, the heating of truncated electron distribution function proceeds faster than predicted by the classical formulae.

Summarizing, one can say that for equilibrium conditions the effects of inelastic scatterings on the electron distribution function in a hot iron plasma are negligibly small; for non-equilibrium conditions (shock–heating) this has to be checked in future calculations.

References

Arnaud, M. and Rothenflug, R.: 1985, Astron. Astr. Suppl. **60**, 425

Brinkmann, W.: 1990, submitted to Astron. Astr.

Brinkmann, W., Fink, H.H., Smith, A., and Haberl, F., 1989, Astron. Astr. **221**, 385

Itoh, H., Masai, K., and Nomoto, K.: 1988, ApJ. **334**, 279

Mewe, R., 1990, in: *Physical Processes in Hot Cosmic Plasmas*, W. Brinkmann, A. Fabian, F. Giovannelli (eds), Kluwer Acad. Publ., Dordrecht, p.999

Mewe, R. and Schrijver, J.: 1978, Astron. Astrophys. **65**, 99

Nomoto,K., Thielemann, F.K., and Yokoi, K.: 1984, Ap.J. **286**, p. 644

Rosenbluth, M.N., MacDonald, W.M., and Judd, D.L.: 1957, Phys. Rev. **107**, 1

Shklovsky, I.S.: 1961, Sov. Astron. **6**, 162

SPARTAN 1 X-RAY OBSERVATION OF DIFFUSE EMISSION AT THE GALACTIC CENTER

W.A. Snyder, R.G. Cruddace, M.P. Kowalski, G.G. Fritz
E.O. Hulburt Center for Space Research
Naval Research Laboratory, Washington, D.C. 20375-5000

and

E.E. Fenimore
Los Alamos National Laboratory
Los Alamos, New Mexico 87545

Abstract

The 1-10 keV X-ray detectors on SPARTAN 1 scanned the central 2 degree region of the galactic center during 20-21 June 1985. Four strong point sources and extended emission ~1° in diameter centered near the galactic center were detected [1]. We have re-analyzed the diffuse component data following the discovery [2] of intense Fe line emission. The radial distribution of this component is examined by grouping the data into radial bins centered on Sgr A West, after first removing known point sources. We confirm that this component is an intense source of Fe line radiation, but find that the Fe line flux is more sharply peaked and narrowly distributed than previously reported. A Gaussian surface brightness distribution with FWHM of 49-92 arcmin (90% confidence limits) is consistent with the SPARTAN 1 data. The spectrum can be approximated with either a single power-law or thermal form, but the high χ^2 suggests a more complex shape. The radial distribution of the Fe line and continuum emission is presented and used to address the nature of the diffuse component.

I. Observations

SPARTAN 1 was released in orbit on 20 June 1985 by the shuttle Discovery. The experiment package consisted of two finely collimated proportional counters, each with 1-12 keV bandpass, 16% energy resolution at 6 keV, 660 cm^2 effective area, and a collimated 5 arcmin x 3 degree FWHM rectangular field-of-view. The mission lasted 2 days and observations were made of the galactic center [1] and the Perseus cluster [3,4,5]. Observations were made by sweeping the 5 arcmin field-of-view across the source at many different position angles. As the field-of-view traverses the source, a count rate is observed which is the convolution of the collimator response with the intrinsic surface brightness of the observed object. The intrinsic surface brightness can be reconstructed by modeling the object with a set of extended and point sources, or by using a non-linear image reconstruction algorithum like maximum entropy [6].

A schematic of the maximum entropy map of the region [1] is shown in Figure 1. Five prominent point sources were detected. They are marked in Figure 1 with the exception of GX3+1 which lies outside the mapped region. A point source at the position of Sgr A West [7,8] was not detected, although a region of extended or confused emission located near the galactic center was evident. This patch of emission is an area of maximum surface brightness of a more extended region seen in the sweep data. This extended feature could be approximated with a two-dimensional Gaussian surface brightness distribution and was identified with the diffuse emission seen in the *EINSTEIN* IPC image of the region [7], although of larger size (see also [8]). The 1/e intensity contour of this extended component is marked as a dotted ellipse in Figure 1.

Using the large-area proportional counters on Ginga, Koyoma et al. [2] discovered that the diffuse galactic center emission was a source of intense 6.7 keV Fe line radiation. The intensity of the emission was at least 10X larger than observed from the galactic ridge. We have re-examined the diffuse component data and have determined its radial properties by binning the galactic center sweep data into 4 arcmin radial bins centered on Sgr A West after first removing known point sources (e.g., GX3+1, A1742-294, 1E1743.1-2843, 1E1740.7-2942 and SP1744.2-2959) and background radiation.

A fit of the radially binned spectra to simple exponential or power-law forms indicates that the radially averaged exponential temperature (or power-law index) does not change significantly over the 60 arcmin radius studied (at 10 Kpc, 1 arcmin = 2.9 pc). The average exponential temperature is ~5 x 10^7 ^{o}K. Exponential and power-law spectra fit the radially binned data equally well, although the distribution of χ^2 is larger than expected, ranging from 4 to 22 for 9 degrees-of-freedom. This indicates that the radially binned spectra must generally have a more complicated shape. The observed 6.7 keV Fe line flux as a function of radius is shown in Figure 2. The expected Fe line flux obtained by Koyama et al. [2] is also indicated. Assuming that the Fe line surface brightness is a radial Gaussian function we obtain a FWHM of 49-92 arcmin (90% confidence range). This is somewhat smaller than the value of 102-114 arcmin obtained with Ginga. The Ginga field-of-view is larger than SPARTAN 1, however, and these results are understandable if the surface brightness is not exactly Gaussian but either flattens at large radii and/or narrows at small radii.

The radially binned spectra have also been fitted to an optically thin thermal plasma [9]. Ninety-percent confidence limits to the Fe abundance of 0.6-1.7 times the cosmic value are set by

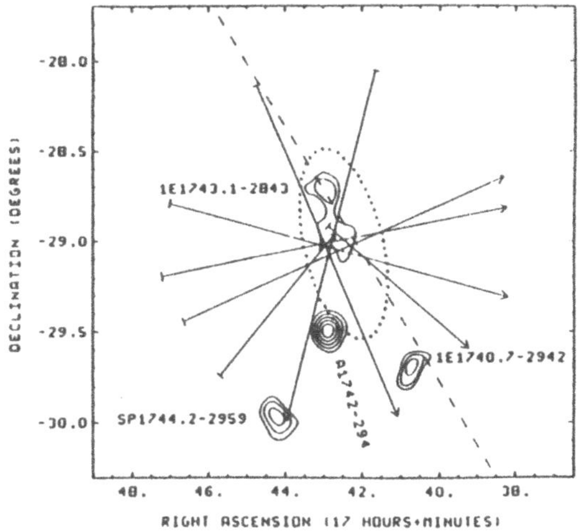

Figure 1. SPARTAN 1 map of the Galactic center region. The sweep pattern for the observation is shown. Each line is the locus of the collimator view axis as it traversed the region.

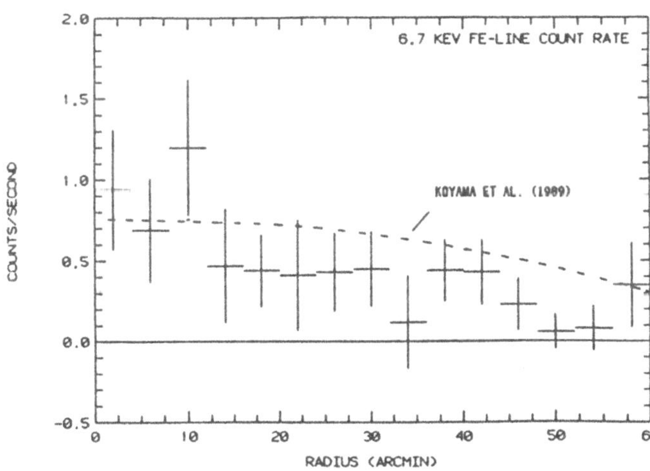

Figure 2. The radially averaged 6.7 keV Fe line count rate observed from the diffuse Galactic center component. The expected intensity observed by Ginga [2] is indicated.

our data. For these models, the average temperature is higher than that derived using an exponential, with temperatures of ~10^8 °K usually obtained.

II. Discussion

Koyoma et al. [2] noted that the intense Fe line flux argues for a diffuse thermal origin for the emission, and suggested that it was caused by a Galactic center explosion. It is unlikely that the X-ray emitting gas is gravitationally or magnetically confined. If the gas temperature is 5 x 10^7 °K, then a total mass of 5 x 10^{10} $M_\odot$ within 180 pc of the Galactic center would be required to bind it. This is much larger than the value of 2-5 x 10^9 $M_\odot$ inferred from HI velocities [10]. With a 2-10 keV luminosity of 41 x 10^{35} erg s^{-1}, we calculate an average electron density of 0.08 cm^{-3} (assuming a 1° x 0.5° x 0.5° ellipsoidal volume), a cooling time for the gas of 5 x 10^8 yr, and an energy content of 6.6 x 10^{52} erg. A 0.2 mG field could bind this gas, but this is larger than the 10 μG field thought to be present [11]. The long cooling time explains the absence of spectral change as a function of radius, because gas expelled from the Galactic center would require only 2 x 10^5 yr to traverse the 180 pc distance, assuming an expansion velocity of 10^3 km s^{-1}. It should be noted, however, that a single massive explosion is not the only mechanism for powering the thermal emission. Multiple supernovae can also satisfy the energy requirments, with a Type I (II) rate of only 10^{-7} (10^{-5}) yr^{-1} (much smaller than observed [12]) needed. The lack of spherical symmetry and the elongated nature of the source may, however, argue against many smaller random events.

References

[1] Kawai, N., et al. 1988, Ap.J., **330**, 130.

[2] Koyama, K., et al. 1989, Nature, **339**, 603.

[3] Ulmer, M.P., Cruddace, R.G., Fenimore, E.E., Fritz, G.G., and Snyder, W.A. 1987, Ap.J., **319**, 118.

[4] Kowalski, M.P., et al. 1991, this conference.

[5] Snyder, W.A., et al. 1990, Ap.J., in press.

[6] Willingale, R. 1981, MNRAS, **194**, 359.

[7] Watson, M.G., Willingale, R., Grindlay, J.E., and Hertz, P. 1981, Ap.J., **250**, 142.

[8] Skinner, G.K., et al. 1987, Nature, **330**, 544.

[9] Raymond, J.C., and Smith, B.W. 1977, Ap.J.Suppl., **35**, 419.

[10] Townes, C.H. 1989, IAU Symp. no. 136, The Center of the Galaxy, ed. Mark Morris (Dordrecht: Kluwer Academic Publishers), p. 1.

[11] Sofue, Y. 1989, IAU Symp. no. 136, The Center of the Galaxy, ed. Mark Morris (Dordrecht: Kluwer Academic Publishers), p. 213.

[12] Allen, C.W. 1973, Astrophysical Quantities, (London: The Athlone Press), p. 221.

IRON LINE DIAGNOSTICS IN ELLIPTICAL GALAXIES AND CLUSTER COOLING FLOWS

Craig L. Sarazin and Michael W. Wise
Department of Astronomy, University of Virginia
P. O. Box 3818, Charlottesville, Virginia U.S.A.

ABSTRACT. X-ray iron line observations of clusters of galaxies have shown that the X-ray emission in clusters is mainly thermal emission from hot diffuse gas. Much of this gas has probably come out of stars and been ejected from galaxies in the cluster. In elliptical galaxies, iron and other X-ray lines can establish whether the X-ray emission is due to diffuse thermal emission, which is particularly uncertain for the fainter galaxies. X-ray line observations of clusters indicate that large amounts of gas are cooling in these clusters. The total luminosities of low ionization X-ray lines (like those of Fe XVII) can give the total cooling rates in these "cooling flow" clusters. The surface brightness of low ionization lines can determine where the gas is cooling. High resolution spectral observations of X-ray lines may detect the inflow in the gas, although this will be difficult if the gas in clusters is turbulent. The hot interstellar gas in elliptical galaxies is expected to have very high heavy element abundances, and the Fe L lines should be very strong. The abundance of iron is predicted to be three to ten times solar in these objects. Measurements of the iron abundance can be used to determine the supernova rate and heating rate of the gas in ellipticals. Other elements (such as oxygen) should have large interstellar abundance gradients. The form of these gradients can determine whether gas is flowing inward, expanding outward, or cooling locally in elliptical galaxies.

1. INTRODUCTION

In the X-ray study of clusters of galaxies, X-ray spectroscopy of Fe lines has played an essential role. The discovery of iron K lines in the spectrum of the Perseus cluster (Mitchell *et al.* 1976), followed by their discovery in the Coma and Virgo clusters (Serlemitsos *et al.* 1977), established immediately that the emission mechanism of clusters was thermal emission. This implied that all of the apparently empty space between the galaxies in clusters was actually filled by diffuse, hot gas with typical temperatures and electron densities of $T \sim 7 \times 10^7$ K and $n_e \sim 10^{-3}$ cm^{-3}. The total mass of this gas was found to somewhat exceed the total mass of all of the galaxies in the clusters. The strength of the iron K lines indicated that the iron abundances in these three clusters were about half of the solar value. Since the only known sources for the production of heavy elements are in the interiors of stars, and the only significant populations of stars are located in galaxies, this indicated that a significant portion of the intracluster gas had come out of stars and

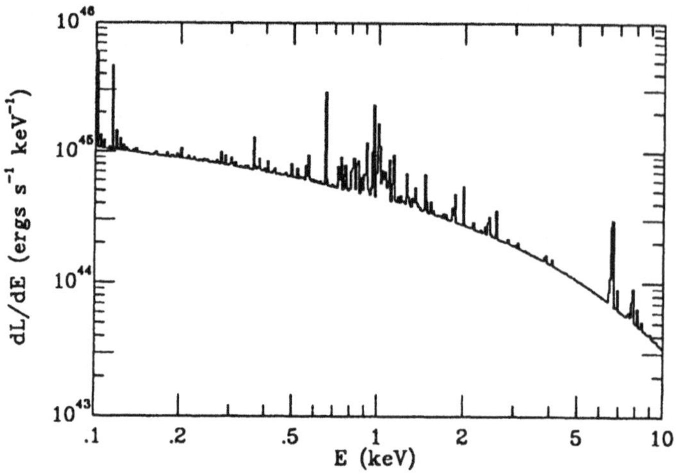

Figure 1. The predicted X-ray spectrum of a cluster cooling flow with a total cooling rate of $\dot{M} = 300\, M_{\odot}\, \mathrm{yr}^{-1}$. The cluster model is the homogeneous model in Figure 2.

been ejected from galaxies. Since the present rates of stellar mass loss in galaxies are generally thought to be at least two orders of magnitude too small to have produced this gas, this suggested that galaxies may have had much more star formation and mass loss in the past (*e.g.*, Sarazin 1988).

X-ray Fe line spectroscopy also led to the detection of low ionization X-ray lines from the centers of some clusters of galaxies (Canizares *et al.* 1988; Mushotzky 1984). This discovery, made with the *Einstein* X-ray Observatory, indicated that there are considerable amounts of cooler gas ($T \sim 10^6 - 10^7$ K) at the cluster centers. It is thought that this gas originates in "cooling flows" of gas moving in from the intracluster medium. While there is considerable evidence supporting the cooling flow picture from X-ray images of clusters, alternative theories have been proposed to explain the observations. The X-ray spectroscopy, however, remains the most difficult piece of evidence to reconcile with these alternative theories.

2. IRON LINES FROM CLUSTER COOLING FLOWS

In many clusters of galaxies, the intracluster gas densities in the central regions of the cluster are sufficiently large that the gas can cool within the lifetime of the cluster. At a typical point in these "cooling flow" clusters, the cooling time is shorter that the age of the cluster (estimated to be 10^{10} years), but is much longer than the dynamical time (or sound–crossing time) for the central regions of the cluster. Under these conditions, the gas will remain nearly in hydrostatic equilibrium as it cools. As gas in the center of the cluster cools, its density will increase as it is crushed by the weight of the exterior gas. This compression produces a slow, subsonic inflow of the gas. Once gas cools significantly below typical intracluster gas temperatures of $T \sim 7 \times 10^7$ K, the cooling accelerates rapidly due to the compression of the gas and the increase in the cooling function below $\sim 3 \times 10^7$ K. As a result, the gas cools rapidly out of the X-ray emitting temperature

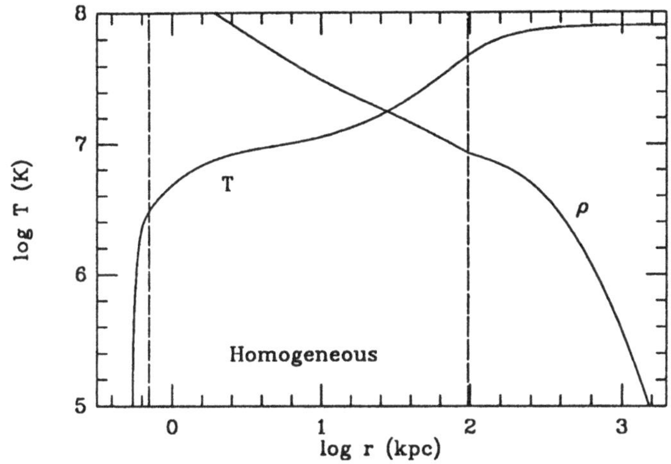

Figure 2. The variation of the gas density and temperature with radius in a homogeneous cluster cooling flow model. Gas is cooling at a rate of 300 $M_\odot$ yr^{-1}. The outer vertical dashed line marks the "cooling radius," the radius at which the cooling time is equal to the Hubble time. The "cooling flow" is the region interior to this radius. The inner dashed line marks the sonic radius r_s of the inflow, at which the inflow becomes transonic.

range to temperatures of $T \ll 10^6$ K. Thus, a characteristic feature of cooling flows is the emission of low ionization X-ray lines, such as the Fe L lines. These lines have been detected from several nearby cooling flow clusters with the *Einstein* Observatory. Fe XVII and O VIII lines have been detected from several clusters with the FPCS by Canizares *et al.* (1988). The SSS was used to observe the Fe L line complex as well as the K lines from Mg, Si, S, and Ca by Mushotzky (1984).

Figure 1 shows the integrated X-ray spectrum for a theoretical model of a cooling flow cluster with a total cooling rate of $\dot{M} = 300\, M_\odot$ yr^{-1}. This spectrum shows the strong Fe L line complex around 1 keV, the Fe L $\Delta n = 0$ lines near 0.1 keV, and the O VIII K line. The Fe K lines near 7 keV originate primarily in the cluster as a whole, rather than the central cooling flow region. Measuring the luminosity of X-ray lines produced at low temperatures (such as Fe XVII) allows one to directly determine the total cooling rate in the cluster. Under the coronal conditions which apply in the intracluster gas, the emissivity of any line can be written as

$$\epsilon_{line} = \rho^2 \Lambda_{line}(T) , \tag{1}$$

where ρ is the gas density and $\Lambda_{line}(T)$ is a function of the temperature and atomic physics. Similarly, the total radiative cooling rate of the gas can be written as

$$\epsilon_{total} = \rho^2 \Lambda_{total}(T) . \tag{2}$$

If one assumes that the gas cools isobarically from a temperature T_o to a very low temperature, the luminosity of any line produced by cooling is given by (*e.g.*, White and Sarazin

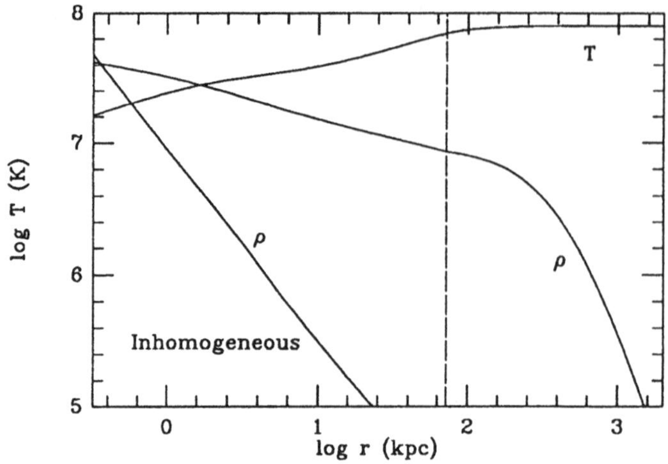

Figure 3. The variation of the gas density, temperature, and gas loss rate with radius in an inhomogeneous cluster cooling flow model. The rate of inhomogeneous cooling is given by equation (4) with $q = 4$. The cluster model and total cooling rate are the same as in Figure 2. The vertical dashed line marks the cooling radius; there is no sonic radius.

1987)

$$L_{line} \approx \frac{5}{2}\frac{\dot{M}}{\mu m_p} k \int_0^{T_o} \frac{\Lambda_{line}(T)}{\Lambda_{total}(T)}\, dT. \tag{3}$$

Here, $\mu \approx 0.62$ is the mean mass per particle in the gas in units of the proton mass m_p. If one considers an X-ray line produced at temperatures which are below those of the ambient intracluster gas, then the luminosity is independent of the upper limit on the integral T_o. This expression indicates that the line luminosities of cooling flows are independent of many of the details of how the gas cools.

For example, if the density of the gas is increased, both the rate of line emission and the cooling rate are increased by the same factor, and the amount of line emission per gram of cooling gas remains unchanged. Under some circumstances, the gas may cool at constant density, but this only changes the coefficient in front of the integral to 3/2, and reduces the line luminosity by at most 40%. Similarly, the line luminosity is nearly independent of the overall abundance of heavy elements. This result occurs because the total cooling rate is dominated by heavy elements, so changing the heavy element abundance affects Λ_{line} and Λ_{total} by almost the same factor. The luminosities of individual X-ray lines are affected by the ratios of abundances of heavy elements, however.

The result is that the luminosities of low ionization X-ray lines (such as Fe XVII and other Fe L lines) are determined almost exclusively by the total cooling rate. These lines, therefore, provide a direct means of determining these rates. Applying this technique to the rather small set of existing data leads to typical cooling rates of $\sim 100\, M_{\odot}\, \mathrm{yr}^{-1}$.

The surface brightness of the soft X-ray lines can show us where the gas is cooling below X-ray emitting temperatures. There are two general ways in which the gas might cool. If the gas is homogeneous, then all of the gas at a given radius will have the same temperature and density, and all gas will cool at the same rate. In this case, the gas

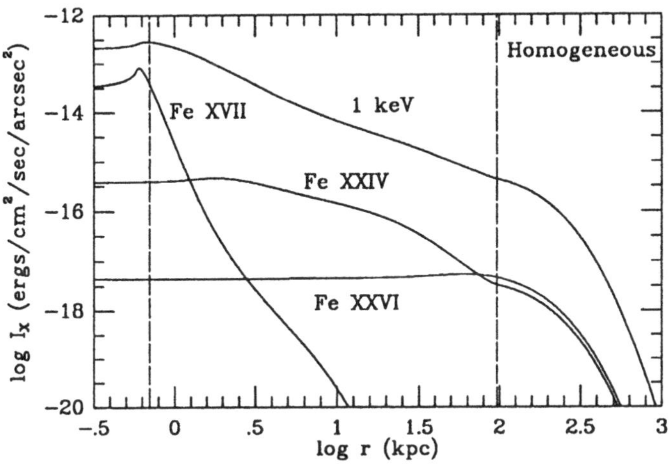

Figure 4. The surface brightness of several X-ray features as a function of projected radius in the homogeneous cluster cooling flow model of Figure 2. The curves labeled by ions give the strongest emission line from each ion. The curve labeled "1 keV" gives the continuum surface brightness at a photon energy of 1 keV. The vertical dashed lines mark the same positions as in Figure 2.

cools and slowly flows towards the center of the cluster. Because the inflow is subsonic throughout most of the flow, the gas is nearly in hydrostatic equilibrium, and remains fairly hot. However, once the inflow becomes transonic, the gas cools catastrophically (Soker and Sarazin 1988). Typically, the radius at which the gas becomes transonic is $r_s \lesssim 1$ kpc. Figure 2 shows the variation of the gas density and temperature in a homogeneous cluster cooling flow model. Note the relatively slow decrease in the gas temperature outside of the sonic radius r_s, and the rapid cooling within this radius.

A second possibility is that the gas in cluster cooling flows is very inhomogeneous. In this case, it is likely that the denser, cooler lumps of gas will cool more rapidly than the diffuse, hotter gas. If the variations in density are large, some of the gas may cool below X-ray emitting temperatures while the remainder of the gas stays hot. One way to approximate this complex situation is to treat the gas as a multiphase medium, assuming that all of the gas at any given radius has the same pressure and velocity (Thomas *et al.* 1987). Another way to describe inhomogeneous cooling flows is to divide the gas into two components: rapidly cooling condensates, and an ambient medium (White and Sarazin 1987). We have adopted the latter approximation. Following White and Sarazin, we assume that a portion of the gas cools rapidly to low temperature, and that the rate at which gas cools is given by

$$\dot{\rho} = q\frac{\rho}{t_{cool}} = \frac{2}{5}q\frac{\rho^3\Lambda_{total}(T)}{P} \tag{4}$$

where t_{cool} is the isobaric cooling time, P is the pressure, and q is the gas loss efficiency. Figure 3 shows a model for a cluster cooling flow with the same parameters as in Figure 2, but assuming inhomogeneous cooling with $q = 4$. In this case, the gas cools below X-ray emitting temperatures throughout the cooling flow. Analyses of the images of cluster

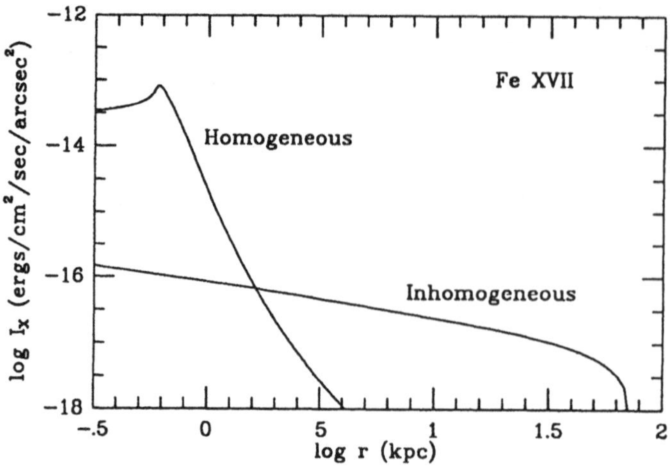

Figure 5. The surface brightness of the Fe XVII line as a function of projected radius in the homogeneous and inhomogeneous cluster cooling flow models (Figure 2 and 3, respectively).

cooling flows from the *Einstein* Observatory suggest that the cooling in clusters is very inhomogeneous, and that most of the gas cools below X-ray emitting temperatures at relatively large radii $r \sim 30 - 100$ kpc (Thomas *et al.* 1987; White and Sarazin 1988).

Figure 4 shows the surface brightness of several X-ray features from the homogeneous cluster cooling flow model of Figure 2. Note that the Kα line of iron is not concentrated in the cooling flow, while the softer X-ray lines and continuum are. Most of the Fe XVII line emission comes from a limb-brightened shell just within the sonic radius of the inflow. The lower energy X-ray lines are produced by gas which is cooling out of the X-ray emitting range, so the surface brightness of these lines shows us where the gas is cooling. Figure 5 compares the surface brightness of the Fe XVII line from the homogeneous cooling flow model (Figure 2) with the inhomogeneous model of Figure 3. In the homogeneous model, the gas cools catastrophically just within the sonic radius, producing a limb-brightened shell. In the inhomogeneous model, gas cools throughout the central region of the cluster, and the Fe XVII surface brightness is much more extended.

If substantial improvements in spectral resolution can be achieved, spectral profiles of iron X-ray lines can be used to measure inflow velocities in the cooling flow gas. If the only source of local line broadening is thermal motion, lines from iron will be rather narrow in the emitted frame ($FWHM \approx 50$ km s^{-1}). However, it is likely that the gas will be significantly turbulent. If the gas has barely subsonic turbulent motions, lines will tend to be rather broad ($FWHM \lesssim 2000$ km s^{-1}). Figure 6 shows the integrated line profiles of the Fe XVII line from the homogeneous and inhomogeneous cooling flow models (Figures 2 and 3). In the homogeneous model, the Fe XVII line is produced in a transonically infalling shell, and the line has a characteristic square profile. In the inhomogeneous model, the line is mainly produced at large radii where the inflow velocities are small. If there is no turbulent broadening, the line is very narrow. However, the gas at large radii in the inhomogeneous models is fairly hot (Figure 3), and the line will be very broad if there is transonic turbulence.

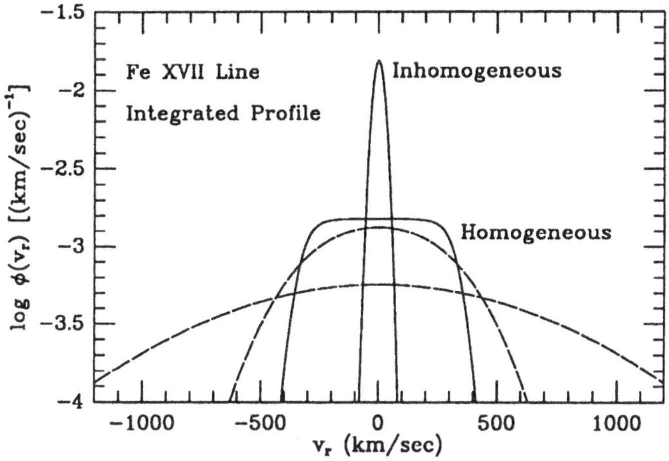

Figure 6. The line profiles of the Fe XVII line for homogeneous (Figure 2) and inhomogeneous (Figure 3) cluster cooling flow models. The line profile function $\phi(v_r)$ integrates to unity over the profile. The observed photon energy is converted to an equivalent radial velocity v_r. The solid lines assume only thermal line broadening; the dashed lines assume just transonic turbulence.

3. IRON LINES FROM ELLIPTICAL GALAXIES

Observations with the *Einstein* X-ray Observatory have established that normal elliptical and S0 galaxies are generally strong X-ray sources (Forman, Jones, and Tucker 1985; Trinchieri and Fabbiano 1985). The X-ray emission indicates that these galaxies contain extensive atmospheres of hot, diffuse interstellar gas. Prior to this discovery, it was generally believed that early-type galaxies were gas-poor systems. In the brighter X-ray ellipticals, the surface brightness of the X-ray emission indicates that the cooling time in the gas is short, and in these galaxies the gas may form a cooling flow. The X-ray properties of elliptical have been reviewed recently by Fabbiano (1989) and Sarazin (1990).

The gas in elliptical galaxies is believed to originate as mass loss from the visible stars. These stars also provide the energy in the gas, since the gas is thought to be heated by a combination of stellar motions and Type Ia supernovae. The heavy element abundances in stars in the central regions of elliptical galaxies are rather high, and Type Ia supernovae may further enhance the abundances. As a result, we expect the hot interstellar gas in ellipticals to have very high heavy element abundances. The combination of these high abundances with relatively cool gas (ambient gas temperatures of $\sim 10^7$ K), and possibly cooling gas as well, suggest that elliptical galaxies should be very strong sources of X-ray line emission. Figure 7 shows the predicted spectrum for a model for elliptical galaxies. Note particularly the very strong Fe L lines near 1 keV.

The line emission in ellipticals is so strong that it may be difficult to determine the hydrogenic continuum, except in the 3–6 keV region. This difficulty could complicate the determination of abundances.

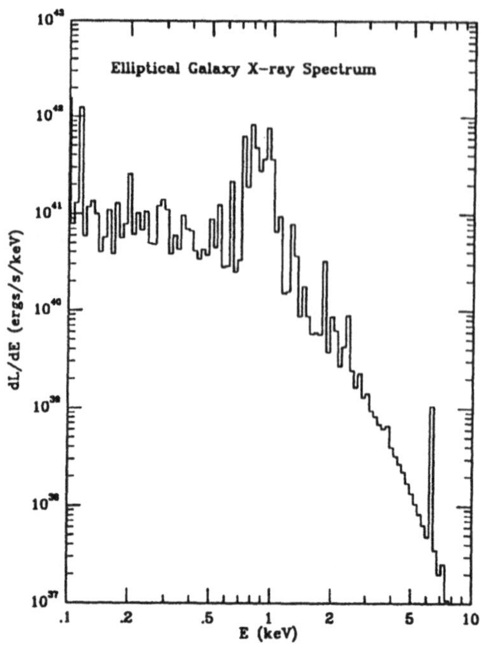

Figure 7. The model X-ray spectrum for an elliptical galaxy. The galaxy has an optical luminosity of $L_B = 10^{11}\,L_\odot$. The model assumes stellar abundances taken from Franx and Illingworth (1990), and a supernova rate of $r_{SN} = 0.075$.

There are two main sources for the heavy elements in the interstellar gas in elliptical galaxies: Type Ia supernovae, and normal stellar mass loss. The rate of normal stellar mass loss is well-determined by stellar evolution and is $\dot{M}_* \approx 1.5(L_B/10^{11}L_\odot)\,M_\odot\,\mathrm{yr}^{-1}$ (Renzini 1990), where L_B is the blue optical luminosity of the galaxy. The rate of Type Ia supernovae, however, is much less certain. Recently, Evans *et al.* (1989) have found a rate of $r_{SN} \approx 0.075(L_B/10^{10}L_\odot)$ per century. This rate is about three times lower than the previous rate from Tammann (1982), and it is probably fair to treat this difference as an indication of the uncertainty in the rate. If each Type Ia supernova produces a mass M_{Fe} of iron, then the interstellar iron abundance will be approximately

$$\frac{(\mathrm{Fe/H})_{gas}}{(\mathrm{Fe/H})_\odot} \approx \left[\frac{(\mathrm{Fe/H})_*}{(\mathrm{Fe/H})_\odot} + 2.4\left(\frac{M_{Fe}}{0.6\,M_\odot}\right)\left(\frac{r_{SN}}{0.075}\right)\right], \tag{5}$$

where, $(\mathrm{Fe/H})_*$ is the iron abundance in the normal stellar ejecta. The iron abundances in the stars in elliptical galaxies are believed to be $(\mathrm{Fe/H})_* \approx 3(\mathrm{Fe/H})_\odot$ in the center of the galaxy, and $(\mathrm{Fe/H})_* \approx 1.5(\mathrm{Fe/H})_\odot$ on average. The supernova rate used in equation (5) is at the low end of the suggested rates, so the iron abundances in ellipticals are expected to be $(\mathrm{Fe/H})_{gas} \gtrsim 4(\mathrm{Fe/H})_\odot$, and could reach ten times solar if the higher Tammann supernova rates are correct. Thus, one expects very high interstellar iron abundances. The iron comes primarily from supernovae, so the iron abundance gives a direct means of determining the supernova rate in elliptical galaxies. Since supernovae provide an important source of heating for the interstellar gas in ellipticals, the iron abundance also measures this heating rate.

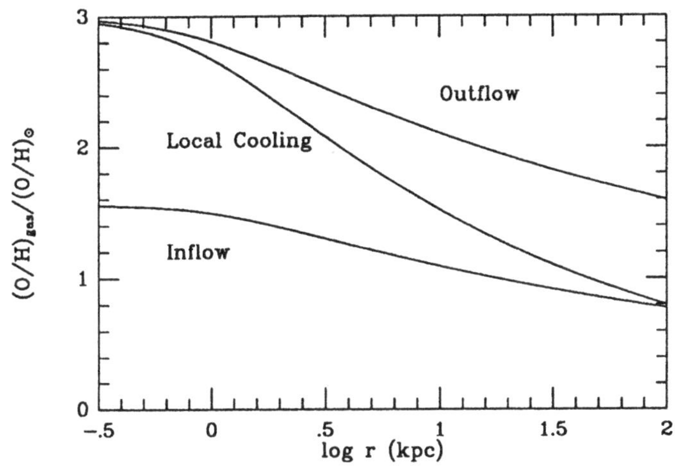

Figure 8. The interstellar oxygen abundance in three models for the dynamics of the gas in elliptical galaxies. In the model labeled "Outflow," the gas undergoes subsonic inflation. The "Inflow" model represents a homogeneous cooling flow, while the "Local Cooling" models is a cooling flow which is so inhomogeneous that most of the gas cools near the location where it is injected by stars.

There are two basic models which have been proposed for the dynamics of the gas in ellipticals. It has been suggested that the gas forms a steady-state cooling flow, which is similar to the situation in clusters (see Sarazin 1990 for a review of the arguments supporting this theory). If the gas forms a cooling flow, it might be homogeneous or inhomogeneous (as was the case with cluster cooling flows). In either case, the total rate of heating of the gas equals the total rate of cooling. Since the gas cools radiatively, the total cooling rate will equal the X-ray luminosity. Alternatively, D'Ercole *et al.* (1989) have suggested that the gas may undergo subsonic inflation. In this model, the heating rate must substantially exceed the X-ray luminosity.

X-ray line derivations of the abundances in the interstellar gas can also be used to determine the dynamical behavior of the gas in ellipticals. Such observations can allow one to distinguish between inflow and outflow models for the gas. The stellar abundance gradients observed in elliptical galaxies will produce gradients in the interstellar abundances, particularly for elements which are not provided mainly by Type Ia supernovae. An example is oxygen; current Type Ia supernovae are estimated to produce only about 5% of the oxygen in the interstellar gas in ellipticals. Since the oxygen comes mainly from normal stellar mass loss, the abundance of oxygen should be strongly affected by stellar abundance gradients. It is also strongly affected by the dynamics of the gas. If the gas in involved in an outflow (D'Ercole *et al.* 1989), the oxygen atoms at any radius will have come from stars interior to that radius. Since stellar abundances increase towards the galaxy center, this will increase the abundance of oxygen. If the gas is involved in a homogeneous inflow, all of the oxygen comes from the outer regions of the galaxy, and the oxygen abundance is depressed. If the gas involves an inhomogeneous cooling flow, much of the gas will cool below X-ray emitting temperatures near the point where it is injected by stars, and the interstellar abundances will mirror the abundances in the stellar ejecta.

Figure 8 shows the predicted interstellar abundances of oxygen for these three models for the dynamics of the gas in ellipticals. The stellar abundance gradient was taken from Franx and Illingworth (1990). Observations of X-ray lines intensities and the abundance information they contain can tell us what the gas is doing, even if the observations lack the spectral resolution to directly detect the Doppler shifts associated with the motions in the gas.

ACKNOWLEDGMENTS

We would like to thank Aldo Treves for all of his efforts to organize this very useful meeting. This work was supported in part by NASA Astrophysical Theory Program Grant NAGW-764.

REFERENCES

Canizares, C. R., Markert, T. H., and Donahue, M. E. 1988, in *Cooling Flows in Clusters and Galaxies*, ed. A. C. Fabian (Kluwer: Dordrecht), p. 63.

D'Ercole, A., Renzini, A., Ciotti, L., and Pellegrini, S. 1989, *Ap. J. (Letters)*, **341**, L9.

Evans, R., van den Bergh, S., and McClure, R. D. 1989, *Ap. J.*, **345**, 752.

Fabbiano, G. 1989, *Ann. Rev. Astr. Ap.*, **27**, 87.

Forman, W., Jones, C., and Tucker, W. 1985, *Ap. J.*, **293**, 102.

Franx, M., and Illingworth, G. 1990, *Ap. J. (Letters)*, **359**, L41.

Mitchell, R. J., Culhane, J. L., Davison, P. J., and Ives, J. C. 1976, *M.N.R.A.S.*, **175**, 29p.

Mushotzky, R. F. 1984, *Phys. Scripta*, **T7**, 157.

Renzini, A. 1990, in *Window on Galaxies*, ed. by G. Fabbiano, J. S. Gallagher, and A. Renzini, p. 255.

Sarazin, C. L. 1988, *X-ray Emission from Clusters of Galaxies*, (Cambridge Univ.: Cambridge).

Sarazin, C. L. 1990, in *The Interstellar Medium of External Galaxies*, ed. by H. A. Thronson, Jr. and J. M. Shull (Dordrecht: Kluwer), 201.

Serlemitsos, P. J., Smith, B. W., Boldt, E. A., Holt, S. S., and Swank, J. H. 1977, *Ap. J. (Lett.)*, **211**, L63.

Soker, N., and Sarazin, C. L. 1988, *Ap. J.*, **327**, 66.

Tammann, G. A. 1982, in *Supernovae: A Survey of Current Research*, ed. by M. J. Rees and R. J. Stoneham (Dordrecht: Reidel), p. 371.

Thomas, P. A., Fabian, A. C., and Nulsen, P. E. J. 1987, *M.N.R.A.S.*, **228**, 973.

Trinchieri, G., and Fabbiano, G. 1985, *Ap. J.*, **296**, 447.

White, R. E., and Sarazin, C. L. 1987, *Ap. J.*, **318**, 621.

White, R. E., and Sarazin, C. L. 1988, *Ap. J.*, **335**, 688.

Iron Line Emission Features in Clusters of Galaxies

Katsuji KOYAMA

Department of Astrophysics, School of Science, Nagoya University
Furo-cho, Chikusa-ku, Nagoya, 464-01

Abstract

The Ginga satellite discovered large scale X-ray emission extending about 12 ° from the Virgo cluster. From the scan profile and the temperature structure, we suggest that the intra-cluster gas of Virgo has two components around the large elliptical galaxies, M87 and M49. This two-subclustering indicates that the Virgo cluster is now under strong evolution.

We demonstrate that the simple evolutionary scenario can roughly explain the relation between X-ray luminosity and gas temperature from many clusters of galaxies observed with the Ginga satellite.

The iron abundance of the outer part of the Virgo cluster is only about 0.1-0.2 of cosmic value. We therefore interpret this to mean that the major components of intra-cluster gas are primordial. However the iron abundance of the Virgo cluster shows central concentration near M87, which indicates that the central part was contaminated by the gas ejected from each galaxy.

The iron abundance from individual clusters show a trend of anti-correlation with the gas temperature. One possible reason for this anti-correlation is that the galaxy formation rate is smaller in clusters of hotter intra-cluster gas.

1. Introduction

The diffuse X-ray emission from clusters of galaxies is one of the most important discoveries of X-ray astronomy. The best X-ray spectrum of a cluster above 2 keV was obtained with the Gas Scintillation Proportional Counters (Koyama et al. 1984) on board the Temma satellite (Tanaka et al. 1984). The X-ray spectra from three bright clusters; the Perseus, Coma and Ophiuchus clusters, are given in the figure 1 by Okumura et al. (1985). We see clear peaks at the energies of 6.7 keV and 8 keV, which are attributable to $K\alpha$ and $K\beta$ lines from highly ionized iron atoms in a thin hot plasma. In fact, the X-ray spectrum can be fitted well with the model of a thermal emission from a thin hot plasma of a temperature of about 10 keV.

In this paper, we will discuss the iron line features of Clusters of galaxies based on the recent observations with the Ginga satellite.

2. Large Scale Structure of the Virgo Cluster

The Virgo cluster is the nearest rich cluster (about 20 Mpc distance) and has a large optical extension of about 12° along the major axis of the cluster. However the previous X-ray instrument had obtained X-ray emission within the small region of less than 2 deg radius from the cluster center M87.

We have conducted X-ray scanning observations with the large area proportional counters (LAC) on board the Ginga satellite (Turner et al. 1989, Makino et al. 1987) along the cluster major axis which runs over two major elliptical galaxies M87 and M49, and two parallel paths about 3 degrees away.

From the profile along the major axis, we found a large scale extended X-ray emission of about 12° covering the whole Virgo cluster (Takano *et al.* 1989). To the scan profile, we tried model fitting of the surface brightness with the standard model;

$$S(r)= S(0)\ \{1+(r/a)^2\}^{-3\beta+0.5} \qquad (r<r_{cut})$$

Figure 1 shows the fitted curve with this model. We used the same parameters as those determined with the Einstein IPC observation of the surface brightness within the small region near M87. We find that the large scale extended emission can be understood as a smooth extrapolation from the Einstein IPC result. Furthermore, we find a more extended component toward the southern part of the Virgo cluster.

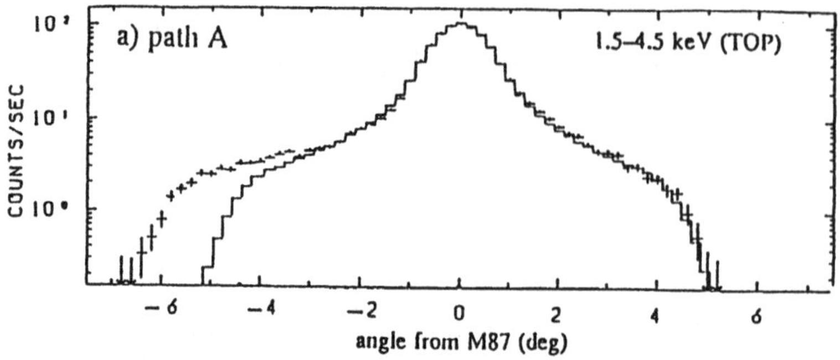

Fig. 1 The fitting curve with the standard model, where the parameters are the same as those determined from the Einstein IPC observation.

Near the central region, the X-ray spectrum was found to exhibit two components: a thermal emission and a power-law component (Takano and Koyama , 1991). Therefore we have carried out two component fitting and determined the temperature. We found that the power-law component disappears at an angle larger than 1 degree from M87. Thus the distribution of the power-law component is the same as the Ginga collimator response, indicating that the power-law component comes from very near M87. Using independent scans near the cluster center, Takano and Koyama (1991) found that the power-law emission is due to two AGNs, M87 and a Seyfert 2 galaxy NGC4388. The angular separation between these two AGNs is significantly smaller than the collimator angular response.

In the outer region, the X-ray spectrum can be fitted by a single temperature thermal emission model with no power-law component; therefore we can estimate the temperature from the single thermal model fitting. If we scan further out to more than about 3 degree from M87, the photon statistics become very poor; therefore we determined the temperature using the two energy band intensity ratio, the flux in the 4.5-9.2 keV band divided by the flux in the 1.5-4.5 keV band. This flux ratio was converted to the temperature assuming single component thermal emission. Figure 2 shows the temperature distribution of the Virgo cluster along the major axis. Since the field of view of the LAC instrument is as large as about 1 degree, we could not find any sign of low temperature cooling flow at the cluster center M87. Instead, we found two clusterings of the temperature; one is a 2-keV component around M87 and the other is a 3-keV component around M49.

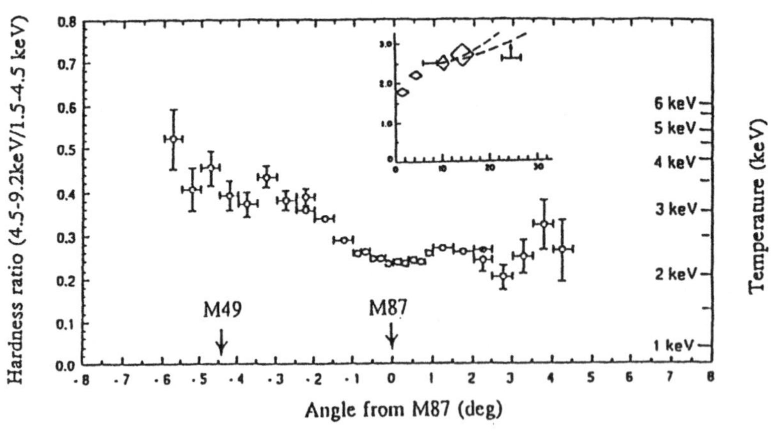

Fig. 2 The temperature distribution of the Virgo cluster along the major axis together with the central part by Fabricant *et al.* (1980).

The two temperature sub-groups of the X-ray emitting region of the Virgo cluster are illustrated in figure 3 overlaid on the optical galaxy map, where each circle shows member galaxies with larger ones corresponding to brighter galaxies. This sub clustering indicates that the Virgo cluster is not relaxed yet. In fact, from the optical observations, Vaucouleurs (1961) already pointed out two major concentrations of the member galaxies around M87 and M49. Therefore the present discovery of two temperature sub-clustering provides us with the independent suggestion that the Virgo cluster is undergoing strong evolution at the present time.

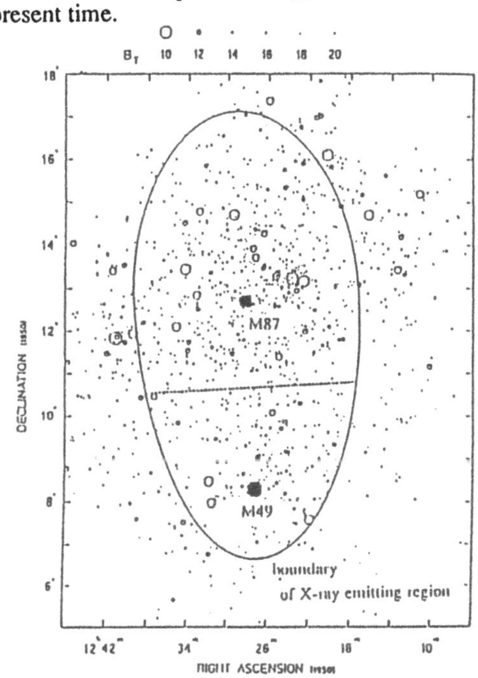

Fig. 3
The two temperature subgroups of the X-ray emitting region of the Virgo cluster on the optical map of the Virgo cluster, where each circle shows a member galaxy.

As the Virgo cluster becomes more evolved, the two sub-clusters will merge with each other. Then, the size of the cluster becomes small and, hence the gas density becomes large. Assuming that total mass is conserved throughout the evolution,the gas temperature (T) would be proportional to the depth of the gravitational potential, or the inverse of the radius (R),

$$T \propto R^{-1}.$$

The X-ray luminosity is proportional to the square density (n), volume ($\propto R^3$) and Gaunt factor ($\propto T^{0.5}$). Therefore we get following equation:

$$L_X \propto n^2 R^3 T^{0.5} \propto T^{0.5} R^{-3}.$$

Then we can get the luminosity and temperature evolution with the function,

$$L_X \propto T^{3.5}.$$

The relation of the temperature and luminosity of many clusters observed by Hatukade (1989) with the Ginga satellite is given in figure 4. The solid line is the $L_X \propto T^{3.5}$ relation which is derived by the simple estimation. Although the best-fit power-law relation gave the index of 3.2, you see that the simple estimation assuming that the cluster is evolving to become more condensed can roughly explain the Ginga results.

The luminosity evolution of cluster of galaxies is also found in the log*N*-log*S* relation. Edge *et al.* (1990) found a clear deviation from the Euclidean distribution even at small cosmological red shifts of about 0.1-0.2. This indicates that the X-ray luminous clusters are undergoing strong luminosity evolution even at this moment.

Fig. 4
The relation of the temperature and luminosityof many clusters observed with the Ginga satellite. The solid line is $L_X \propto T^{3.5}$ relation .

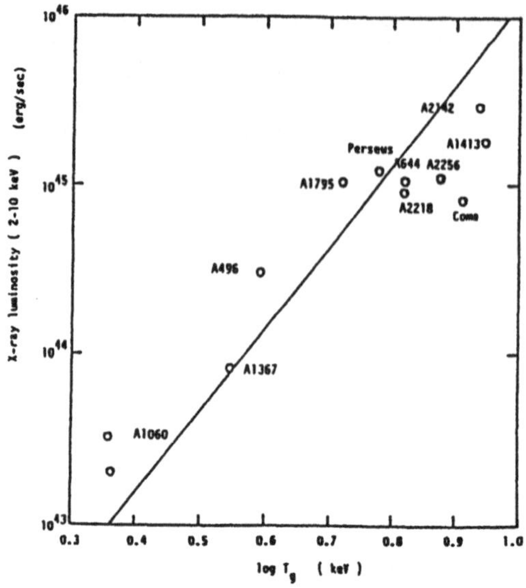

3. Iron Distribution in the Virgo Cluster

Since heavy elements such as iron are not primordial but originates in stars, the iron line is a powerful tool to study the origin of the intra-cluster gas. The K_α and K_β energies were well determined with the Temma satellite (see figure 3 by Okumura *et al.* 1987) . From the figure, we see K_α and K_β energies are roughly consistent with the theoretical values in the ionization equilibrium. The energy resolution of the Ginga satellite is not as good as the Temma satellite. However the effective area is ten times larger, thus enabling us to obtain better statistics. Hatsukade (1989) have determined the line energy with a single emission line. This model found the line energy to be significantly larger than the K_α line energy. Therefore he added another line which represents the K_β line. Figure 8 shows the relation between the gas temperature and iron line K_α and K_β energy by Hatsukade (1989). The data fit well to the theoretical lines in which we have assumed ionization equilibrium. Therefore we can conclude that the hot plasma in the cluster of galaxies is now in ionization equilibrium. Then we can easily estimate the iron abundance from the observed iron line intensity .

The distribution of the iron line intensity along the major axis of the Virgo cluster was observed by Takano (1990). Possible contamination from the power-law sources, M87 and NGC4388, is estimated to be 10 % assuming that the equivalent widths of the iron lines from these AGNs are about 200-300 eV (see Awaki et al. 1991). Therefore we will neglect this contamination in this paper. We then obtained the equivalent width of the iron line using the continuum intensity of the thermal emission. Since the equivalent width is not a steep function of the temperature in the range of 2-3 keV (ex. see Rothenflug and Arnaud, 1985), we can reliably estimate iron abundance in the units of the cosmic value. Koyama *et al.* (1990) showed the distribution of iron abundance along the major axis of the Virgo cluster, and found a clear peak of iron abundance near the cluster center at M87. We can conclude that the iron abundance from the central region (within 1° from M87) is about 0.5 cosmic value. The iron abundance at the outer region of the cluster (larger than 1° from M87) is 0.1~ 0.2 cosmic value.

Since the iron abundance of the outer part of the Virgo cluster is only 0.1-0.2 of cosmic value, we can assume that the intra-cluster gas of Virgo mainly originated from the primordial gas. The outer region of the Virgo cluster has not yet been contaminated by the gas ejected from each galaxy. By contrast, the central region of the cluster has been contaminated by a more efficient mass-loss from galaxies due to a denser galaxy distribution than the outer region of the cluster. Therefore the iron line distribution of the Virgo cluster gives us good example of tracing the hot gas evolution in the cluster. We note here that Ponman *et al.* (1990) also obtained a similar abundance distribution from the Perseus cluster.

4. Iron Abundance and Gas Temperature

The Ginga field of view is too large to investigate the spatial structure other than the Virgo cluster. Therefore we will discuss the mean iron abundance and the gas temperature of the cluster of galaxies. In figure 5, Hatsukade (1989) summarized the iron abundance and the temperature from many clusters of galaxies. We find that the iron abundance is not constant but shows a clear decrease as the gas temperature increases.

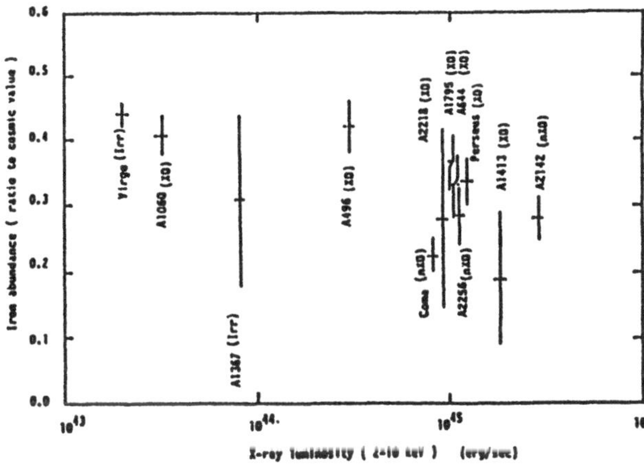

Fig. 5 The iron abundance and the temperature from many clusters of galaxies (Hatsukade 1989).

We will assume that the intra-cluster gas consists of two components: the primordial gas which includes no iron, and the gas ejected from the galaxy with cosmic abundance. Then the iron abundance in the intra-cluster gas is a simple function of the ratio of the primordial gas and the ejected gas. From the observation, we found when the gas temperature is high, iron abundance is small. If we simply assume that the temperature is proportional to the depth of the gravitational potential of the cluster, then a cluster of deeper potential tends to have a smaller amount of ejected gas. This may be interpreted that the formation rate of the galaxies is small in a cluster of deep gravitational potential. In fact David et al. (1990) reported that the ratio of gas mass and galaxy mass positively correlated with the gas temperature.

The present observation was provided by a non-imaging device and hence, the energy resolution is not good. I believe that further systematic observations of the temperature, the mass and the iron abundance in the hot gas from samples of many clusters with fine spatial and energy resolution is essential to investigate the formation and evolution of the clusters. This will be a major subject of the next Japanese satellite ASTRO-D.

References

Awaki, H. *et al.* 1991, submitted to *Publ. Astron. Soc. Japan.*

David, L. P. *et al.* 1990, *Astrophys. J.*, **356**, 32.

Edge,A.C, *et al.* (1990), *M.N.R.A.S.*, **245**, 559.

Fabricant,D.. 1980, *Astrophys. J.*, **241**, 552.

Hatukade, I. (1989), Ph .D. Thesis, Osaka University

Koyama, K. *et al.* 1984, *Publ. Astron. Soc. Japan*, **36**, 659.

Koyama, K. *et al.* (1990), submitted to *Nature*

Makino, F. and the Astro C team 1987, *Astrophys. Letters Commun.*, **25**, 223.

Okumura, Y. *et al.* (1988), *Publ. Astron. Soc. Japan*, **40**, 639.

Ponman, T. J., *et al.* (1990), *Nature* **347**, 450.

Rothenflug,R., and Arnaud,M. (1985) , *Astr. Astrophys.* 144, 431

Takano, S. (1990). Ph. D. Thesis, The Tokyo University

Takano, S. and Koyama, K.(1991) *Publ. Astron. Soc. Japan*, **43**, (accepted).

Tanaka, Y. *et al.* 1984 *Publ. Astron. Soc. Japan*, **36**, 641.

Turner, M.L.J. *et al.* 1989, *Publ. Astron. Soc. Japan*, **41**, 345.

SPARTAN-1 X-RAY OBSERVATIONS OF IRON LINE EMISSION IN THE PERSEUS CLUSTER

M.P. Kowalski, W.A. Snyder, R.G. Cruddace, G.G. Fritz
E.O. Hulburt Center for Space Research
Naval Research Laboratory, Washington DC 20375-5000

E.E. Fenimore
Los Alamos National Laboratory
Los Alamos, New Mexico 87545

and

M.P. Ulmer
Northwestern University
Evanston, Illinois 60208

I. Introduction

The detection of iron K-line emission in the Perseus cluster of galaxies has established the fact that the dominant source of X-rays is a hot intracluster gas emitting thermal bremmstrahlung radiation [1,2]. Averaged over the whole cluster, this gas has an iron abundance about half that of the solar value. However, these observations were made with instruments which had wide fields-of-view and little or no spatial resolution. Observations of iron L-lines in the core of Perseus yielded an iron abundance of ~0.4 solar [3]. Although the Perseus cluster has been mapped in broad bands by imaging instruments [4,5], until recently no spatially resolved spectral observations have been made. Such measurements hold the key to determining the structure and evolution of the cluster gas as derived from the density, temperature, and abundance distributions.

SPARTAN-1 was one of the first instruments designed to provide spatially resolved spectral observations. In an earlier paper [6], we reported on the average spectra of 2 regions centered on NGC 1275: a region of radius 5 arcmin (0.16 Mpc for z=0.0183, H_0=50 km s^{-1} Mpc^{-1}) surrounding NGC 1275 and dominated by the cluster cooling flow, and an external shell having inner and outer radii of 6 and 20 arcmin (0.2-0.6 Mpc). More recently, we published results obtained by comparing the predicted count-rates produced from a 3-dimensional model to the broad-band surface brightness and hardness ratio [7]. In that work, we extended the analysis to the limit of the data, a radius of 50 arcmin, but a constant abundance of half-solar was assumed. Here we have examined the abundance-radius relation with nonconstant models. For SPARTAN-1, the abundance is driven almost entirely by the iron K-line emission.

II. Observations, Analysis, and Results

SPARTAN-1 was deployed in orbit by the shuttle Discovery on 20 June 1985 [8]. The free-flying payload contained two co-aligned X-ray proportional counters, each with a 1-12 keV bandpass, 660 cm^2 effective area, 16% energy resolution at 6 keV, and a rectangular collimator (0.5 arcmin x 3° FWHM field of view with triangular response). During the 2-day mission observations were made of the galactic center [9,10] and the Perseus cluster. Observations were accomplished by scanning the collimator field at a uniform rate across the source in a direction perpendicular to the long collimator axis. To obtain uniform angular coverage the scan angle was changed each orbit, and NGC 1275 was at the center of the scan pattern [7]. About 11 ksec of good data on Perseus exist after screening.

The data were grouped radially in bins about NGC 1275, and two distinct methods of analysis were used. In the first method each radial bin spectrum was fit independently by a single temperature thermal model [11] with temperature, abundance, and amplitude as free parameters. The galactic absorption was held fixed at our best-fit value of 2.8 x 10^{21} cm^{-2} [6]. Figure 1 is a summary of the best-fit abundances and the 1-sigma limits for a single interesting parameter [12]. The result for the innermost bin is taken from our early analysis [6], while points at greater radii are this work. In the innermost bin, a power-law point source was added to the thermal model to account for NGC 1275. There is a clear abundance gradient in this figure as there is no evidence for iron beyond a radius of ~20 arcmin. Except for the innermost bin there is excellent agreement between all our values and those obtained by the Spacelab-2 coded-aperture imaging experiment [13]. The disagreement in the innermost bin is about a factor of 2, and efforts to resolve this discrepancy have not been successful [14].

Therefore, a second method of analysis was undertaken, in which 3-dimensional models were constructed and fit to the SPARTAN-1 data (details given in [15]). A number of different analytic forms were investigated for each of the density, temperature, and abundance distributions. Different forms for the abundance distribution produced results that were in general agreement with each other, so we present results here only for a linear distribution. A summary of the best-fit model and results is given in Table 1. We can exclude a zero abundance gradient at 90% confidence, thus the abundance gradient suggested by us in [6] is now established firmly by both SPARTAN-1 and Spacelab.

III. Discussion

An important consequence of an abundance gradient in clusters is that the amount of iron required to produce the observed K-line strength as measured by wide-field instruments is reduced greatly [16]. To account for such a gradient, a number of mechanisms have been proposed in which the metal-enriched gas is separated from the galaxies [17]. Unfortunately, the predicted gradients are not well-defined and similar gradients are predicted for different mechanisms, therefore a more detailed comparison between the SPARTAN-1 data and these models is not warranted. However the SPARTAN-1 result is consistent with ram pressure models which predict that stripping should occur at a density of ~5 x 10^{-4} cm^{-3} [18]. In our best fit model (Table 1), the density drops to this value at a radius of ~24 arcmin which is in good agreement with the radius (22 arcmin) at which the linear abundance model goes to zero.

IV. References

[1] Mitchell, R.J., et al. 1977, MNRAS, **181**, 25p.

[2] Serlemitsos, P.J., et al. 1977, Ap.J. (Letters), **211**, L63.

[3] Mushotzky, R.F., et al. 1981, Ap.J. (Letters), **244**, L47.

[4] Branduardi-Raymont, G., et al. 1981, Ap.J., **248**, 55.

[5] Fabian, A.C., et al. 1981, Ap.J., **248**, 47.

[6] Ulmer, M.P., et al. 1987, Ap.J., **319**, 118.

[7] Snyder, W.A., et al. 1990, Ap.J., in print.

[8] Cruddace, R.G., et al. 1989, NRL Report 9207.

[9] Kawai, N., et al. 1988, Ap.J., **330**, 130.

[10] Snyder, W.A., et al. 1990, this conference.

[11] Raymond, J.C., and Smith, B.W. 1977, Ap.J. Suppl., **35**, 419.

[12] Malina, R., et al. 1976, Ap.J., **209**, 678.

[13] Ponman, T.J., et al. 1990, Nature, in print.

[14] Ponman, T.J. 1990, private communication.

[15] Kowalski, M.P., et al. 1990, Ap.J. (Letters), in preparation.

[16] Fabian, A.C., and Pringle, J.E. 1977, MNRAS, **181**, 5p.

[17] Sarazin, C.L. 1986, Rev. Mod. Phys., **58**, 1.

[18] Gunn, J.E., and Gott, J.R., III. 1972, Ap.J., **176**, 1.

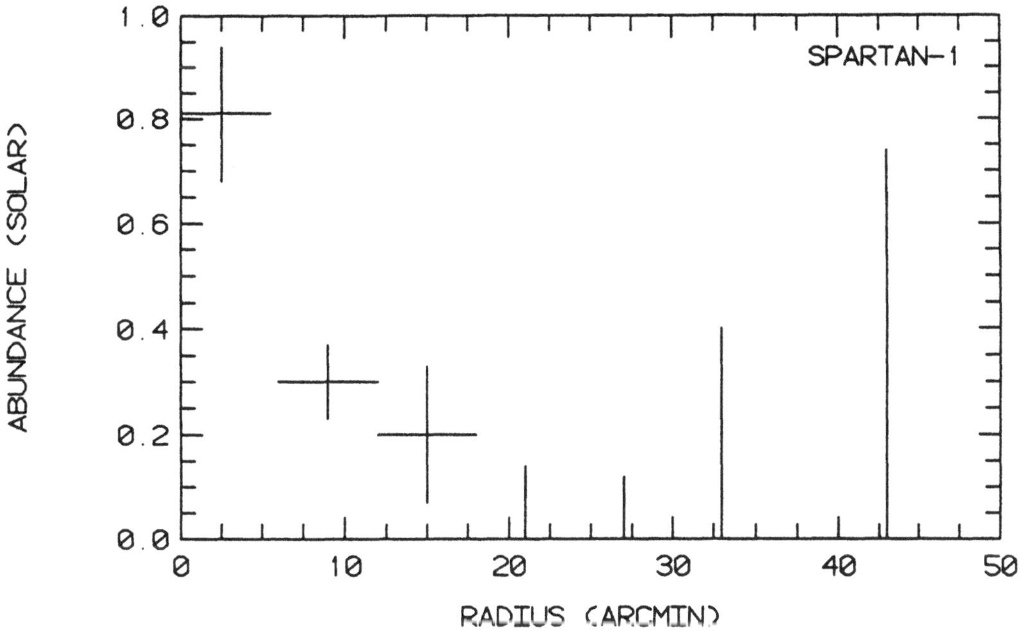

Figure 1. Abundance of elements as a function of radius. Error bars and upper limits are 1 sigma.

Table 1

MODEL AND BEST FIT PARAMETER VALUES

Density	
inner component (cooling flow): $\rho \propto (1 + r^2/a_c^2)^{-Y}$	a_c=0.20 Mpc, Y=1.89
outer component (cluster emission): $\rho \propto (1 + r^2/a^2)^{-X}$	a=0.41 Mpc, X=1.09
critical radius between inner and outer components:	r_{cr}=0.28 Mpc
density at critical radius	ρ_{cr}=1.76x10^{-3} cm^{-3}
Temperature	
inner component (cooling flow): $T \propto r^m$	m=0.54
outer component (cluster emission): $T \propto \rho^{n-1}$	n=1.0
critical radius between inner and outer components:	r_{cr}=0.28 Mpc
temperature at critical radius:	T_{cr}=7.95x10^7 K
Abundance[1]	
single component: $A = A_o + A_1 r$	A_o=0.77±0.40, A_1=-1.1±0.9 Mpc^{-1}

[1] Errors are 90% confidence for a single interesting parameter [12].

Error calculations for temperature and density parameters in progress (see also [7]).

THE DISTRIBUTION OF IRON IN GALAXY CLUSTERS – RESULTS FROM THE SPACELAB 2 X-RAY TELESCOPE

T.J.Ponman, M.P.Watt, P.A.Knight, D.Bertram, M.J.Church
C.J.Eyles, G.K.Skinner and A.P.Willmore

School of Physics and Space Research, University of Birmingham,
Edgbaston, Birmingham B15 2TT, England

ABSTRACT

Hard X-ray spectral imaging observations of a number of galaxy clusters were obtained by the coded mask X-ray telescope (XRT) on Spacelab 2. The energy band of the telescope encompasses the iron K complex, so that the spatial distribution of the iron line emission can be mapped. Here we present results for the distribution of iron abundance which can be inferred in the three brightest X-ray clusters: Perseus, Coma and Ophiuchus.

1. INTRODUCTION

The Spacelab 2 XRT[1] used coded mask imaging, in conjunction with a position sensitive proportional counter, to produce X-ray spectral images with spatial resolution of $12'$ (for the 'coarse' telescope). For the present analysis the data have been further subdivided into statistically independent subpixels $3'$ square, and binned along the spectral dimension into 42 energy channels spanning the range 3.5–25 keV.

During a shuttle flight in July–August 1985 observations of a number of clusters were obtained, including exposures of 24600 s, 27800 s and 3000 s for the Perseus, Coma and Ophiuchus clusters respectively. Hard X-ray radial surface brightness profiles for these clusters are shown in Fig.1. The process of background subtraction[2] leaves systematic residuals at the level of $\sim 1 \times 10^{-4}$ ct s^{-1} arcmin^{-2}, which is marked in the figure. In the case of Perseus and Coma, statistical errors are of the same order, and the X-ray emission can be traced out to $r \gtrsim 2$ Mpc from the cluster centre. In the case of Ophiuchus, the poorer statistics limit the detection to $r \lesssim 1$ Mpc.

2. THE PERSEUS CLUSTER

Details of the discovery of a strong iron abundance gradient in Perseus have been presented elsewhere[3], and are confirmed by the results from Spartan 1 (Kowalski, this workshop). An initial indication of this gradient can be obtained by extracting spectra from concentric annuli and fitting them with hot plasma models[4]. The result, shown in Fig.2, actually underestimates the abundance gradient, since the effects of projection along the line of sight and blurring by the instrumental point spread function are not taken into account. To allow for these, a parameterised model of the emission as a function of cluster radius was set up and fitted to the data[2]. This model included a power law point source of X-radiation arising from NGC1275, a surrounding cooling flow component based on results from earlier Einstein observations[5,6] and parameterised distributions for the density, temperature and metallicity of the hot intracluster gas.

The results, for a variety of simple models for the iron abundance distribution, indicate that (i) the abundance is low ($\lesssim 0.1$ solar) outside the cluster core, and

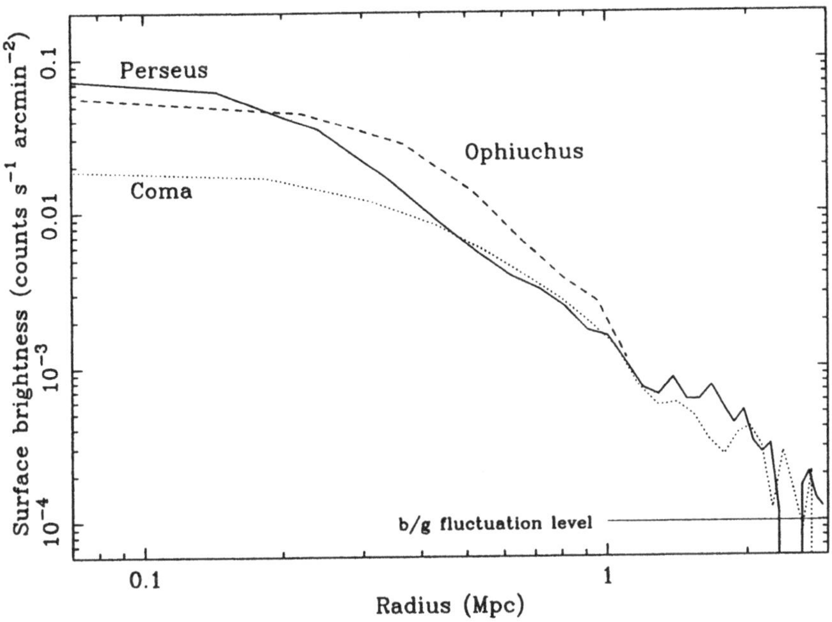

Figure 1. Hard X-ray surface brightness distributions for the Perseus, Coma and Ophiuchus clusters. Conversion of radial distances to Mpc assumes $H_o = 50$ km s^{-1} Mpc^{-1}.

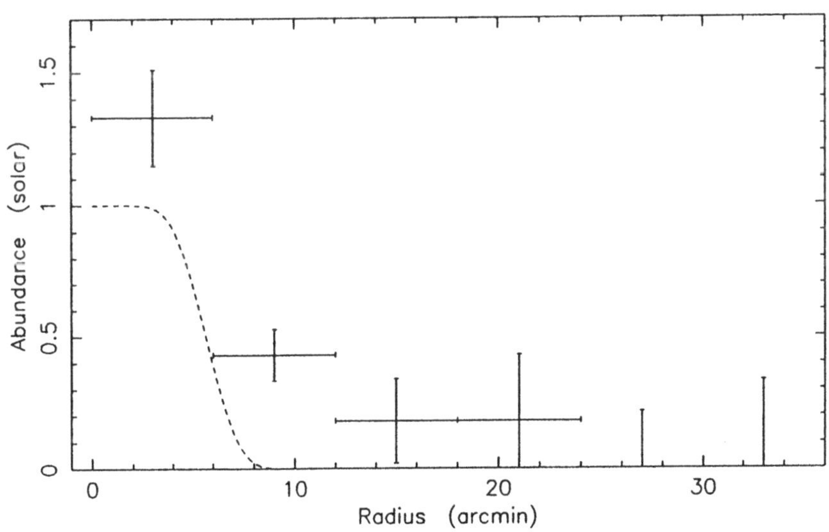

Figure 2. Iron abundance in the intracluster gas as a function of projected radius for the Perseus cluster. A power law component corresponding to emission from NGC1275, as determined from the 3D fitting, has been included in the spectral model for the innermost annulus. The dashed line shows the half-width of the instrument point spread function.

(ii) it reaches a high value ($\sim$ 2 solar) at the cluster centre. Since the work published in ref.3, we have carefully reexamined the effects of various sources of systematic error on our results and have discovered that the effects of uncertainties in our spectral response at low energies were previously underestimated. Correcting this reduces the central metallicity in our fits by about 15%. For example, with a top-hat model for the metallicity, our best fit gives $Z = 1.76$ within a radius $R_Z = 9.1'$, and $Z = 0$ (90% confidence upper limit $Z = 0.1$) outside R_Z.

The central iron abundance can be reduced somewhat further by including a 6.4 keV fluorescent iron line in the model for NGC1275. The best fit line strength of 9.4×10^{-4} photon cm^{-2} s^{-1} gives an improvement in the model fit which an F-test shows to be significant at 95% confidence. This line corresponds to an equivalent width of 1.25 keV in the spectrum of the central power law component, reminiscent of the exceptionally strong line seen in NGC1068 (ref.7), and it reduces the central metallicity in the intracluster gas to $Z = 1.54$. However, this is still well above the value obtained by the Spartan 1 group. The major source of error in our result, corresponding to perhaps 20%, is uncertainty about the properties of the cooling flow component, and an attempt to find a consistent solution for the temperature and density in the flow, using both Einstein and Spacelab XRT data, is currently underway.

3. THE COMA AND OPHIUCHUS CLUSTERS

These clusters are simpler to analyse than Perseus, since they have no central active galaxy. However the statistical quality of the XRT data are poorer, due to the lower surface brightness of Coma, and the short exposure time on Ophiuchus. The results for the metallicity as a function of projected radius, from hot plasma fits to data from concentric annuli, are shown in Fig.3. In the case of Coma, there is no obvious sign of any abundance gradient, and 3D model fitting, as described above for Perseus (but in this case without any point source or cooling flow contribution) shows that a uniform metallicity of $Z = 0.21$ is consistent with the data. This is in good agreement with the results obtained by Hughes, from observations with EXOSAT, Tenma and Ginga (ref.8 and this workshop)

A rather wide range in iron abundance has been reported from fits to the integrated spectrum of Ophiuchus[9]. We obtain $Z = 0.47^{+0.33}_{-0.26}$ (1 sigma limits), in good agreement with the Tenma result[10]. Fig.4 shows that investigation of the spatial distribution of the iron is limited by the statistical quality of the data. We can only fit the spectral data reliably out to $20'$, and it appears from the 3D fitting performed to date (which does not yet include a component for the cooling flow believed to be present in this cluster) that a uniform abundance cannot be ruled out.

4. DISCUSSION

It now seems clear that the distribution of iron is very different in the two clusters, Perseus and Coma, for which reasonable data are available. What can be the cause of this, and which is the more typical?

The intracluster gas in Coma is unusual in that it apparently contains no cooling flow[11] . The binary nature of the core of Coma, which contains two giant elliptical galaxies, suggests that it might have resulted from the merger of two smaller clusters. The resulting hiatus in the intracluster gas might be expected to disrupt preexisting cooling flows and also to spread any central metal concentration more widely within the merged cluster.

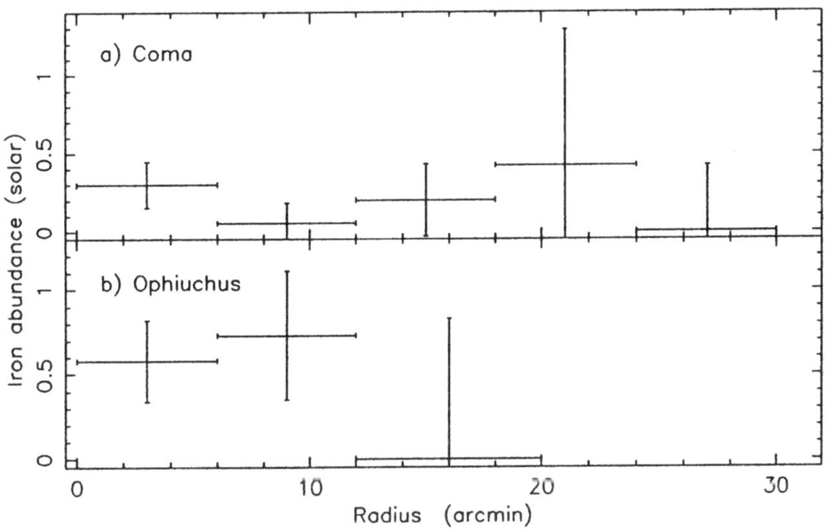

Figure 3. Iron abundance versus projected radius from spectral fits to data from concentric annuli for (a) Coma and (b) Ophiuchus.

The case of Ophiuchus is now very important. If future observations (for example with BBXRT) demonstrate that the iron is centrally concentrated, this would be consistent with a picture[3] in which the bulk of the intracluster gas is primordial, but ram pressure stripping of galaxies leads to deposition of highly metallic gas in the cluster core. Absence of a strong metallicity gradient would then denote a cluster in which considerable mixing of the intracluster gas has occurred, probably due to subcluster merging.

REFERENCES

1. Skinner G.K., *et al. Astr.Lett.Comm.* **27**, 199 (1988)
2. Eyles C.J, *et al. Astrophys.J.* (submitted)
3. Ponman T.J. *et al. Nature* **347**, 450 (1990)
4. Raymond J.C. & Smith B.W. *Astrophys.J.Suppl.* **35**, 419 (1977)
5. Fabian A.C, Hu E.M., Cowie L.L. & Grindlay J. *Astrophys.J.* **248**, 47 (1981)
6. Branduardi-Raymont G. *et al. Adv.Space Res.* **5**, 133 (1985)
7. Koyama K. *et al. Publs.Astr.Soc.Japan* **41**, 731 (1989)
8. Hughes J.P., Gorenstein P. & Fabricant D. *Astrophys.J* **329**, 82 (1988)
9. Arnaud K.A. *et al. Mon.Not.R.astr.Soc.* **227**, 241 (1987)
10. Okumura Y. *et al. Publs.Astr.Soc.Pacific* **40**, 639 (1988)
11. Fabian A.C., Nulsen P.E.J. & Canizares C.R. *Ann.Rev.Astr.Astrophys.* (in press)

THE SPATIAL DISTRIBUTION OF IRON IN THE COMA CLUSTER OF GALAXIES

John P. Hughes
Harvard-Smithsonian Center for Astrophysics
60 Garden Street, Cambridge, MA 02138 USA

1. Introduction

The discovery of K-shell emission lines from highly ionized atoms of iron in the X-ray spectra of clusters of galaxies (Mitchell *et al.* 1976, Serlemitsos *et al.* 1977) confirmed the thermal origin of the X-ray emission and led to a number of important realizations. It became clear that (1) the hot gas which was the source of this emission constituted as much matter as the optically visible galaxies, (2) this same hot gas was a sensitive probe of the total mass binding the cluster, and (3) the electrons in the hot plasma, through inverse-Compton scattering, would distort the microwave spectrum of the cosmic background radiation (the Sunyaev-Zel'dovich effect). Over the ensuing decade, considerable observational effort has gone into studying the hot gas in clusters of galaxies along just these lines of research.

The discovery of iron line emission also posed a number of simple questions, which, for mostly observational reasons, have been rather more difficult to answer. Perhaps the most basic questions are: what is the total mass of iron and how is it spatially distributed throughout the cluster? Since the iron luminosity is proportional to $\int n_{Fe}\, n_H\, dV$, while the mass in iron is proportional to $\int n_{Fe}\, dV$, a highly concentrated iron distribution can satisfy the observed luminosity, but yield an integrated mass in iron orders of magnitude less than the uniform case. Consequently theories of the origin and evolution of the hot gas in clusters of galaxies have been largely unconstrained by observational data on the spatial distribution of iron.

Recently, data with the potential to investigate the spatial distribution of iron have become available for at least two clusters of galaxies: Perseus and Coma. In this volume M. Kowalski and T. Ponman discuss results related to the Perseus cluster (see also Ponman *et al.* 1990). My study will concentrate on the Coma cluster.

2. Analysis

The data came from three different satellite experiments: the *Tenma* gas scintillation proportional counters, the EXOSAT medium energy proportional counters, and the *Ginga* large area proportional counters. All were mechanically collimated with fields of view of roughly 45′ (EXOSAT), 1° × 2° (*Ginga*), and 3° (*Tenma*). There were two

EXOSAT datasets: one centered on the cluster and one off-centered from this by about 45′. A summary of results from these observations is given in Table 1 (errors are one-sigma). The various datasets all yielded different values for best-fit temperatures when analyzed with isothermal models, yet a global isothermal model can be rejected with high confidence (> 99.95%). Since the different experiments viewed different regions of the cluster, the measured temperatures contain information about the spatial temperature distribution in Coma. The four datasets are in fact consistent with a large class of temperature distributions with negative radial temperature gradients; details are in Hughes *et al.* (1990). In contrast to the temperature values, the agreement of the measured abundance values with a global value of 0.239 ± 0.013 seemed remarkable and warranted further study. One technical point: since the abundance values were somewhat sensitive to the shape of the continuum and the intrinsic variation of iron line intensity with temperature, the abundance values in the table were determined from fits to the best-fit nonisothermal model. In general the abundances determined this way were some 2% higher than those determined using isothermal models.

TABLE 1

Temperature and Abundance Measurements of the Coma Cluster

Instrument	kT[a] (keV)	FOV (°)	Abundance[b]
Tenma	7.50 ± 0.12	3	0.237 ± 0.017
Ginga	8.21 ± 0.15	1×2	0.247 ± 0.024
EXOSAT (centered)	8.50 ± 0.3	0.75	0.231 ± 0.032
EXOSAT (off-center)	7.39 ± 0.7	0.75	$0.344^{+0.081}_{-0.146}$

[a] Using isothermal model.

[b] Using best-fit nonisothermal model; relative to cosmic: $[\mathrm{Fe}]/[\mathrm{H}] = 4 \times 10^{-5}$.

3. Iron Abundance Models

I have introduced a simple parameterization for the spatial distribution of iron in the cluster:

$$n_{Fe} = n_{Fe,0}[1 + (R/R_{Fe})^2]^{-3\beta_{Fe}/2}.$$

This is the same form as the isothermal-β model used to describe the radial variation of gas density and which, for appropriate choice of β, can describe the distribution of galaxies as well. I assume that emission from hydrogen and helium dominates the X-ray surface brightness, which allows one to determine R_H and β_H from the ***Einstein*** imaging data (using the same functional form as above). Hughes *et al.* (1988) give values of 7.′6 for R_C and 0.63 for β. With this information it is possible to determine the projected iron and continuum (hydrogen and helium) emission, convolve with the various collimator beam patterns, and predict an iron abundance for each instrument for a given set of $n_{Fe,0}$, R_{Fe}, and β_{Fe} values.

The best-fit values for R_{Fe} and β_{Fe} correspond very nearly to a spatially uniform metallicity. When R_{Fe} and β_{Fe} are allowed to vary freely, I find that there is no constraint on the value of R_{Fe}. Values from 0.′01 to 100′ (the limit of values considered) are all allowed. However the data do constrain β_{Fe} to lie within a well-defined band, for example, with $R_{Fe} = 10'$, $\beta_{Fe} = 0.52^{+0.32}_{-0.19}$. Over the allowed range of parameter space, the integrated mass of iron in the cluster is strongly bounded. In Table 2, for three different integration radii and the various values of R_{Fe} used, I give the minimum allowed mass of iron (relative to hydrogen, relative to the cosmic ratio). Remarkably, the current data on Coma require that the mass of iron in the cluster within 3 Mpc (about 10 core radii) be no less than about 70% of the value calculated assuming a uniform abundance distribution (*i.e.*, a value of 0.239).

TABLE 2

Minimum Integrated Iron Mass Contained Within Various Radii[a]

R_{Fe} (′)	1 Mpc	2 Mpc	3 Mpc
0.01	0.18	0.19	0.18
0.1	0.18	0.19	0.18
1	0.18	0.19	0.19
10	0.19	0.22	0.19
100	0.20	0.27	0.17

[a] Ratio to hydrogen, relative to the cosmic value.

Using the above formulation, it is possible to begin investigating the origin of the iron in the Coma cluster. I consider three simple models which address this question: (1) iron distributed like the gas density (*i.e.*, uniform metallicity), (2) iron distributed like the number density of galaxies, and (3) iron distributed like the product of the gas and galaxy density distributions. Note that in all cases I consider the iron sedimentation timescale to be large and neglect the effects of gravitational settling. The first case roughly corresponds to situations in which the iron was primordial or was produced by an early generation of stars, or if all the X-ray emitting gas (including hydrogen and helium) was lost by the cluster member galaxies. The second situation assumes that the iron was lost from galaxies in a manner which did not depend on the galaxy's position or velocity, *e.g.*, from a strong galactic wind. Here and in the next case, hydrogen and helium are considered to be largely primordial. The last case approximates a scenario in which all of the observed iron was ram-pressure stripped from cluster galaxies. Then the iron distribution can be represented as the product of a source term (the galaxy distribution) and a term to represent the efficiency of stripping, which goes as the product of the gas density and a power of the galaxy's velocity through the intracluster medium (Gaetz, Salpeter, and Shaviv 1987). For simplicity I assume that the cluster galaxy velocities near the core are approximately independent of position. My conclusions concerning the stripping scenario would be weakened only if the velocity distribution had a strong positive radial gradient, which seems unlikely. The more

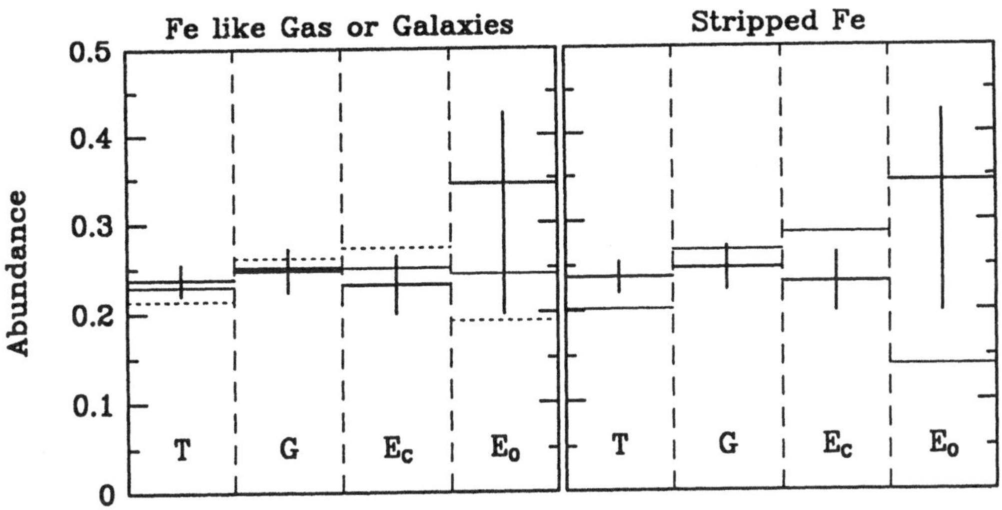

FIGURE – The left panel compares the abundance data (solid dark crosses) with models for the iron distributed like the gas (thin solid line) or like the galaxies (thin dotted line); the right panel compares the data with a model for the distribution of iron assuming a pure stripping scenario. In each panel the fields of view decrease from left to right, that is, from *Tenma* (T) to *Ginga* (G) to EXOSAT (E).

reasonable situation of a negative radial velocity gradient would tend to produce a more sharply concentrated iron distribution and hence a stronger constraint.

The results are shown in the figure. Both cases in the left panel (iron distributed like the gas or galaxies) are acceptable (at the 90% confidence level). However the stripping scenario case in the right panel can be rejected at about the 99% confidence level. There just is not enough iron in the outer parts of the cluster (*i.e.*, beyond about 1 Mpc) in the stripping scenario. These results imply that either ram-pressure stripping is not the dominant source of the iron in galaxy clusters, or that there is a significant amount of radial mixing in the intracluster gas which can efficiently homogenize the spatial distribution of iron.

4. References

Gaetz, T. J., Salpeter, E. E., and Shaviv, G. 1987, *Ap. J.*, **316**, 530.

Hughes, J. P., Butcher, J. A., Stewart, G. C., and Tanaka, Y. 1990, *Ap. J.*, submitted.

Hughes, J. P., Yamashita, K., Okumura, Y., Tsunemi, H., and Matsuoka, M. 1988, *Ap. J.*, **327**, 615.

Mitchell, R. J., Culhane, J. L., Davison, P. J. N., and Ives, J. C. 1976, *M.N.R.A.S.*, **176**, 29.

Ponman, T., J., Bertram, D., Church, M. J., Eyles, C. J., Watt, M. P., Skinner, G. K., and Willmore, A. P. 1990, *Nature*, **347**, 450.

Serlemitsos, P. J., Smith, B. W., Boldt, E. A., Holt, S. S., and Swank, J. H. 1977, *Ap. J. (Letters)*, **211**, L63.

II. X-ray Binaries

IRON LINES FROM X-RAY BINARIES

T. R. Kallman
Laboratory for High Energy Astrophysics
NASA/Goddard Space Flight Center

I Introduction

Iron lines have a unique importance in aiding understanding of X-ray binaries. Their strengths depend on both the conditions in the emitting gas and on the broad band properties of the continuum from the accreting object. Furthermore, the energies of the iron K lines and the ratio of the L lines of iron and the K lines of medium Z elements to the iron K lines are probes of the ionization state in the gas near the compact object. In X-ray binaries (hereafter referred to as XRB) these studies are facilitated by the brightness of the sources, and by the fact that the timescales affecting both the source of continuum and the line emitting gas are within the range accessible to most observations.

This review will discuss the formation of iron line features in X-ray binaries. The emphasis will be on the intrinsic efficiency of the line formation, and on the energies of the emitted lines. These will be explored first in the case of idealized models for the line emitting region, and then will be discussed in the context of more realistic models for XRB emitting regions.

II Photoionization Models for Iron Line Emission

The relative importance of the various processes affecting iron line emission can be illustrated by considering the line emission spectra from an idealized model for the emission line region in XRB. The model that we consider consists of a slab of gas of with constant density which is heated by ionizing radiation from the compact object with prescribed spectrum and flux. The line emission arises as a consequence of the absorption and reprocessing by this gas. The physical processes affecting the heasting, ionization, and excitation of the gas are all assumed to be in a steady state, and the elemental abundances are solar. The results of such photoionization models depend on the shape of the ionizing continuum, the column density of the slab, and on the ratio of the continuum flux at the illuminated face of the slab to the gas density or pressure, the ionization parameter. We adopt the following definitions for the ionization parameter: $\xi=4\pi F/n$ (Tarter, Tucker, and Salpeter 1969) if the gas density n is prescribed (we define F as the radiation energy flux between 1 and 1000 Ry). At large ionization parameter the gas is more ionized and hotter, and conversely (e.g. Kallman and McCray, 1982).

1) K lines

Iron is the unique among the most abundant elements in the relatively high efficiency for inner shell fluorescence. This is due to the abundance of iron, and to the fact that the transition rate for decay of the multiply-excited level produced by innershell ionization scales more rapidly with the nuclear charge than does the rate for the competing process, Auger ionization (e.g. Bambynek, et al., 1972). The fluorescence yield ω is described as the probability that a K line photon is emitted following the removal of a K shell electron by photoionization, and ranges from 0.34 in FeI to 0.49 in FeXXIII (Bambynek, et al. 1972; Jacobs and Rozsnyai 1986). In Fe XXIII the yield drops to ω=0.11, and in Fe XXIV it is ~0.75 (Krolik and Kallman, 1987). The energy $\epsilon_{K\alpha}$ of the $K\alpha$ photon is $\simeq$6.4 keV in FeI thru Fe XXVII, increasing at higher ionization stages up to nearly 7 keV in FeXXVI (Makishima, 1986). The rate of line emission also depends on the photoionization cross section and its behavior as a function of ionization state. The energy of the edge moves from 8.49 KeV to 8.76 KeV and the threshold cross section moves from 2.19×10^{-20} cm^2 to 2.02×10^{-20} cm^2 between Fe XXIII and Fe XXV.

For Fe XXVI recombination from Fe XXVII is likely to be the dominant emission process in a photoionized gas, since there is no fluorescence in this ion and since the temperature in a photoionized plasma is likely to be too low to allow efficient collisional excitation. The relevant recombination rate can be derived from cascade matrix calculations familiar from the study of nebular hydrogen emission lines (e.g. Osterbrock, 1974). In a steady state the recombination into Fe XXVI is balanced by photoionization, so the line emission rate per Fe XXVI ion will be proportional to the photoionization rate coefficient. The constant of proportionality is equivalent to a fluorescence yield, and is in the range $\omega_{FeXXVI} \simeq 0.3 - 0.7$ depending on temperature and assumptions about the nature of the recombination cascade process. Iron emission features can also be formed by scattering of line photons into the observer's line of sight, as pointed out by Band *et al.* (1989). The efficiency for this process depends formally on the oscillator strengths for the various lines involved. Since various sum rules guarantee that the summed line oscillator strengths will be comparable in magnitude to those for the continuum, i.e. the photoionization cross section, it follows that this process will have intrinsic efficiency comparable to K shell fluorescence. Both processes will have decreases in macroscopic efficiency at large column densities when the emission or scattering region is shielded from the incident continuum. The difference between the two lies in the fact that scattering saturates at lower columns owing to the greater bandwidth, and hence greater cross section, for line absorption as opposed to continuum processes. Another important difference lies in the fact that scattering of K lines can only occur for ions with unfilled L shells, i.e. Fe XVIII – Fe XXVI. For ions Fe XVIII - Fe XXIV scattering works with comparable efficiency for both L lines and K lines.

A convenient way of characterizing the efficiency of reprocessing continuum photons into iron lines is in terms of the equivalent width. A simple estimate of this quantity for fluorescence from an optically thin slab is

$$EW = \varepsilon_{line} \frac{L_\varepsilon |_{\varepsilon=\varepsilon_{th}}}{L_\varepsilon |_{\varepsilon=\varepsilon_{line}}} \sigma_{th} \omega x y N \tag{1}$$

where EW is the line equivalent width, ε_{line} is the lineenergy, $L_\varepsilon |_{\varepsilon=\varepsilon_{th}}$ is the ionizing specific luminosity evaluated at the threshold energy of the emitting ion $L_\varepsilon |_{\varepsilon=\varepsilon_{line}}$ is the ionizing specific luminosity at the line energy, σ_{th} is the K shell photoionization cross section at the threshold energy, ω is the fluorescence yield, x is the ionization fraction of the emitting ion, y is the elemental abundance of iron, and N is the total column of the reprocessor. If the spectrum is flat between the line and the K edge, x=1, $y = 3\times10^{-5}$ and we take the cross section for neutral iron, then the equivalent width will be approximately $EW \simeq 2.3$ KeV N_{24} where N_{24} is the cloud column density in units of 10^{24} cm^{-2}. These assumptions are only correct when the dominant line production mechanism is inner shell fluorescence. When the dominant line emission mechanism is recombination the abundance of the emitting ion, x, is a decreasing function of ionization parameter

At column densities greater than about 10^{24} cm^{-2} the approximate formula (2) breaks down, owing to the effects of continuous absorption of the continuum photons as they they enter the slab, and of line photons as they leave. Both these effects cause the reprocessing efficiency to decrease at large columns relative to the estimate of equation (2) (Makishima, 1986, and references therein). This decrease will depend on the detailed radiation transfer processes which can, for example, redistribute ionizing photons into or out of the ionizing continuum of the various iron ions. It will also depend on the geometry of the reprocessor and the incident continuum. For example, a slab seen in reflection is likely to present a smaller column to the emitted line photons, since they will traverse a 'bleached' region as they escape, than it will to line photons if it is viewed in transmission.

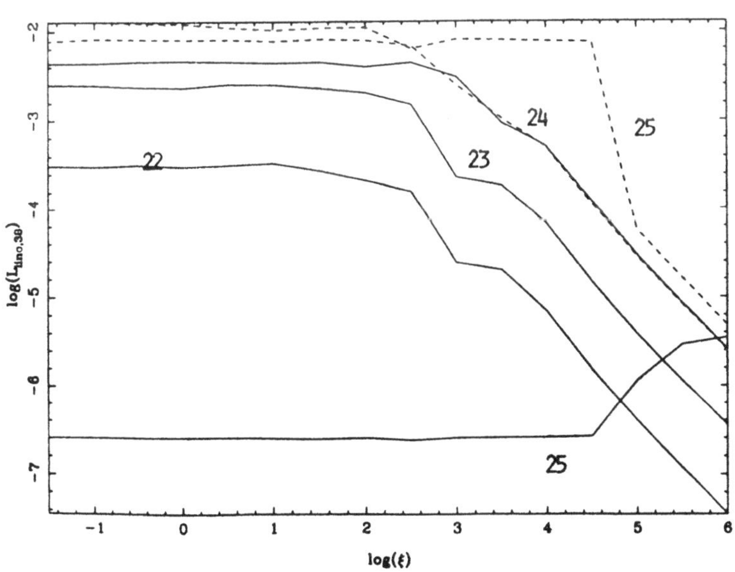

Figure 1: Luminosities of reflected (dashed) and transmitted (solid) iron K lines for constant density slabs of varying column density. Curves are labelled by the log of the slab column density.

A more accurate estimate of the efficiency of the reprocessing is given by the numerical photoionization model results shown in Figure 1. This shows the line luminosity in units of 10^{38} erg s^{-1} (which we denote by $L_{line,38}$) emitted by model slabs (which may be optically thick) as viewed in transmission (solid lines) and in reflection (dashed lines) as a function of ionization parameters for various column densities N=10^{22} cm^{-2}, 10^{23} cm^{-2}, 10^{24} cm^{-2}, and 10^{25} cm^{-2}. The line luminosity corresponds to what an observer would see from an ensemble of such slabs arranged into a spherical shell completely surrounding the source of ionizing continuum. The two sets of curves are overlapping, or nearly so, whenever the clouds are optically thin. The continuum is assumed to be a bremsstrahlung spectrum with temperature kT_{rad}=10 KeV and a total ionizing luminosity of 10^{38} erg s^{-1}, so that the equivalent width of these lines, calculated relative to the incident ionizing continuum, is related to the line luminosity by

$$EW = 10KeV L_{line,38} e^{\varepsilon_{line}/10KeV} \quad (2)$$

Since any given model can emit a complex of lines from several ionization states, we define the line energy ε_{line} to be the luminosity-weighted average of the various components. The values of this quanitity are shown in Figure 2.

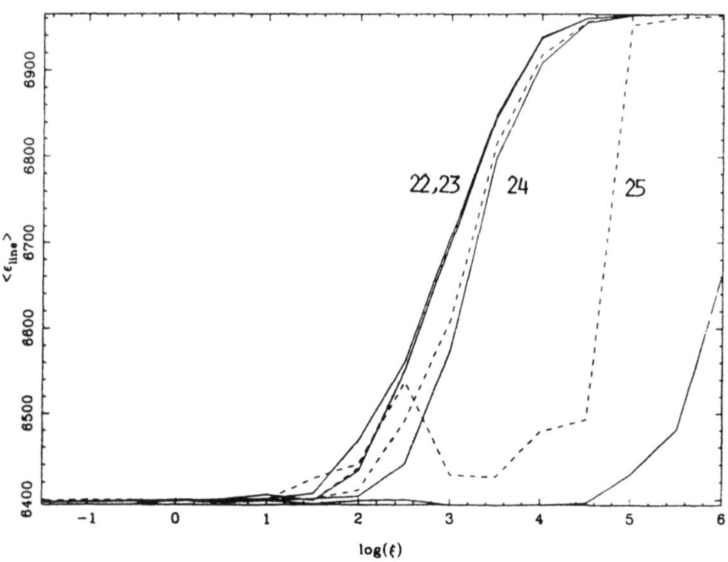

Figure 2: Mean energies of reflected (dashed) and transmitted (solid) iron K lines for constant density slabs of varying column density. Curves are labelled by the log of the slab column density.

The range of ionization parameters and column densities for which the clouds are optically thin at the energy of the iron K lines is approximately log(ξ)$\geq$2.5 for a column of 10^{24} cm^{-2}, and the entire range spanned by the models for smaller column densities. At N=10^{25} cm^{-2} the clouds are optically thick at all values of ionization parameter. Within this optically thin range, the estimates of equation (1) are crudely consistent with the numerical results. The line luminosity is proportional to the cloud column density, and the maximum attainable line equivalent width is approximately 95 eV at a column of 10^{23} cm^{-2} when emission from both faces of the slabs are included.

At column densities of 10^{24} cm^{-2} and above the optically thin approximation breaks down for the K lines. For this column density at low ionization parameter the line luminosity is approximately a factor of 2 smaller than that predicted by equation (1). For a column of 10^{25} cm^{-2} the discrepancy with equation (1) is approximately a factor of 30, and we would expect that further increases in column density would not affect the line strengths.

The line equivalent width decreases with increasing ionization parameter at large ionization parameter owing to the depletion of ions with any bound electrons. In the limit of fully stripped iron we expect the emitted luminosity to scale proprotional to ξ^{-1}, in accord with the numerical results. Also apparent in Figure 1 is a plateau in the emitted luminosity near logξ=3, which is due to the enhanced fluorescence efficiency for Fe XXIV. The monotonic decrease between logξ=2 and logξ=3 is due to the decrease in fluorescence efficiency from Fe XXIII, and to the decreased K shell cross section and the increased threshold energy from Fe XVIII - Fe XXIII.

Departures from the optically thin approximation are also apparent from the asymmetry between the luminosities ot the reflected and transmitted line components. This is most obvious at low values of ionization parameter (logξ $\leq$2.5) for a cloud column density of 10^{24} cm^{-2}, where the ratio between the two components is approximately 2:1. At large ionization parameters for this choice of column the cloud is sufficiently highly ionized that absorption of the line photons as they escape is unimportant. However, effects of absorption for logξ $\geq$2.5 and column density 10^{24} cm^{-2} can be discerned from the difference between the energies of the reflected and transmitted lines. This is because the line photons emitted by more highly ionized material near the illuminated face of the cloud escapes preferentially from that face, while the photons escaping the unilluminated face of the cloud are more likely to be emitted far from the illuminated face. Absorption within the cloud *of the incident ionizing continuum photons* causes the material farther from the illuminated face to be less highly ionized, and hence to emit a lower energy line, than material close to the illuminated face.

2) L lines

A consequence of the screening of nuclear charge by the K electrons is that fluorescent yields for the emission of L lines are never greater than 10^{-2} for near-neutral iron ions (e.g. Bambynek, et al, 1972; Jacobs and Rozsnyai, 1988). Therefore efficient L line emission can only come from ions Fe XVI and higher. Fe XVI can emit via L shell fluorescence owing to the absence of non- radiative decay channels of the excited ion left after the L shell ionization of Fe XV. Fe XVII - XXVI can emit collisional excitation or recombination.

Estimates of L line intensities are complicated by the many possible line transitions, and by the variety of physical processes which can lead to line emission (e.g. Kahn, 1990, Liedahl, et al. 1990). Also, the greater number of potentially observable lines and the larger spectral band they cover makes simple estimates necessarily much more crude. For either recombination or collisional excitation the line equivalent width can be written in terms of an effective 'fluorescence yield' using equation (1), where the yield is defined in the same sense as it is for the recombination emission if K lines from Fe XXV and XXVI. That is, the yield is the fraction of recombinations into an ion which result in L line emission. Collisional excitation may also be included in this description if the yield includes the ratio of the rates of excitation to photoionization.

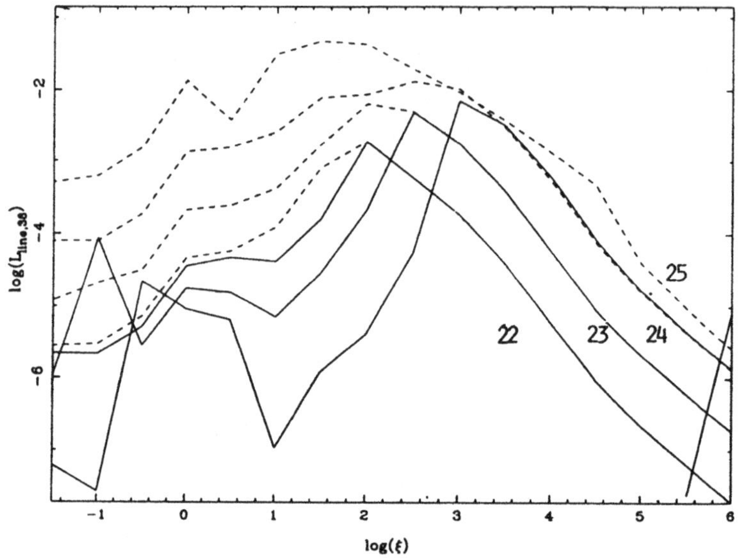

Figure 3: Luminosities of reflected (dashed) and transmitted (solid) iron L lines for constant density slabs of varying column density. Curves are labelled by the log of the slab column density.

Rather than present analytic estimates for these rates, we show in Figures 3 and 4 the results of numerical calculations of the L line intensities and energies from the same model slabs described in the previous subsection. In calculating the iron L line intensities we employ the rate data of Raymond and Smith (1976) and Smith, et al. (1986) for collisional excitation, and of Kallman and McCray (1982) for recombination (see Kallman, Vrtilek and Kahn, 1988 for a fuller description of the calculation). We consider these results to be accurate to within a factor of 2; sufficient for comparison with medium resolution ($\Delta\varepsilon/\varepsilon \simeq 10$) observations such as those obtained with the solid state detectors on the *Einstein* (HEAO-2) spacecraft. Also, since such observations cannot distinguish between line from the L shell of iron and K shell lines of medium Z elements such as nitrogen, oxygen, and silicon, we include in our calculation all the lines with wavelengths longer than 2 Å and shorter than 20Å. We will refer to these collectively in what follows as 'L lines'; of these the contributions from the true iron L lines are roughly comparable to the contributions from the lines of other elements.

The scaling behavior of L lines is qualitatively similar to those of the K lines: The emitted luminosities are approximately proportional to cloud column density at large ξ, when the optically thin approximation obtains. The luminosities also decrease proportional to ξ in this region, owing to the depletion of ions with L shell electrons. For $\xi \sim 2$ the luminosity passes through a maximum, with the decrease at lower ξ due to the fact that ions with M shell electrons predominate. In this range of ionization parameter the optically thin assumption is invalid, resulting in luminosities which do not scale with column density, and in large asymmetries between the reflected and transmitted components.

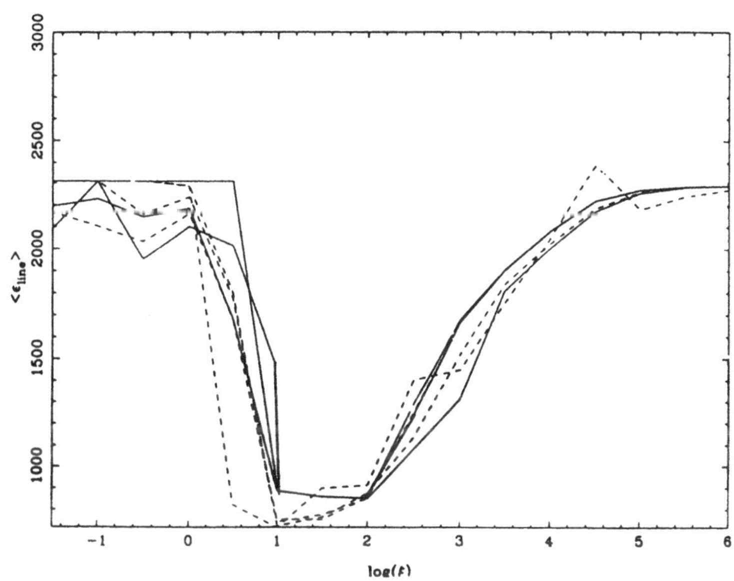

Figure 4: Mean energies of reflected (dashed) and transmitted (solid) iron L lines for constant density slabs of varying column density. Curves are labelled by the log of the slab column density.

Figure 4 shows the weighted mean energy of the L lines. At large ξ this quantity approaches 2.3 KeV, corresponding to the weighted average of the K line energies of of the hydrogenic species of all the various medium Z elements. At small ξ the mean energy also approaches this value, owing to the effect of photoelectric absorption which suppresses the lower energy lines. In the intermediate region, where the L luminosity peaks, the mean energy is 0.8 – 1 KeV.

Of particular interest to current observations of XRB's is the ratio of the L line luminosities to those of the K lines. This is because, even if the two sets of lines are emitted from two (or more) separate regions with differing conditions, the total L/K ratio cannot lie outside of the range of ratios spanned by the various regions. Figures 1 and 3 show that there is no obvious lower limit on the L/K ratio, but that the upper limit (near $\xi \simeq 3$ by the 10^{24} cm^{-2} slab seen in reflection) is approximately 1.6. An example of the reflected spectrum obtained from such a model slab is shown in figure 5. This maximum L/K ratio value may be regarded as uncertain to within 50%, say, owing to the coarseness of the model grid employed here and to other uncertainties in the models themselves including the uncertainties in the atomic rate data employed for the L line emission calculation. Note also that under the conditions needed for this maximum L/K ratio the efficiency of making these lines is high, but is not the maximum value found in the model grid. The K line equivalent width is 150 eV under these conditions. The ratio of the equivalent widths of the L and K lines is also smaller than the ratios of the luminositites by approximately a factor 1.9 due to the lower continuum flux in the vicinity of the K line for our choice of ionizing spectrum.

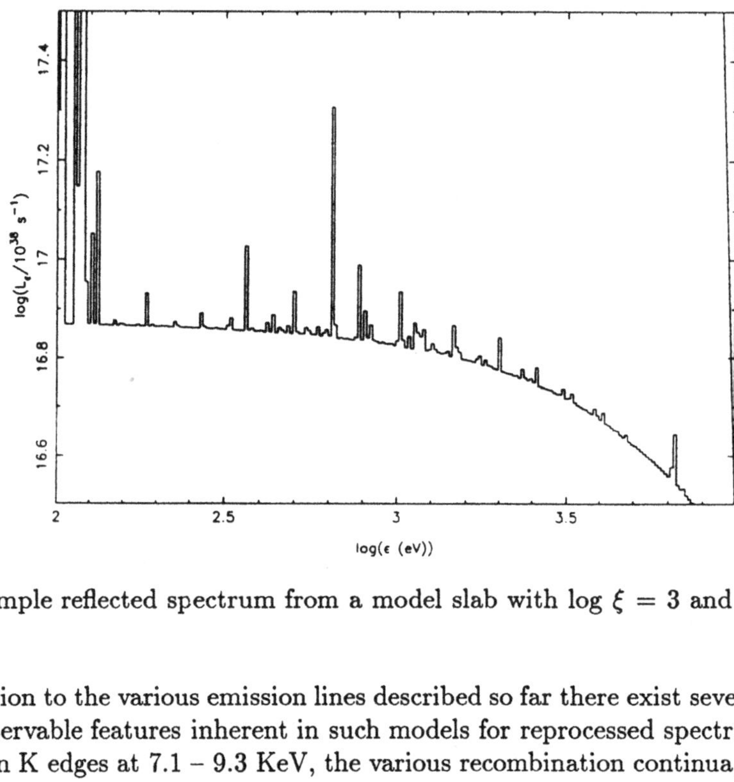

Figure 5: A sample reflected spectrum from a model slab with log ξ = 3 and log N = 24.

In addition to the various emission lines described so far there exist several other potentially observable features inherent in such models for reprocessed spectra. These include the iron K edges at 7.1 – 9.3 KeV, the various recombination continua, and the absorption at low energies produced by photoionization of electrons bound with energies 0.01 – 0.6 KeV. Although the behavior of these features can be examined in the same way as has been done for the emission lines, comparison of the results with observations depends more sensitively on the instrumental details than do the emission lines, and so we will not present them here.

At this point it is worthwhile to summarize what the idealized slab models tell us about iron line production efficiency, and the mean energy of the emitted line. In order to get a high energy K line, i.e. $\varepsilon \geq 6.5$ KeV, all that is needed is that the ionization parameter of the reprocessor be greater than $\log\xi \simeq 2$. This applies to both low (10^{22} cm^{-2}) and high (10^{24} cm^{-2}) column density clouds, although line energies characteristic of hydrogenic iron can only be obtained at high column density if $\log\xi \geq 4$. The converse may also be emphasized: that high ξ material cannot emit a low energy (6.4 KeV) iron line with significant efficiency. This fact has been pointed out in the context of low mass X-ray binaries by Kallman and White (1988), and in the context of active galaxies by Begelman and deKool (1990). In order to obtain high reprocessing efficiency, as measured either by the line luminosity or the equivalent width, the ionization parameter must be $\xi \leq 2$ and the column must be $\geq 10^{24}$ cm^{-2}. For any given column density, the greatest efficiency is obtained at $\xi \leq 2$. To get large L/K ratios the ionization parameter must be near $\xi \sim 2-3$. The greatest such ratios are obtained at columns 10^{22} cm^{-2}–10^{23} cm^{-2}, but with significantly reduced efficiency.

IV Application to X-ray Binaries

We can now apply the result of the idealized models to what is known about the reprocessing environments in and near X-ray binaries. The various possible sites for reprocessed emission include an accretion disk and its associated X-ray induced corona, the companion star, accretion stream, and winds emanating from any of the above sites. For each of these we can estimate the range of likely ionization parameter and column density. It must also be emphasized that the model results for the line luminosity given in the previous section assumed that the reprocessor completely surrounded the continuum source, and that these results must be corrected for partial covering when comparing them with observations.

In the case of an accretion disk, the vertical column is likely to be greater than 10^{23} cm^{-2} (Shakura and Sunyaev, 1972), although the exact value depends on the distance from the source of gravity, influence of X-ray heating, viscosity parameter, mass accretion rate, etc. An accretion disk corona (e.g. White and Holt, 1982, Begelman, McKee and Shields, 1982; Begelman and McKee, 1982) will have a large ionization parameter ($\log\xi \geq 4$) and can have a column up to 10^{24} cm^{-2} (e.g. Kallman and White, 1988). If so, the disk/corona combination must always emit a high energy iron line with relatively low efficiency, and a very low L/K ratio. Low energy iron lines require the corona to have low column density (less than 10^{23} cm^{-2}), and low ionization parameter ($\log\xi \leq 2$). The division between these two regions occurs near radii of 10^{8} cm for conditions likely to occur in low mass X-ray binaries. Large L/K ratios can only be obtained near this radius. This fact is potentiall in conflict with the observed L/K ratio in Sco X-1 and Cyg X-2 , which is near 2 (e.g. White, 1990), together with the observed iron line centroid energies which are greater than 6.4 KeV (e.g. White, Peacock, and Taylor, 1984). One major uncertainty in modelling disk reprocessing is the geometry; we do not fully understand the flaring angle of the disk or the transfer of photons in the corona, so that the effective 'covering fraction' of the disk, and the ionizing flux incident on it, can only be derived indirectly (e.g. Kallman, 1990; Vrtilek, et al., 1990).

In the case of winds, those familiar from the study of massive X-ray binaries have ionization parameters spanning the range from $\log\xi \simeq 0$ (e.g. 4U1700-37) to $\log\xi \simeq 1$ (e.g.

4U0900-40) to log$\xi \geq 4$ (e.g. SMC X-1). The various column densities decrease inversely with the increased ionization parameter, with the consequence that intermediate objects such as Cyg X-1 (and 4U0900-40 to a lesser extent) are likely to produce the greatest L/K ratio (see, e.g. Barr and Van der Woerd, 1989). The companion star in all classes of XRB's will have an effectively infinite column density and low ionization parameter, and so can only be the site of 6.4 KeV reflected emission and no L shell lines (Basko, 1976; Hatchett and Weaver, 1977).

REFERENCES

Bambynek, W., Crasemann, B., Fink, R.W., Freund, H.-U., Mark, H., Swift, C.D., Price, R.E., and Rao, P.V., 1972, *Rev. Mod. Phys.*, **44**, 716.

Band, D., Klein, R., Castor, J., and Nash, J., 1989, in, *Proceedings of the 23rd ESLAB Symposium on Two Topics in X-ray Astronomy*,, ed. J. Hunt and B. Battrick (ESA SP-296).

Barr, P., and Van der Woerd, H., 1989, in, *Proceedings of the 23rd ESLAB Symposium on Two Topics in X-ray Astronomy*,, ed. J. Hunt and B. Battrick (ESA SP-296).

Basko, M., 1978, *Ap. J.*, **223**, 268.

Begelman, M., and deKool, M., 1989, in, *Proceedings of Atlanta Workshop on Variability in Active Galactic Nuclei*,, ed. H.R. Miller and P. W. Wiita.

Begelman, M. C., McKee, C. F., and Shields, G.A., 1983, *Ap. J.*, **271**,70.

Begelman, M. C., and McKee, C.F., 1983, *Ap. J.*, **271**, 89.

Jacobs, V., and Rosznyai, B. F., 1986, *Phys. Rev. A.*, **34**, 216.

Hatchett, S. P., and Weaver, R., 1977, *Ap. J.*, **215**, 285 .

Kahn, S., 1990,,, this volume.

Kahn, S., Seward, F.D., and Chlebowski, T., 1984, *Ap. J.*, **283**, 286 .

Kallman, T. R. 1989, in, *Proceedings of the 23rd ESLAB Symposium on Two Topics in X-ray Astronomy*,, ed. J. Hunt and B. Battrick (ESA SP-296).

Kallman, T.R., and McCray, R. A., 1982, *Ap. J. Supp.*, **50**, 263.

Kallman, T. R., Vrtilek, S. D., and Kahn, S. M., 1988, *Ap. J.*, **345**, 498.

Kallman, T., and White, N.E., 1988, *Astrophys. J.*, **341**, 955.

Krolik, J.H. and Kallman, T.R., 1984, *Astrophys. J.*, **320**, L5.

Liedahl, D., Kahn, S.M., Goldstein, W., and Osterheld, A., 1990, *Ap. J.*, **350**, L37.

Osterbrock, D. E., 1974, *Astrophysics of Gaseous Nebulae* ,, (San Francisco: W.H. Freeman and Co.).

Raymond J.C., and Smith, B. H., 1977, *Ap. J. Supp.*, **35**, 419.

Shakura, N.I., and Sunyaev, R.A., 1973, *Astr. and Ap.*, **24**, 337.

Smith, B., Raymond, J., Cowan, R., and Mann, J., 1985, *Ap. J.*, **298**, 898.

Tarter, C. D., Tucker, W., and Salpeter, E. E., 1967, *Ap. J.*, **156**, 943.

Vrtilek, S. D., et al. 1989, in, *Proceedings of the 23rd ESLAB Symposium on Two Topics in X-ray Astronomy*,, ed. J. Hunt and B. Battrick (ESA SP-296).

White, N.E., and Holt, S.S., 1982, *Ap. J.*, **257**, 318.

White, N.E., Peacock, A., and Taylor, B.G. 1985, *Ap. J.*, **296**, 475.

BLACK HOLE CANDIDATES IN BINARIES

Yasuo Tanaka
Institute of Space and Astronautical Science
3-1-1 Yoshinodai, Sagamihara, Kanagawa-ken 229, Japan

ABSTRACT.

The X-ray properties of the galactic black hole candidates are briefly reviewed. These sources show either of the two types of energy spectrum: (1) a hard spectrum of approximately a single power-law form, or (2) a composite of an ultrasoft component and a hard, power-law type tail. The power-law component of either spectral class exhibits distinct spectral features associated with the iron K-absorption edge, which is qualitatively different from that expected for an absorption due to forground matter. These features are discussed in relation to the accretion disk structure.

1. INTRODUCTION

Several galactic black hole candidates have so far been known (see e.g.[1], and references therein). For some of these systems, mass lower limits of the compact objects have been determined, among which the limits for Cyg X-1 [2] and A0620-00 [3] are considered to be most reliable.

In addition to those with the mass lower limit exceeding 3 $M_\odot$, those sources which exhibit either of the following two sets of X-ray properties as summaraized in Table 1 are considered as black hole candidates. (I) The **[hard]** state: The energy spectrum is approximately of a single power-law form. Intensity varies rapidly and erratically ("flickering") on various time scales down to millisconds with large amplitudes. (II) The **[ultrasoft]** state: The energy spectrum is ultrasoft (much softer than those of binary neutron star sources) and accompanied by a high-energy tail of approximately a power-law form. Intensity varies relatively slowly (little flickering). Examples of the spectrum of the [hard]-state sources are shown in Fig. 1 (Cyg X-1, GX339-4, GS2023+33), and those of flickering together with the power-density spectra

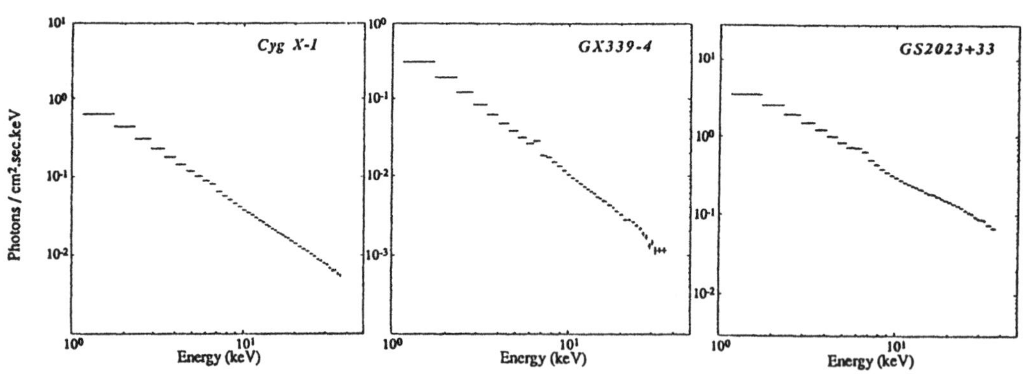

Fig. 1. Energy spectra of the [hard]-state sources.

Table 1. X-ray properties of the black hole candidates

Spectrum	Time Variation	Sources
[Hard]-state		
Approximately a single power-law	Flickering	Cyg X-1*
		GX339-4*
		GS2023+33
		GS1826-24
[Ultrasoft]-state		
Ultrasoft + hard tail	Little flickering	Cyg X-1*
		GX339-4*
		LMC X-3
		LMC X-1
		A0620-00
		GS2000+25

* Occasionally switches between the [hard]- and [ultrasoft]-states.

Table 2. Power-law slopes

Sources	Photn index
Cyg X-1*	1.6 - 1.8
GX339-4*	1.7 - 1.8
GS2023+33*	~1.4
GS1826-24*	~1.7
GX339-4+	~2.5
GS2000+25+	2.0 - 2.2
AGNs	1.3 - 2.8 Mean~1.7

* [Hard] state
+ [Ultrasoft] state

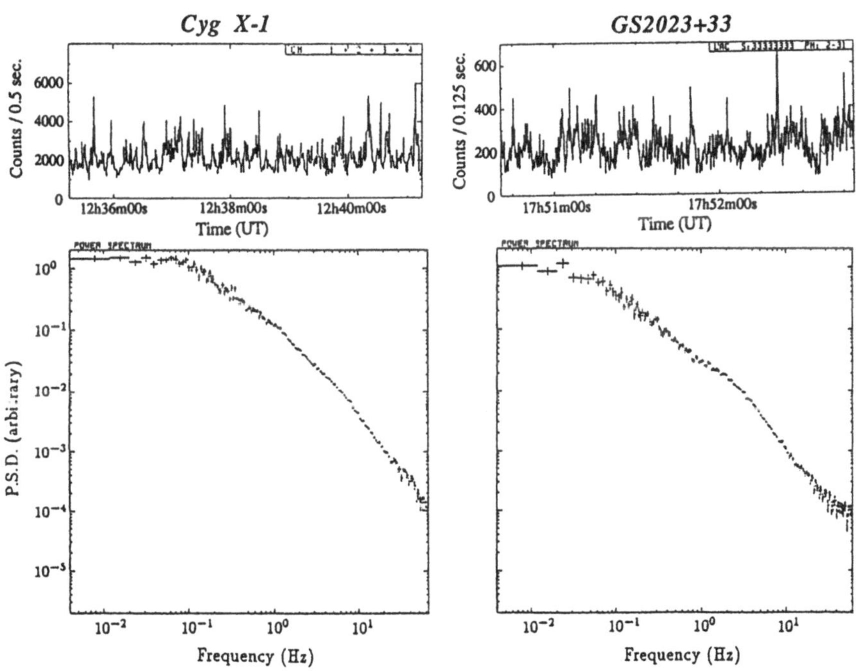

Fig. 2. Flickering of Cyg X-1 and GS2023+33, and the power density spectra.

in Fig. 2. Figure 3 shows examples of the spectrum of the [ultrasoft]-state sources (GX339-4, LMC X-3, GS2000+25). Cyg X-1 and GX339-4 are known to switch occasionally between these two states. While these X-ray properties have not as yet been proven to be genuine black hole signatures, they are quite distinct from those of X-ray binaries known to include a neutron star. Based on this classification, two extremely bright transients discovered from Ginga, GS2000+25 and GS2023+33 may also be considered as the black hole candidates [1].

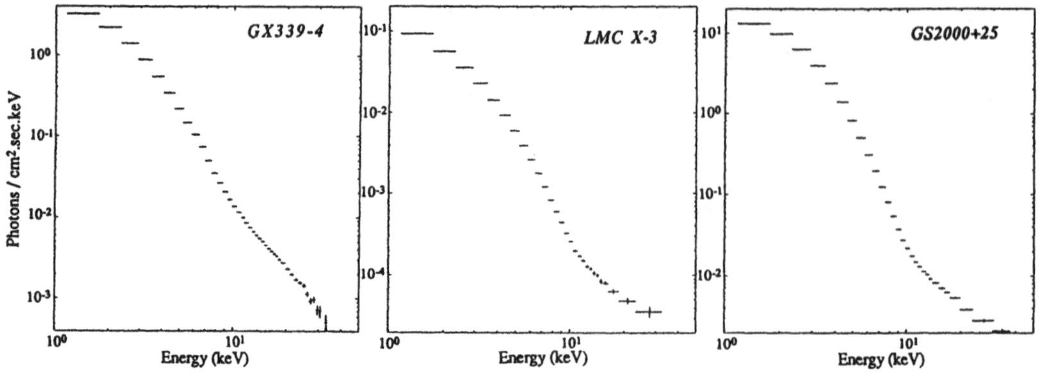

Fig. 3. Energy spectra of the [ultrasoft]-state sources.

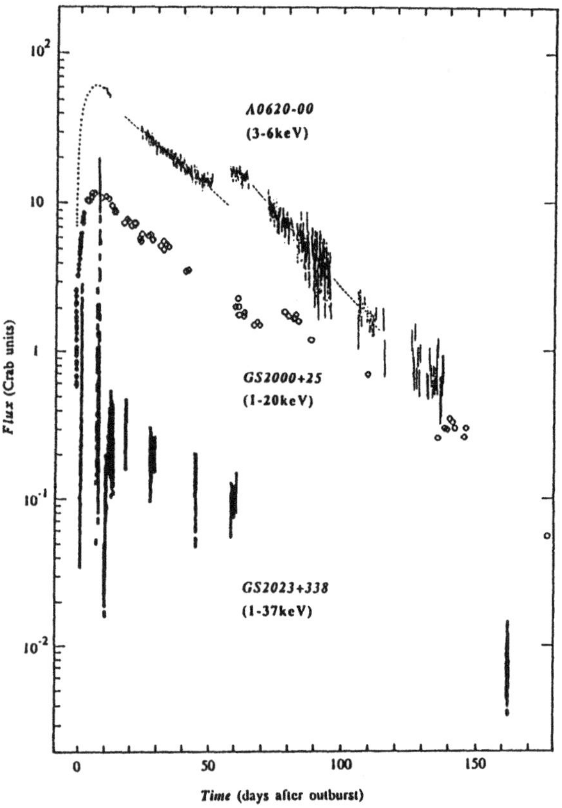

Fig. 4.

Light curves of GS2000+25 and GS2023+33. Light curve of A0620-00 [4] is also shown for comparison.

Figure 4 shows the light curves of these two Ginga transients in comparison with that of A0620-00 [4]. As a matter of fact, the decay times for all three sources are strikingly similar to each other. Although what determines the decay time is not well understood, yet the result may suggest close similarity of these systems.

2. THE ULTRASOFT COMPONENT

The most likely origin of the ultrasoft component is the thermal emission from the accretion disk. With this interpretation, the bolometric luminosity of the ultrasoft component is expected to be proportional to the accretion rate $\dot{M}$. It has been shown that the spectrum of the ultrasoft component was very well fitted with a multicolour blackbody disk model. This multicolour blackbody disk model [5] qualitatively represents the emission spectrum from an optically-thick accretion disk. Basic assumptions are that (1) gravitational energy release is dissipated locally into thermal emission and (2) the continuity of accretion flow holds throughout the disk. This model includes two parameters corresponding respectively to $r_{in}(\cos\theta)^{1/2}$ and kT_{in}, where r_{in} is the innermost radius of the optically-thick accretion disk, T_{in} the temperature at r_{in}, and θ the inclination angle of the disk.

An important conclusion obtained from a comparison between GS2000+25 and GS2023+33 is that the presence or absence of the ultrasoft component is not governed by $\dot{M}$. A previous speculation was that an optically-thick accretion disk switches into optically thin when accretion rate decreases below some critical value. However, it is not the case. The estimated peak luminosities of both GS2000+25 (2 - 3 kpc; [6]) and GS2023+38 (1.5 - 2 kpc) are well in excess of 10^{38} ergs/cm^{2}s at the peak, and perhaps near the Eddington limit. For GS2000+25, the ultrasoft component was persistently present throughout the decay over three orders of magnitude in the measured intensity. Whereas, the spectrum of GS2023+33 was always a single power-law form right from the beginning, and the ultrasoft component was always absent. As a matter of fact, there are cases for both Cyg X-1 and GX339-4 in which total energy flux observed is not very much different between their [ultrasoft] and [hard] states.

Another important finding is that the innermost radius of an optically-thick disk, whenever present, remains essentially constant against large changes in the accretion rate. Figure 5 shows the evolution of energy spectrum of GS2000+25 [7]. It is evident that these spectra are composed of an ultrasoft component and a hard tail. In fact these spectra are well fitted with a composite of a multicolor blackbody disk model and a power-law, except for a significant structure noticeable around 7 - 8 keV which we shall discuss in Section 4. As shown in Fig. 6, the bolometric flux of the ultrasoft component decreases monotonously and smoothly with time over two orders of magnitude. Such a gradual decrease of $\dot{M}$ would be natural in case of a sporadic

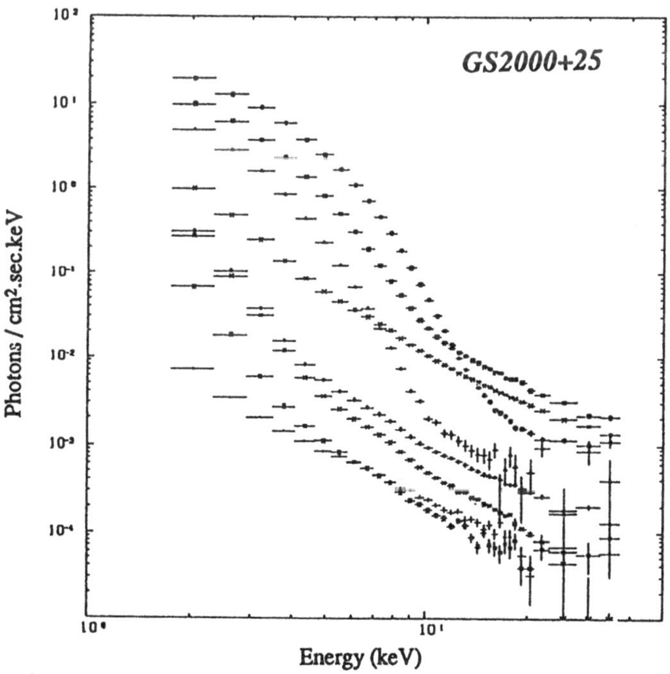

Fig. 5.

Energy spectra of GS2000+25 observed from Ginga. The days of observations (1988) are indicated.

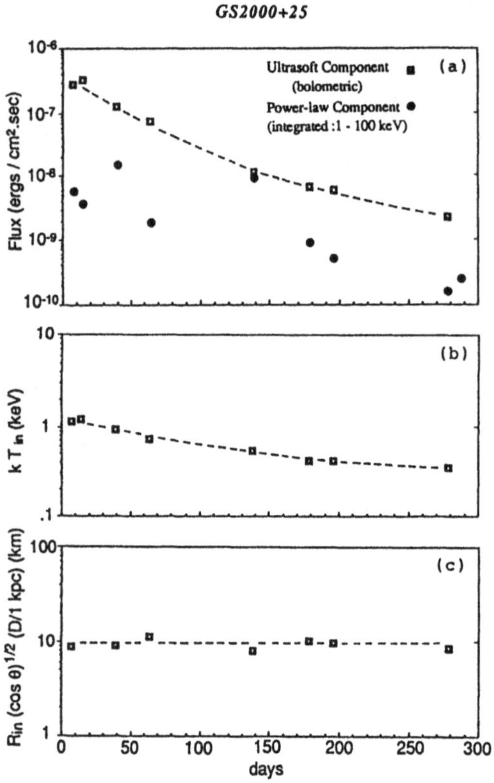

Fig. 6. Time histories of parameters for GS2000+25 spectra: bolometric flux of the ultrasoft component and the integrated flux of the power-law component, kT_{in}, and $r_{in}(\cos\theta)^{1/2}$ (see text for notations).

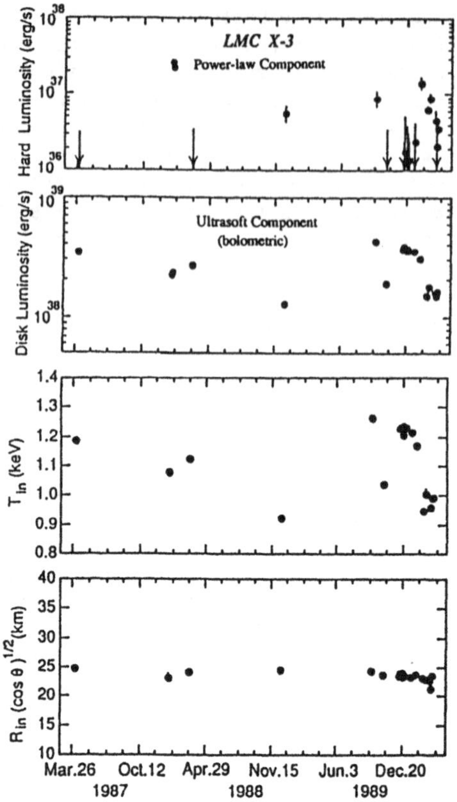

Fig. 7.

Time histories of parameters for LMC X-3 spectra.

accretion in the transient sources. On the other hand, the best-fit values of $r_{in}(\cos\theta)^{1/2}$ are found to be strikingly constant at about 25 km for an assumed distance of 2.5 kpc [6]. The decay of the ultrasoft component is the result of gradual cooling of the disk. A similar result is slao found from LMC X-3 as shown in Fig. 7 [8]. Over a period of their observations, LMC X-3 changed its intensity by a factor of four. Again, the innermost disk radius remained constant at about 24 km for a known distance of ~50 kpc, and the intensity changes of the ultrasoft component is solely due to changes in the disk temperature. It is also interesting to note that the values of $r_{in}(\cos\theta)^{1/2}$ obtained for GS2000+25 and LMC X-3 are roughly equal. Similarly, LMC X-1 at a distance of ~50 kpc also gives a comparable value of 30~35 km for the same parameter [9].

Although the face values of the parameter, $r_{in}(\cos\theta)^{1/2}$, should not be taken seriously, its invariance against a large change in M must carry some important meaning. The most plausible interpretation of this result would be that the size of the accretion disk remains unchanged for a large range of M. Since the only size conceivably independent of M would be the radius of the innermost stable Keplerian orbit, we may conclude that an optically-thick accretion disk extended to the neighbourhood of $3r_g$ from the compact object persists stably over a large range of M.

3. THE HARD COMPONENT

Behavior of the hard power-law component of the [ultrasoft] sources is quite remarkable. As shown in Fig. 6 and 7, for both GS2000+25 and LMC X-3, the intensity of the hard component which is approximated by a single power-law changes dramatically on a time scale of a day or possibly hours, independently of the intensity of the ultrasoft component, or $\dot{M}$. The intensity of the power-law component relative to that of the ultrasoft component varies over two orders of magnitude from ~ 1 to 10^{-2} in GS2000+25 and more than an order of magnitude in LMCX-3. Similar changes of the hard component uncorrelated with the ultrasoft component was also observed from LMC X-1 [9], and GX339-4 [8] when it was in the [ultrasoft] state.

This outstanding property of the hard component of the [ultrasoft] sources is a great puzzle. In the case of GS2000+25, a quasi-stationary accretion flow seems to have continued throughout the decay, as represented by a smooth decrease of the ultrasoft component. What governs the power in the hard component, totally independent of $\dot{M}$? It is most likely to be related to the yet unknown mechanism of generating power-law component. The fact that the power-law component extends beyond several 100 keV [10] or even as high as 1 MeV [11] strongly suggests that the site of formation of the power-law component is near the compact objects. If the gravitational energy released within $3r_g$ were somehow converted into the power-law component, varying power-law component would imply that either the emission is anisotropic or a part of gravitational energy released does not go into X-rays.

It has been known that the intensity of the [hard]-state sources flickers. In contrast, the hard component in the [ultrasoft]-state sources exhibits little flickering. This fact may indicate that an optically-thick accretion disk extended to the vicinity of the compact object has an effect of stabilizing the mass flow.

As regrds the spectrum of the hard component of the black hole candidates, it is also remarkable that the slope of the power law remains fairly constnat against large changes in intensity. The constancy of the power-law slope independent of intensity holds for both the flickering [hard]-state sources and the hard tail of the [ultrasoft]-state sources. In Table 2, the observed power-law slopes of the black hole candidates are listed. One notices that the range of slopes is essentially the same as that for the active galactic nuclei. While these two classes of source are totally different in scale and power, we suspect that the same kind of mechanism would be responsible for generating the power-law components.

4. SPECTRAL FEATURES

The energy spectrum of the hard component of the black hole candidates, the [ultrasoft]- as well as [hard]-state sources, has so far been considered to be expressed by a single power law. However, accurate measurements from <u>Ginga</u> with excellent photon statistics have revealed a significant deviation from a single power law due to presence of a unique spectral feature. In this Section, we shall discuss this spectral structure.

4.1. The [Hard]-State Sources

For an example, the spectrum of Cyg X-1 is shown in Fig. 8. One clearly notices a feature starting at ~ 7 - 8 keV and extending towards higher energies, which appears like a shallow and broad absorption. In order to visualize the shape of the structure, we take the ratio of the observed spectrum to a model spectrum. In this case, the model spectrum is a power law which fits to the observed spectrum in both low-energy (< 6 keV) and high-energy (> 25 keV) parts, disregarding the range in between.

The results for Cyg X-1, GS2023+33 are shown in Fig. 9. The shape of the structure is essentially the same for both sources. From the shape of the structure, it is obvious that this structure is not due to photoelectric absorption by foreground matter.

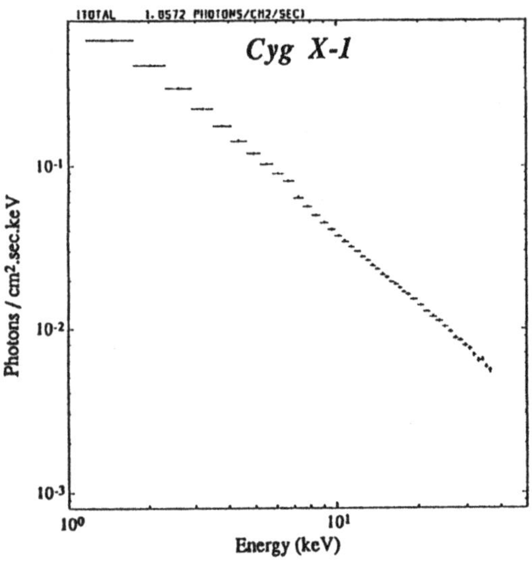

Fig. 8. Energy spectrum of Cyg X-1.

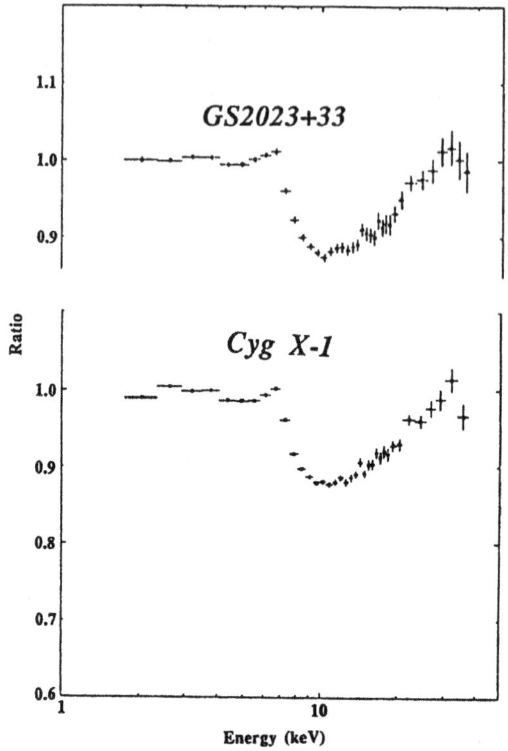

Fig. 9. Ratio of the observed to the model power-law spectra (see text) for Cyg X-1, GS2023+33.

We then looked for a model spectrum which reproduces the observed structure, and found that the following model (here after refered to as the partial reflection model) gives a satisfactory fit to the observed spectrum: a composite of [a single power law] + [its reflection from nearby matter] + [fluorescent iron line]. The basic idea behind this model is illustrated in Fig. 10. Suppose the X-ray beam from the central source hits the surface of matter which is assumed to be optically thick for Compton scattering, some photons are photo-absorbed but the rest will be reflected back. Effective reflection coefficient as a function of energy depends on the ionization state of the scattering medium [12]. The fluorescent iron line is expected as a result of the iron K-absorption. The result of the fit to the observed spectrum of Cyg X-1 with the above model is shown in Fig. 11, in which the effective reflection coefficient for essentialy unionized matter of cosmic abundances was employed [13] (nearly equal to the case of ξ_0=0 or 0.02 of Ref [12]). The best-fit energy of the absorption edge turned out to be 7.2 keV, and that of the line to be 6.4 keV with negligible line width. These values are indeed in excellent agreement with those of the K-edge and the fluorescent line of neutral iron atoms. In addition, the amount of the reflected component requires that the scattering material covers roughly half the total solid angle viewed from the central source.

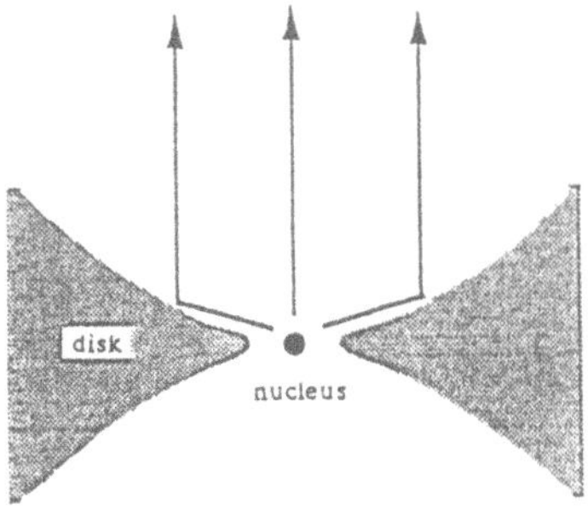

Fig. 10.

Concept of the partial reflection model. The observed X-rays comprise the direct beam, partially reflected beam, and associated fluorsecent iron line (dashed line).

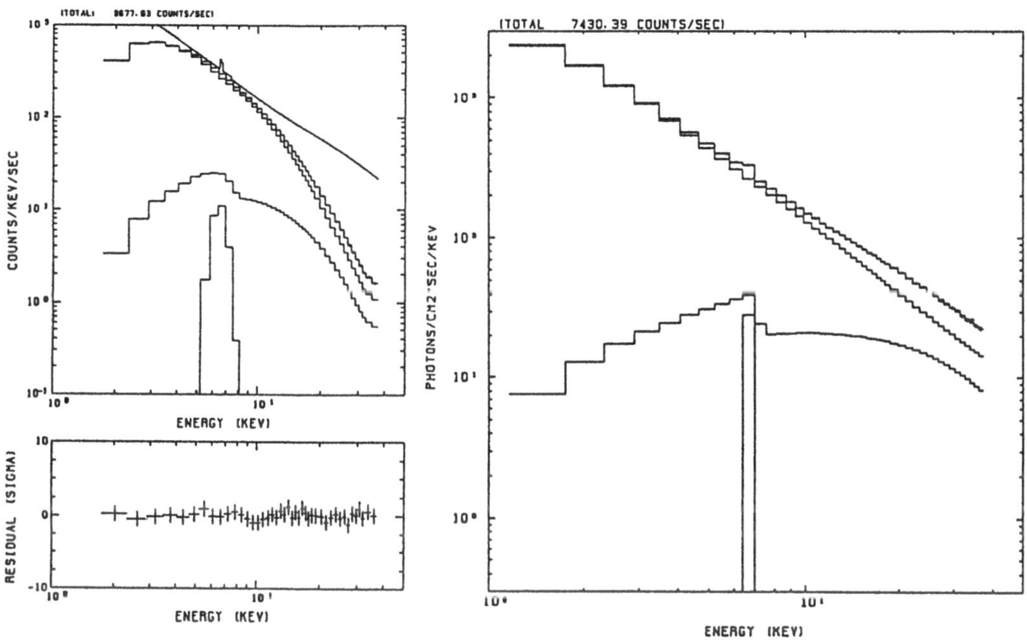

Fig. 11. Fitting of the partial reflection model to the observed spectrum of Cyg X-1; the pulse-height spectrum (left), and the incident photon spectrum (right).

At this point, it is important to note that the above structure underlying in the continuum is crucial for the determination of properties of the iron line. As a matter of afct, if one simply assumes a single power law for the continuum without taking into account the above structure and performs fitting over an insufficiently wide energy range (e.g. 2 - 15 keV), one will obtain an apparently broadened line (FWHM ~1 keV) at some lower energy ("redshifted" to ~6.1 keV), as shown in Fig. 12. This is obviously a false result, hence consequent interpretations are irrelevant.

As regards the above partial reflection model, it has been shown that this model satisfactorily explains the observed spectra of Seyfert 1 galaxies [14]. However, the above reflection model, if plausible, is no more than a mathematical tool for reproducing the observed spectral shape, unless its reality is demonstrated. Fortunately, GS2023+33 verified the actual presence of the reflected component. Intrinsically, GS2023+33 appears to be a power-law source, as seen in Fig. 13. It exhibited, however, a variety of spactral shapes in the early phase, as displayed in Fig. 14 (pulse height spectra). Among these spectra, a certain case (spectrum (f) in Fig. 14) can only be explained satisfactorily by means of reflection of a power-law

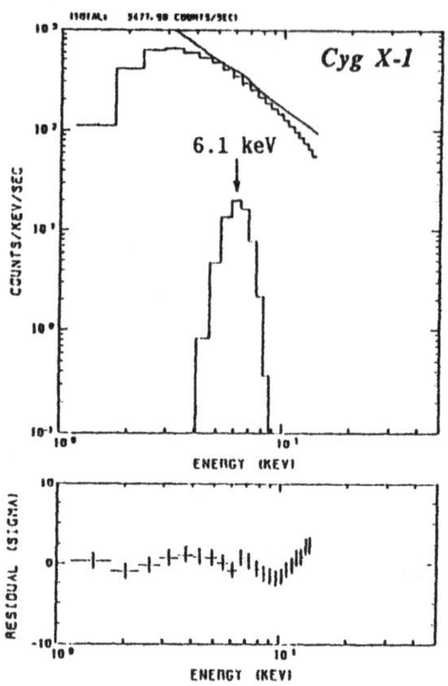

Fig. 12.

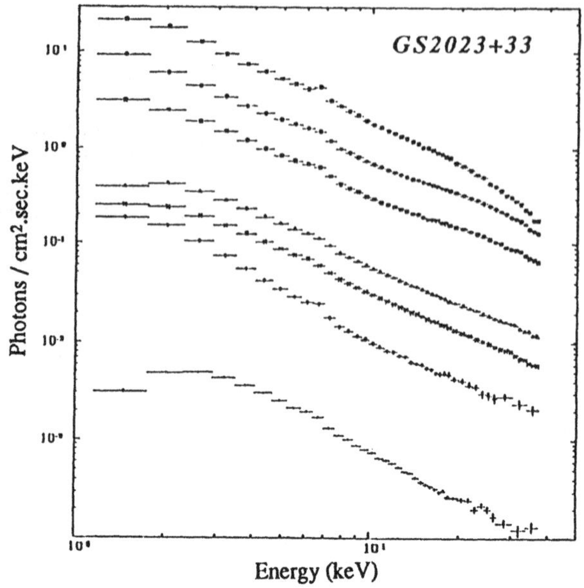

Fig. 13.

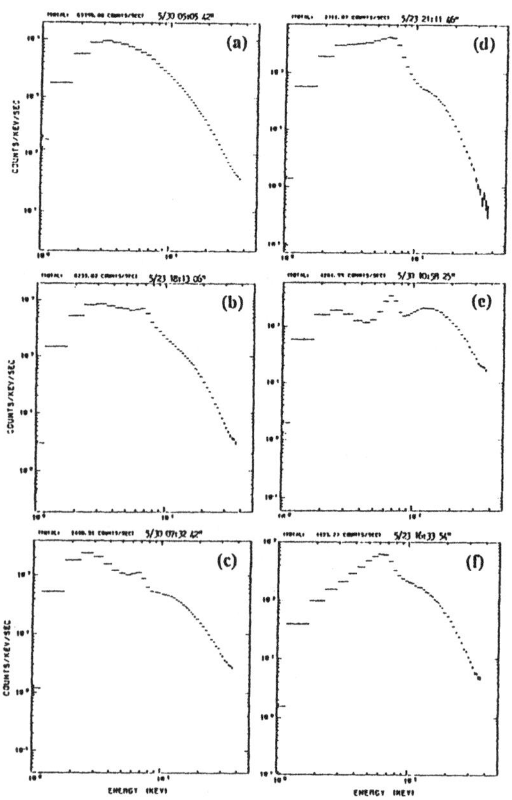

Fig. 14.

Fig. 12.

Fitting of a simple model; [a single power law + a line], to the observed spectrum of Cyg X-1.

Fig. 13.

Energy spectra of GS2023+33 observed from Ginga at various intensity levels.

Fig. 14.
Vrious shapes of the pulse-height spectra of GS2023+33 observed from Ginga.

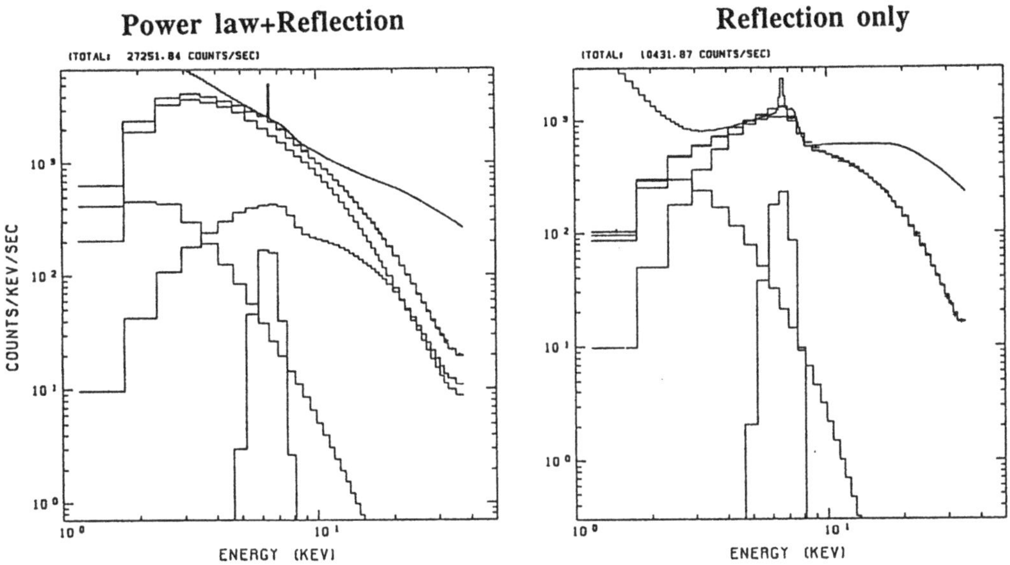

Fig. 15. Fitting of the partial reflection model to Spectrum (a) in Fig. 14 with an incident spectrum of $E^{-1.63}$ (left), and of the reflection only model to Spectrum (f) in Fig. 14 with an incident spectrum of $E^{-1.57}\exp(-E/38\ \mathrm{keV})$ (right).

component. The fit of the reflected power-law model to an observed spectrum is shown in Fig. 15, in which a weak persistent soft component (not an ultrasoft component) is also present. Thus, we conclude that the reflecting component is a real existence. In fact, GS2023+33 exhibited the same spectral shape quite frequently. Most probably, these are the cases in which the central power-law source is screened by some obstacle on the line of sight, but the reflecting surface is still visible.

What screens the central source is yet unclear. One possibility is to consider that the accretion disk occasionally changes its geometry and its outskirts block the line of sight to the central source. However, the following effect observed from GS2023+33 is not consistent with this possibility: In those events when the reflected component became predominant, we often found that the incident power-law showed such a high-energy turn off as noticeable in Fig. 15. This effect would be difficult to explain, if X-rays from the central source were simply stopped by an outer fringe of the accretion disk. The softening of X-rays from the central source could possibly be due to down-Comptonization by a thick electron cloud (fully ionized gas) near the central source. Yet, how this softening is related to the occultation of the central source is sitll a puzzle. This problem remains to be an important issue.

Incidentally, it is very intersting to compare the spectra observed form GS2023+33 with those of Seyfert 1 and Seyfert 2 galaxies, of which some examples are shown in Fig. 16 [15]. Of particular interest is to note that the shape of the Seyfert 2 spectra look similar to that of the reflected power-law component as seen from GS2023+33. This fact supports the view that Seyfert 1 and Seyfert 2 galaxies are the same class of object (see e.g. [16], and references therein).

4.2. The [Ultrasoft]-State Sources

The hard component of the black hole candidates in the [ultrasoft] state also shows a similar spectral feature to that observed in the spectra of the [hard]-state sources

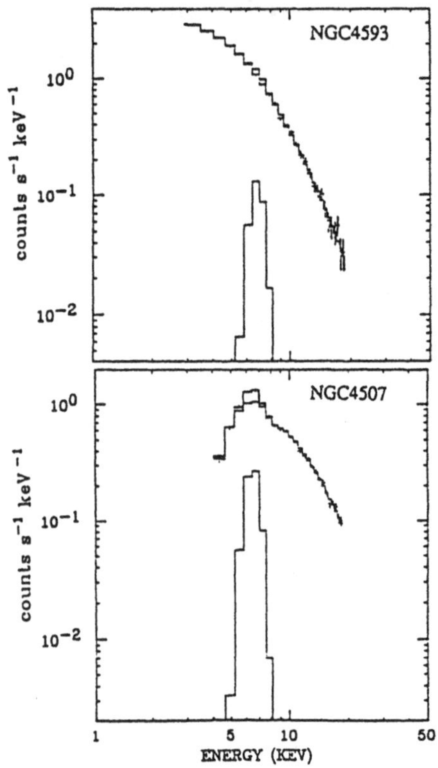

Fig. 16. Typical Seyfert 1 (upper) and Seyfert 2 (lower) spectra (pulse-height spectra) observed from Ginga [15].

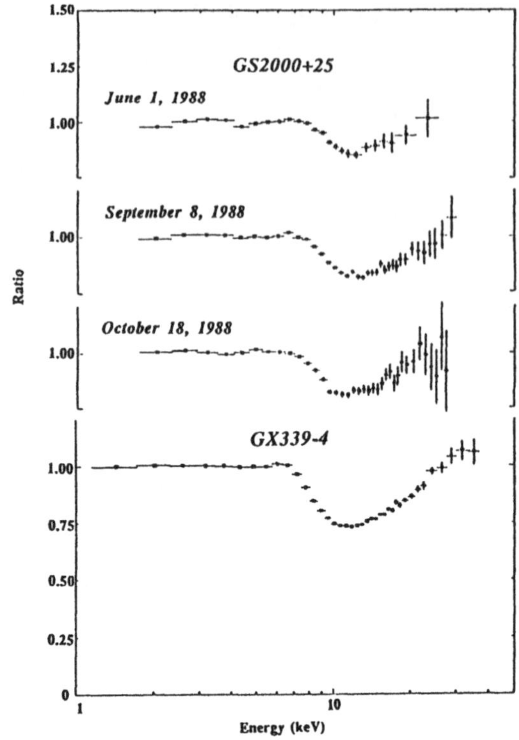

Fig. 17. Ratio of the observed to the model spectra (see text) for GS2000+25, and GX339-4 in the [ultrasoft] state.

discussed above, and hence considered also to be related to the iron K-edge. As we did in the preceeding section, we generate the ratio of the observed spectra to the model spectra in order to discern characteristic shape of the structure. This time, the model spectrum is the sum of a power law and an ultrasoft component for which the multicolour blackbody disk model is employed (see Section 2.). Figure 17 shows the results for GS2000+25 at three different intensity levels (see Fig. 5), and GX339-4 when it was in the [ultrasoft] state. The structures for these two sources are very similar to each other. Furthermore, the structure for the [ultrasoft]-state sources appears significantly different from that for the [hard]-state sources shown in Fig. 9 in the following respects: (1) the structure is significantly deeper, (2) the "edge" is more gradual, trailing to higher energies, and (3) a less pronounced peak due to the iron line.

These charcteristics can be qualitatively understood if the scattering material responsible for the reflected component is highly ionized such that a large fraction of iron atoms are in helium-like and hydrogen-like ions; (1) The K-edges are distributed over the range 7.2 - 9.3 keV. (2) The depth of absorption increases because photoionization opacity below the iron K-edge is much reduced. (3) The fluorescent iron line is a mixture of those at different energies over 6.4 - 6.9 keV, causing an apparent broadening. As a matter of fact, the partial reflection model can satisfactorily reproduce the observed spectra, when we employ the effective reflection coefficient corresponding to a high ionization state similar to the case of ξ_0 = 2 in Ref.[12], as shown for GS2000+25 and GX339-4 in Fig. 18. Also, the result implies that the reflecting surface covers roughly half the total solid angle viewed from the central source for an isotropic emission.

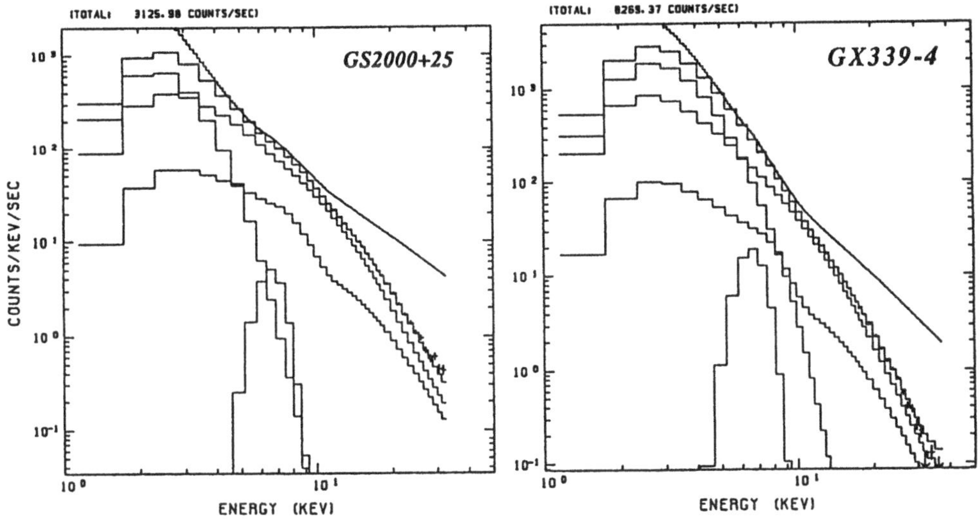

Fig. 18. Fitting of the partial reflection model to the observed spectra of GS2000+25, and GX339-4 in the [ultrasoft] state.

We thus conclude that the difference in the spectral structure that appears in the hard component between the [ultrasoft] state and the [hard] state is due to different ionization states of the scattering material. This is qualitatively consistent with the picture that the accretion disk is responsible for generation of the reflected component for the following reason. In the [ultrasoft] sources, an optically-thick accretion disk is located near the center, and hence a large incident flux density expected will cause a high degree of ionization. Whereas, in the [hard] sources, optically-thick region would be far away from the center, as suggested by the absence of an ultrasoft component, consequently the incident flux density would be low.

The above studies clearly demonstrate the importance of investigation of the spectral features due to iron. The solid angle coverage and the degrees of ionization of the disk surface can be determined from observations of the spectral features, which should provide us with an important clue to elucidate the structure of accretion disks.

Note added in manuscript

In early January, 1991, a new bright transient source was discovered in the Constellation Musca by the Ginga All Sky Monitor [16] and also by the Granat Watch experiment [17]. This source was respectively named GS1124-68 and GRS1121-68. This source was brightened up to about eight Crab flux units at the maximum, and importantly the observed X-ray properties characterize itself to be one of the [ultrasoft] sources in all respects.

ACKNOWLEDGEMENT

The author is indebted to his colleagues for many unpublished results presented in this paper. In particular, the results on GS2000+25 and GS2023+33 are the joint works with K. Yoshida and M. Takizawa, to whom the author is grateful for their collaboration. The results on LMC X-3 and GX339-4 are for the most part due to K. Ebisawa. The author also thanks H. Inoue and K. Mitsuda for useful discussions.

REFERENCES

[1] Y. Tanaka (1989), Proc. 23rd ESLAB Symp. (ESA SP-296), p3.
[2] B. Paczynski (1974), Astron. & Astrophys. **34**, 161.
[3] J.E. McClintock and R.A. Remillard (1986), Astrophys. J. **308**, 110.
[4] M. Elvis et al. (1975), Nature **257**, 656; L.J. Kaluzienski et al. (1977), Astrophys. J. **212**, 203.
[5] K. Mitsuda et al. (1984), Publ. Astron. Soc. Japan **36**, 741.
[6] C. Chevalier and S.A. Ilovaisky (1990), Astron. & Astrophys. **238**, 163.
[7] M. Takizawa (1988), M.S. Thesis; also [1].
[8] K. Ebisawa (1991), Ph.D. Thesis.
[9] K. Ebisawa (1988), M.S. Thesis.
[10] R.A. Sunyaev et al. (1988), Soviet Astron. Letters **14**, 771; V.V. Efremov et al. (1989), Proc. 23rd ESLAB Symp. (ESA SP-296), p15.
[11] J.C. Ling and Wm.A. Wheaton (1989), Astrophys. J. Lettters **343**, L57.
[12] A.P. Lightman and T.R. White (1988), Astrophys. J., **335**, 57.
[13] T. Mihara (1990), private communication.
[14] K.A. Pounds et al. (1990), Nature **344**, 132; Matsuoka et al.(1990), Astrophys. J. **361**, 440.
[15] H. Awaki (1991), Ph.D. Thesis.
[16] H. Awaki et al. (1991), to appear in Publ. Astron. Soc. Japan.
[17] F. Makino et al. (1991), IAU Circular No. 5161; N. Lund et al. (1991), idem.

IRON LINES IN X-RAY PULSATORS

Fumiaki Nagase

Institute of Space and Astronautical Science,
3-1-1, Yoshinodai, Sagamihara, Kanagawa 229, Japan

Abstract. Iron lines in X-ray pulsars are characterized by an emission line at an energy of 6.4 keV and an absorption edge at 7.1 keV. The 6.4-keV emission line is attributed to the fluorescence Kα line due to reprocessed X-rays by cold circumstellar matter surrounding the binary system. Detailed features of iron lines have been measured so far mainly with the use of gas scintillation proportional counters on board the *Tenma* and *EXOSAT* satellites. The line properties are powerful tools to investigate the circumstellar matter in these binary systems, such as the accretion disk, the stellar wind, and the companion star atmosphere. Recent observations with *Ginga* revealed previously unknown complex features in the spectra of several X-ray pulsars, such as Cen X-3, Her X-1, and Vela X-1, especially in their low luminosity states. The results indicate that the temperature distribution and the ionization structure of the circumstellar matter are complicated and that at least two zones of cold matter and hot plasma are required to interpret the spectra observed in the low intensity states of Cen X-3 and Her X-1.

1. Introduction

There are about 30 known X-ray pulsars in close binary systems [1]. The energy spectra of these X-ray pulsars are commonly represented by a power law with a photon index of 0.8-1.5 [2] which is modified at high energies by an exponential cut-off. Emission line features at $\sim$ 6.5 keV were discovered with the use of the proportional counters on board *HEAO 1* and *OSO 8* [2] [3].

Spectroscopic studies using the gas scintillation proportional counters (GSPC) aboard *Tenma* and *EXOSAT* revealed that the iron line emission from X-ray pulsars is predominantly at 6.4 keV [1] [4] [5] [6], unlike the iron line emission from low-mass X-ray binaries which has been observed exclusively at 6.7 keV [7] [8] [9]. The 6.4-keV emission line in X-ray pulsars is attributed to fluorescence Kα emission due to reprocessed X-rays by cold circumstellar matter in the binary systems. Therefore measurements of iron line in the spectra of X-ray pulsars provide very useful information about the circumstellar matter in close-binary systems. The properties of the iron Kα line observed from X-ray pulsars are briefly summarized in section 2.

Using the LAC counters (*L*arge *A*rea proportional *C*ounter) aboard *Ginga*[10] [11], we obtained X-ray spectra with good statistics from several X-ray pulsars, such as Her X-1, Cen X-3, Vela X-1, 4U 1538-52, LMC X-4, and SMC X-1, even during their low states, eclipses or pre-eclipse dips with typical intensities of a few mCrab. The spectra provide an opportunity to examine the temperature, the ionization state, and the distribution of matter surrounding the magnetized neutron star. The new results are described in section 3.

2. Iron Line Properties in X-ray Pulsars

In this section we shall briefly summarize the properties of iron line emission from X-ray pulsars obtained from previous observations. Spectroscopic studies with the *Tenma* GSPC contributed significantly to establish a general understanding of the line emission in X-ray pulsars owing to good energy resolution of the GSPC counters [4] [5] [6].

The iron emission line energies are consistent with 6.4 keV for most X-ray pulsars as tabulated in [1] [4] [5] [6]. The width of the iron line is smaller than the resolution of the GSPC, i.e., 0.5 keV (FWHM) at 6.4 keV. Detailed spectroscopic studies of iron lines were performed with *Tenma* and line energies of 6.42 ± 0.02 keV and 6.46 ± 0.05 keV were obtained for Vela X-1 [12] [13] and for GX 301-2 [14] [15], respectively. The iron K-edge absorption features appeared at 7.24 ± 0.03 keV for Vela X-1 [13] [16] and 7.36 ± 0.05 keV for GX 301-2 [15]. From the energy of emission line, the ionization state of the line-emitting iron is estimated to be between Fe I and Fe XIX [1], whereas from the energy of K-edge absorption, the ionization state of X-ray absorbing iron is estimated to be between Fe V and Fe X [1].

The observed iron line intensity varied significantly from source to source, and from observation to observation. Yet there exists a rough proportionality between the iron line intensity and the continuum intensity above 7.1 keV [12] [15]. This relation and the observed energy of iron line provide firm evidence that these iron lines are the result of the fluorescent reprocessing of X-rays above 7.1 keV by relatively cold matter surrounding the source.

The equivalent width of the iron lines is relatively large (100-1000 eV) and highly variable in the case of massive wind-fed binaries, such as Vela X-1 and GX 301-2. The equivalent width is typically less than 200 eV for low-mass binaries such as Her X-1 and 4U 1626-67 and for Be-star binaries, such as A 0535+26. The relation between the equivalent width and the absorption column density toward the observer (N_H) is complicated and different from source to source [1] [6]. The results cannot be interpreted by a simple model according to which the fluorescence line is produced in spherically symmetric matter surrounding the central source; they require a non-uniform distribution of the reprocessing matter for many systems (see [1] [5] [6] for details). An interesting example is Vela X-1. The observed N_H values for Vela X-1 range widely from 2×10^{22} to 5×10^{23} cm^{-2}, the equivalent width of the iron line lies in the range of 100 to 150 eV, thus independent of N_H. This suggests that matter with $N_H \geq 10^{23}$ cm^{-2} is present near the source even at times that $N_H \simeq 10^{22}$ cm^{-2} along the line of sight to the observer.

Another clue to the distribution of reprocessing matter is the absence of pulsed modulation of the iron line intensity [12] [17]. This suggests that the reprocessing matter covers a fairly large portion of solid angle so that the pulsed modulation is smeared out. We conclude from the absence of the pulsed modulation of the line intensity that either dense and clumpy matter surrounds the neutron star in Vela X-1 unless otherwise the reprocessing matter corotates with the pulsar.

Several reprocessing sites have been discussed in [1] [5] [6] for the origin of fluorescence lines. Significant line emission was detected during the eclipse of Vela X-1 [12] [16], suggesting that both the stellar wind and the companion star atmosphere contribute significantly to the fluorescence emission line. The observed orbital phase dependence of the iron line intensity in Vela X-1 [16] requires an additional local site for fluorescence emission. Matter stagnated around the Alfvén shell [5] could be responsible for the orbital phase dependence. The accretion disk may be the site for the fluorescence line in the case of disk-fed pulsars such as Her X-1.

3. *Ginga* Results of Iron Line Spectroscopy

X-ray pulsars are usually bright sources with intensities greater than a few tens of mCrab in their high-on non-eclipse states. However, X-ray intensities at low-off states, during eclipse or dip phases are often as low as a few mCrab or much weaker, even for such bright sources as Cen X-3, Her X-1, and Vela X-1. The LAC counters aboard *Ginga* have a total effective area of 4000 cm^2 and they allow for good statistics even for sources as weak as 1 mCrab. We can see clear evidence for iron emission in the spectra of the low-off state of Her X-1 and in the eclipse phases of Cen X-3 and Vela X-1. Recent results of such spectral studies are described below.

3.1 Cen X-3

The observations of Centaurus X-3 were performed on March 22.0 to 24.2(UT), 1989 throughout the eclipse. The X-ray light curve during the observations revealed three different states, namely, a high-on non-eclipse phase, an intensity dip just before the eclipse ("pre-eclipse dip") and an eclipse phase. Energy spectra in the three different phases are shown in Figure 1(a). The spectra at the pre-eclipse dip and during the eclipse are significantly different from those of the high-on non-eclipse state. The pre-eclipse dip shows a flat continuum spectrum, whereas the eclipse spectrum shows an excess of soft X-rays although the spectral slope at high energy is the same as that of high-on state. An iron emission line is prominent in both the pre-eclipse dip and the eclipse spectra (see Figure 1(b)). Details of the observations are described by Nagase *et al.* [18].

The flatness at higher energies of the PHA ratio of the eclipse spectrum, i.e., the count-rate ratios for each pulse height channel between the eclipse spectrum and high state spectrum (Figure 1(b)), suggests that the X-rays are due to the scattering by optically thin matter. This is supported by the fact that no pulse modulation has been detected in the X-ray intensity during eclipse. An additional soft component is required to explain the excess of the PHA ratio in the low energies. If the X-rays in the eclipse is due to the scattering origin, soft X-rays are usually subjected by absorption, and thus the PHA ratio at low energies should be smaller than the ratio at higher energies, or at least should not exceed that even if the scatterer is highly ionized plasmas. It is remarkable that a hump is present in the PHA ratio spectrum at about 8 keV in addition to the prominent iron $K\alpha$ line at about 6.5 keV.

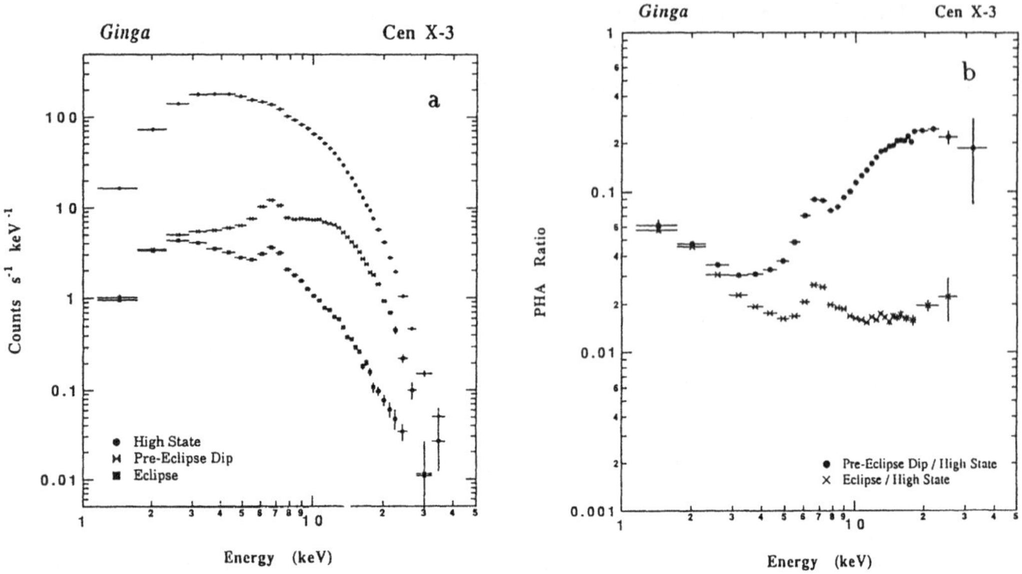

Figure 1: (a). Pulse-averaged pulse-height spectra of Cen X-3 observed with *Ginga* at three different phases: i.e., high-on non-eclipse phase, pre-eclipse dip phase and mid-eclipse phase. (b). The count-rate ratios for each pulse-height channel between the three different states of Cen X-3; the mid-eclipse spectrum divided by the high-state non-eclipse spectrum, and the pre-eclipse dip spectrum divided by the high-state non-eclipse spectrum.

The spectrum during pre-eclipse dip can be interpreted by the sum of a highly absorbed direct component and a soft X-ray component which is comparable with that seen in the eclipse phase (see Figure 1(b)). One should note, however, that the highly absorbed direct component does not reach unity but stays at a level of one-third of the non-eclipse spectrum even at highest energy range. This can be explained in terms of thick scattering matter in the line of sight between the source and observer. This is supported by the fact that the pulse amplitudes in this phase are not only diminished to an invisible level at low energies but also significantly reduced at high energies.

The spectrum of Cen X-3 in the high-state non-eclipse phase has a shape typical for X-ray pulsars and can be fitted by a power law continuum with a high-energy turn-over. To fit the high energy turn-over, we adopted absorption features in the form of a Lorentzian $\times E^2$, resembling the cross sections for the cyclotron resonant scattering [19]. The best-fit values are 1.01 ± 0.03 and $(1.0 \pm 0.1) \times 10^{22}$ cm^{-2}, respectively, for the photon index and the absorption column density. An iron line is required to obtain an acceptable fit. The energy and equivalent width of the line obtained from the fit are 6.50 ± 0.08 keV and 180 ± 23 eV, respectively.

The spectrum observed during mid-eclipse phase requires the inclusion of a soft component in addition to the above model. Hence we adopted a thermal bremsstrahlung spectrum for the soft component. The scattering continuum in the fit is assumed to have the same shape as that of the high state spectrum. The fit is shown in Figure 2(a),

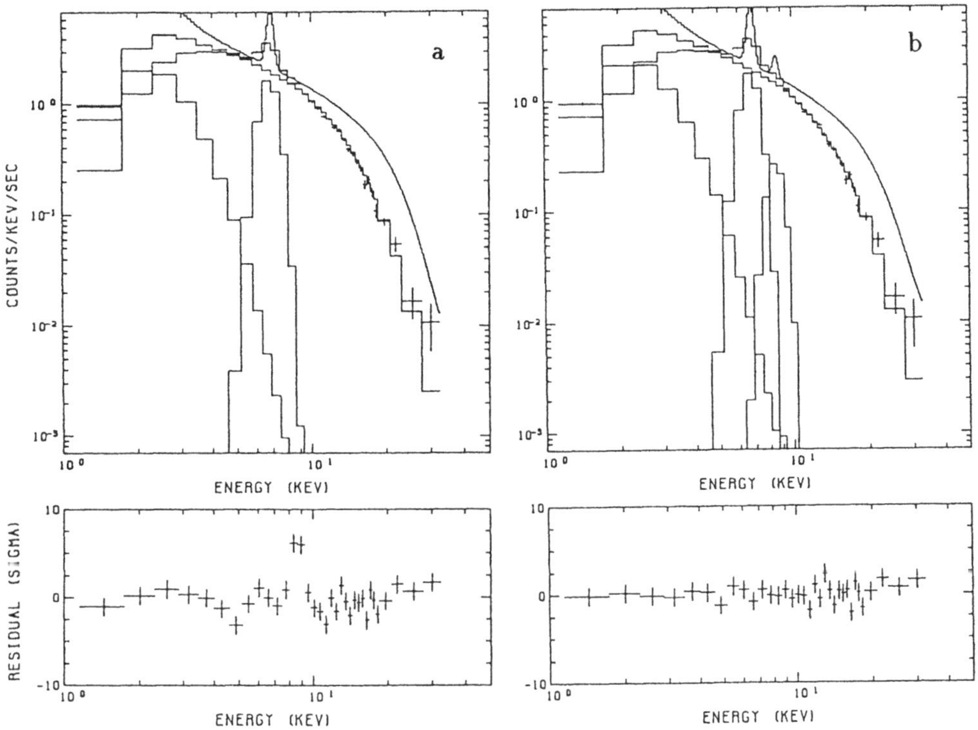

Figure 2: (a). Mid-eclipse spectrum of Cen X-3 fitted by a model described in the text (upper panel). The crosses are the observed data, the histograms are the calculated model spectra. The uppermost solid line represents an incident photon spectrum inferred from the best-fit parameters. The lower panel represents residuals of data from the best-fit model. (b). Same as (a) but with a second line near 8.5 keV in addition to the model in (a).

however, is not acceptable (reduced $\chi^2 = 4.7$ for 28 degrees of freedom). If we include another emission line near 8.5 keV, we obtain an acceptable fit as shown in Figure 2(b) with a reduced χ^2 value of 1.2 (DOF=26). The energies of the two lines are 6.61 ± 0.03 and 8.4 ± 0.2 keV, and their equivalent widths are 1.30 ± 0.08 keV and 0.23 ± 0.04 keV, respectively. The iron Kα line has a significantly larger energy than 6.4 keV, suggesting that it originate from relatively hot (highly ionized) reprocessing matter, unlike what is observed during the high state which is consistent with the line being predominantly produced in cold matter. The line at 8.4 keV is considered to be a blend of emission lines around 8 keV due to highly ionized iron and nickel ions, i.e., Fe Kβ and Ni Kα lines.

The spectrum in the pre-eclipse dip can be fitted by a similar model and taking into the effect of scattering (see [18] for details). It is interesting to mention that the energy of the iron emission line in this phase is obtained at energies between 6.4 keV and 6.7 keV, suggesting that both cold and hot matter are attributing to the iron lines. The iron line parameters during the observations are shown in Figure 3 together with the continuum flux in the energy range of 1-37 keV.

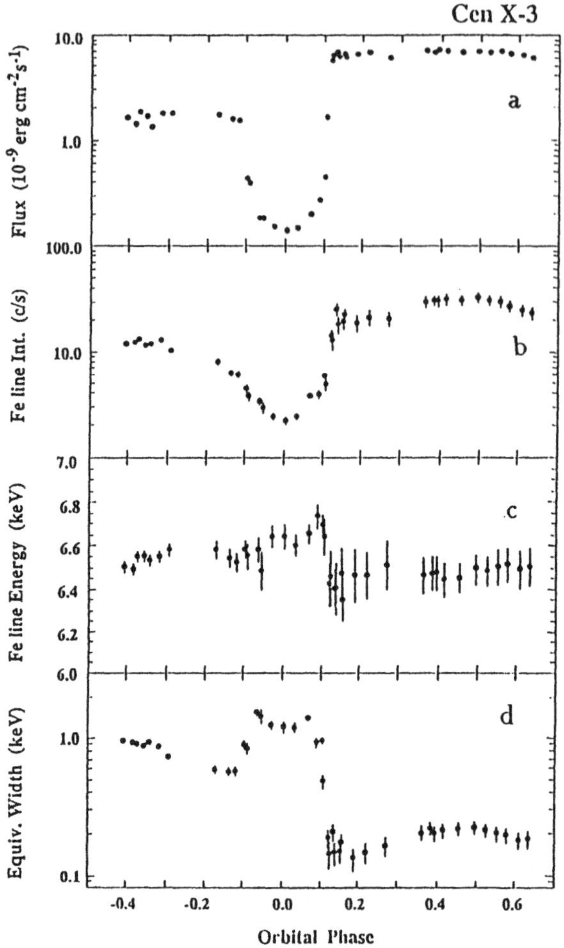

Figure 3: Time dependence of the parameters derived from the fit for individual spectra during the pre-eclipse dip, the eclipse phase and the high-on non-eclipse state of Cen X-3. (a) continuum X-ray flux in the range of 1-37 keV, and (b) line intensity, (c) energy, and (d) equivalent width of the iron line.

From Figure 3 we can deduce several interesting properties. The intensity of the iron line at the eclipse phase decreases to about 10 % of that at the high-on state, whereas the ratio of the continuum fluxes in the two phases is only about 2 %, thus leading to a relatively prominent line feature in the spectrum obtained during the eclipse phase. This feature is similar to that observed from Vela X-1 [16]. The line energy is obtained at 6.5 keV in the high state which is not too much different from other X-ray pulsars. This suggests that the iron line is predominantly due to reprocessing in cold matter, at least in this phase. The line energy increases from 6.5 keV to 6.6 keV during the pre-eclipse dip; the average value of the line energy during the eclipse phase is about 6.65 keV. The equivalent widths of the iron lines are 0.15-0.2 keV during the high-on state, 0.5-1.0 keV during the pre-eclipse dip phase and about 1.5 keV during the eclipse phase. The large equivalent width during the eclipse phase is also consistent with the interpretation that the X-ray emission during this phase is due to scattering [5].

These results suggest the existence of two lines, a fluorescence line near 6.4 keV due to reprocessing in cold matter and one near 6.7 keV which originates from a hot plasma. Hence, two zones of cold matter and a region with hot plasma should exist near Cen X-3. The region of the 6.4 keV line is estimated to be less than 3×10^{10} cm (see [18] for details). Whereas, the region of the 6.7 keV line is larger than the radius of the companion, i.e., 8×10^{11} cm. The stellar wind irradiated by X-rays from the neutron star will be highly ionized and form an extended zone of highly ionized plasma surrounding the neutron star [20] which may be responsible for the 6.7-keV iron line emission.

3.2 Her X-1

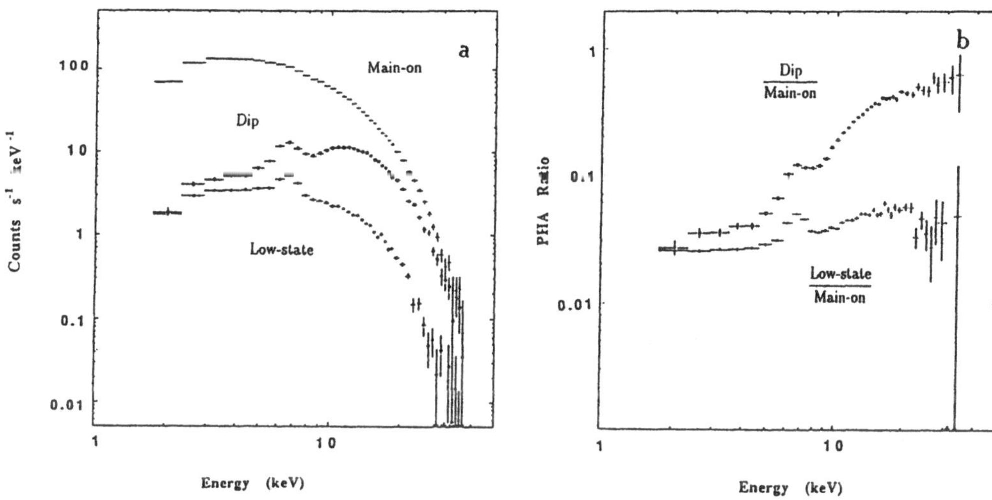

Figure 4: (a). Pulse-average spectra of Her X-1 during the main-on non-eclipse phase and the pre-eclipse dip phase in the high state and in the low state. (b). The count-rate ratios for each pulse-height channel between the three different states of Her X-1; the low-state spectrum divided by the main-on non-eclipse spectrum, and the pre-eclipse dip spectrum divided by the main-on non-eclipse spectrum.

Observations of Her X-1 were made with *Ginga* on August 17 and 19, 1988 during the low-off state of the 35-d cycle. Ten days later on August 28, the turn on to the (main-on) high state occurred. A spectrum of a pre-eclipse dip phase was obtained during this observation. The spectra in the low state and the pre-eclipse dip phase are compared in Figure 4(a) with that of the main-on state. The spectrum of Her X-1 in the low state is due to entirely scattering, as the observed X-rays do not pulsate. An upper limit to the 1.24 s pulse amplitude was 2.4 % of the average flux, unlike the large pulse amplitude of 65 % observed during the main-on non-eclipse phase. Scattering by an optically thin plasma does not fully explain the low state spectrum as it is slightly harder than the incident spectrum (see the ratio of Figure 4(b)). The average X-ray flux in the low state was 3.4×10^{-10} erg cm^{-2} s^{-1} in the range 2-37 keV, which is about 5 % of the flux in the main-on state. If the continuum spectrum during the low state is due to scattering by an optically thin hot plasma, it should have the same shape as the incident spectrum which in turn is assumed to be the same as that during the main-on state non-eclipse phase.

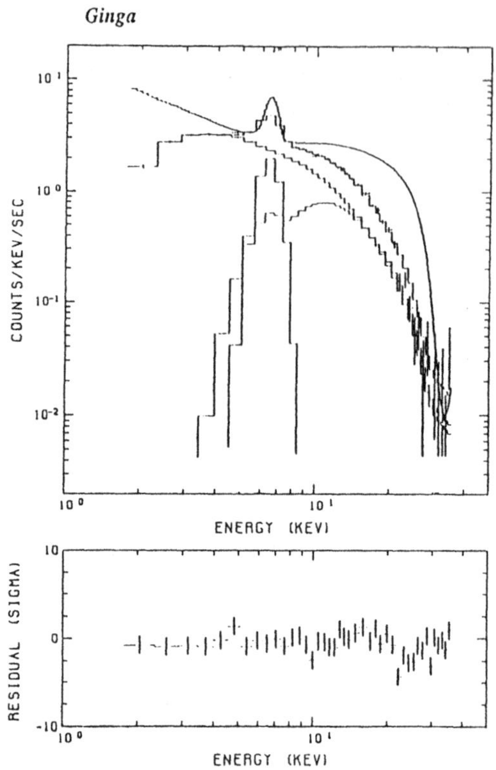

Figure 5: Low state spectrum of Her X-1 fitted by a two component model described in the text (upper panel). Notations of symbols are same as the legends of Figure 2(a). The lower panel represents residuals of data from the best-fit model.

If the spectrum during the low state is due to scattering by optically thick matter, soft X-rays should be absorbed more strongly. The excess in the ratio in Fig. 4(b) from a flat spectrum at higher energies may be partly due to scattering by optically thick matter. Hence, the combination of scattering by an optically thin hot plasma and by optically thick cold matter is a possible explanation for the spectrum of Her X-1 observed during the low state of the 35-d cycle. A fit by such a model, which is shown in Figure 5, gives an acceptable set of parameters. An iron line was required in the fit at an energy of 6.50 ± 0.08 keV with an equivalent width of 1.0 ± 0.1 keV and an intrinsic line width of 0.8 ± 0.4 keV (FWHM). The equivalent width obtained is consistent with the scattering origin of the low state spectrum [5]. An accretion disk corona of radius $r < 5 \times 10^{11}$ cm and an outer edge region of the accretion disk at $r > 3 \times 10^{11}$ cm are likely sites for the two scattering components. The outer edge region of the disk is mainly responsible for the iron line as it is relatively cold, although the large width suggests also a contribution due to reprocessing in a hot scattering region. Details of the observations and modeling of the low state spectrum of Her X-1 are described elsewhere by Mihara *et al.* Ref. [21].

The spectrum of the pre-eclipse dip can be explained in terms of the sum of a highly absorbed incident X-rays and scattering of X-rays on a thin hot plasma (see Figure 4(b)). The pre-eclipse dip spectrum is fitted by the same manner as that of Cen X-3, although the soft component included for the fit for Cen X-3 is not required.

3.3 Vela X-1

The wind-fed pulsar, Vela X-1 (4U 0900-40), was observed with *Ginga* from 16.7 to 24.2 (UT) in February 1988 covering the eclipse and eclipse transitions. The observed pulse-average spectra of Vela X-1 are shown in Figure 6 (ingress and egress). The change of the energy spectrum of Vela X-1 as a function of the orbital phase has been extensively investigated by *Tenma* observations [13] [16] and by *EXOSAT* observations [22] for the non-eclipse phase. In this paper, we shall mainly discuss the spectra observed during the eclipse transitions and eclipse phase. The essential features along the eclipse transitions are the same as those obtained from the *Tenma* observations (i.e., Fig. 3 of [16]), but the observations with *Ginga* are of higher quality (see lowest spectra in Figure 6(a) and (b)).

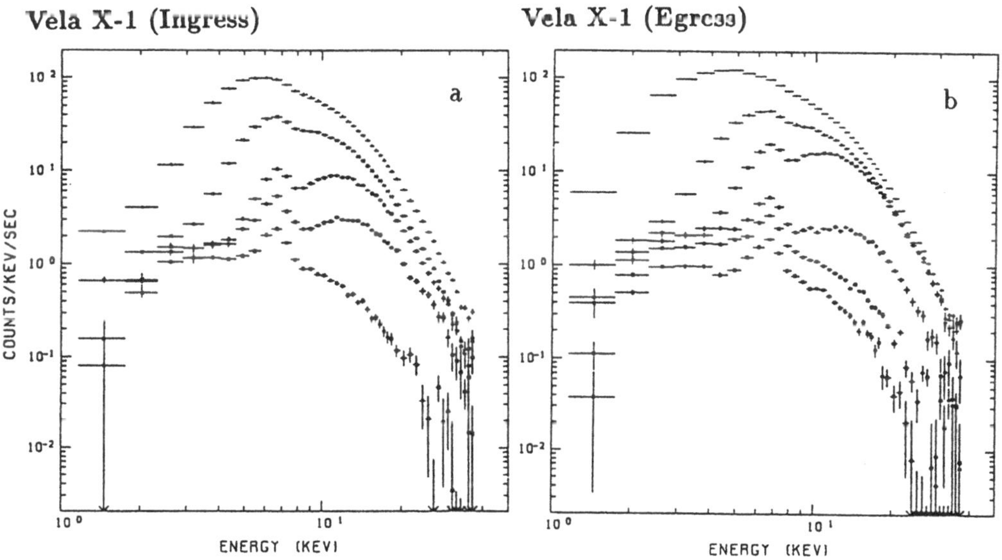

Figure 6: (a). The change of pulse-average energy spectra for Vela X-1 observed with *Ginga* during immersion. The lowest spectrum in the panel is the one obtained during the earlier part of the eclipse. (b). same as (a) but during emersion. The lowest spectrum in the panel is the one obtained during the later part of the eclipse.

An increase of the absorption measure toward the eclipse center during eclipse transitions is evident. The spectra during the eclipse are relatively flat but soft X-ray emission below 4 keV is significant. X-rays during the eclipse do not pulsate at all energy ranges, suggesting that they are due to scattering. It is seen from Figure 6 that the spectra observed during eclipse transitions are a combination of a highly absorbed incident X-rays and a scattering component with a spectrum similar to the eclipse spectrum. Furthermore, iron line features at ~ 6.4 keV are prominent in the spectra during the eclipse and the eclipse transitions. It is noted that the line energy is consistent with 6.4 keV in the case of Vela X-1 even during the eclipse, unlike in the case of Cen X-3 described in section 3.1. These properties could be due to the scattering and reprocessing by the companion star atmosphere and the stellar wind.

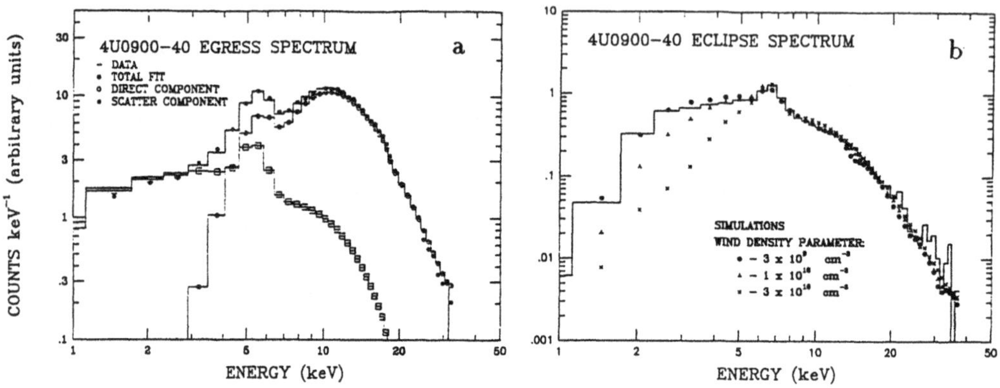

Figure 7: Model spectra for an egress and eclipse of Vela X-1, which are computed using a self-consistent stellar wind model described in the text, are compared with the observed spectrum (histogram).

It was shown by Lewis *et al.* [23] that a self-consistent stellar wind model, involving an *ad hoc* exponential atmosphere surrounding the companion [16] and an analytically connected Castor-Abbott-Klein (CAK) wind [24], can reproduce the observed spectral variation throughout the eclipse transitions, taking into account X-rays scattered by the wind. A crude model for self-consistent ionization zones [22] [25] has also been incorporated in the model calculations. In practice, however, this makes very little difference to the eclipse spectrum. An important finding is that the Kα fluorescence lines of S and Si make very significant contributions to the low energy channels below 3 keV in the eclipse spectrum. The calculated spectra for an eclipse transition and the eclipse are compared with the observations in Figure 7(a) and (b). Lewis *et al.* [23] showed that the entire spectra during eclipse transitions and during the eclipse can be reproduced using the following set of wind parameters, i.e., a scale height of 6.6 ± 2 *lt*-s for the *ad hoc* exponential atmosphere, a density of the CAK wind of 3×10^9 cm^{-3} at the neutron star orbit, and an iron abundance of the wind of of 1.0 ± 0.5 times the cosmic abundance. An insignificant amount of emission of the 6.7 keV line in the Vela X-1 eclipse spectrum, unlike in Cen X-3, may be due to the fact that the X-ray irradiated ionization zone with $\xi(= L_x/nr^2) \geq 3$ is limited to a small region surrounding the neutron star [25], as the X-ray luminosity of Vela X-1 is relatively small ($\sim 5 \times 10^{36}$ erg s^{-1}).

The author would like to acknowledge W. Lewis, T. Mihara, S. Rappaport, and T. Takeshima for preparation of the data and for their valuable comments. He is also grateful to W. H. G. Lewin for careful reading of the manuscript and for useful comments.

References

[1] Nagase, F. 1989, *Publ. Astron. Soc. Japan*, **41**, 1.

[2] White, N. E., Swank, J. H., and Holt, S. S. 1983, *Astrophys. J.*, **270**, 711.

[3] Pravdo, S. H. 1979, in *X-ray Astronomy*, ed. W. A. Baity and L. E. Peterson (Pergamon Press, Oxford), p. 169.

[4] Nagase, F. 1985, *Adv. Space Res.*, **5**, 95.

[5] Inoue, H. 1985, *Space Sci. Rev.*, **40**, 317.

[6] Makishima, K. 1986, in *The Physics of Accretion onto Compact Objects*, ed. K.O. Mason, M.G. Watson and N. E. White (Springer-Verlag, Berlin), p. 249.

[7] White, N. E., *et al.* 1985, *Astrophys. J.*, **296**, 475.

[8] White, N. E., *et al.* 1986, *Mon. Not. Roy. Astron. Soc.*, **218**, 129.

[9] Hirano, T., *et al.* 1987, *Publ. Astron. Soc. Japan*, **39**, 619.

[10] Makino, F. and the ASTRO-C team, 1987, *Astrophys. Lett. Communications*, **25**, 223.

[11] Turner, M. J. L., *et al.* 1989, *Publ. Astron. Soc. Japan*, **41**, 345.

[12] Ohashi, T., *et al.* 1984, *Publ. Astron. Soc. Japan*, **36**, 699.

[13] Nagase, F., *et al.* 1986, *Publ. Astron. Soc. Japan*, **38**, 547.

[14] Leahy, D. A., *et al.* 1989, *Mon. Not. Roy. Astron. Soc*, **236**, 603.

[15] Leahy, D. A., *et al.* 1989, *Mon. Not. Roy. Astron. Soc*, **237**, 269.

[16] Sato, N., *et al.* 1986, *Publ. Astron. Soc. Japan*, **38**, 731.

[17] Leahy, D. A., and Matsuoka, M. 1990, *Astrophys. J.*, **355**, 627.

[18] Nagase, F., *et al.* 1990, in preparation.

[19] Makishima, K., *et al.* 1990, *Publ. Astron. Soc. Japan*, **42**, 295.

[20] Hatchett, S., and McCray, R. 1977, *Astrophys. J.*, **211**, 552.

[21] Mihara, T., *et al.* 1990, submitted to *Publ. Astron. Soc. Japan.*

[22] Haberl, F., and White, N. E. 1990, *Astrophys. J.*, **361**, 225.

[23] Lewis, W. *et al.* 1991, in preparation.

[24] Castor, J. I., Abbott, D. C., and Klein, R. I. 1975, *Astrophys. J.*, **195**, 157.

[25] McCray, R., *et al.* 1984, *Astrophys. J.*, **282**, 245.

IRON LINES IN X-RAY SPECTRA OF HIGH MASS X-RAY BINARIES

Frank Haberl
EXOSAT Observatory, Astrophysics Division, Space Science Department
European Space Agency, Postbus 299, 2200AG Noordwijk, The Netherlands

Abstract

Observations of high mass X-ray binaries with *EXOSAT* have shown that the X-ray spectra can be modeled with absorption by a gas stream from the primary trailing the neutron star. The dense gas stream highly absorbs the direct X-rays from the neutron star at orbital phases when the line of sight passes through the gas stream. A second, low-absorption component becomes visible which can be explained by Thomson scattering of X-rays around the dense gas stream. The spectra of GX301-2 show an iron emission line near 6.4 keV with a wide range of line intensities. Adding the iron line to the low-absorption component of the spectrum the derived Fe column densities to explain the line intensity are consistent with the equivalent H column densities causing the photo-electric absorption of the low-absorption component.

1. Introduction

In supergiant X-ray binaries, SXRB's, an accreting neutron star or black hole is orbiting an OB supergiant companion within a few stellar radii (see White 1985 for a recent review). The X-ray source is viewed through the stellar wind and will suffer photo-electric absorption, principally by the K edges of oxygen, silicon, and sulphur and the K and L edges of iron. X-ray spectra of the SXRB's show variable absorption up to an equivalent of 10^{24} H cm^{-2}, with a prominent iron K absorption edge at 7.1 keV (e.g. Swank *et al.* 1976). The absorption increases around the times of eclipse ingress and egress as the X-ray source passes behind the dense innermost regions of the wind (e.g. Branduardi, Mason and Sanford 1978). An additional absorption component that occurs following the time of superior conjunction of the X-ray source in the 4U1700-37/HD153919 system could be modeled as due to absorption by a dense gas stream from the surface of the primary trailing the neutron star (Haberl, White and Kallman 1989). Also X-ray spectra from Vela X-1 (Haberl and White 1990) and GX301-2 (Haberl 1990) indicate the presence of a gas stream.

The low-energy absorption spectra at orbital phases when the line of sight passes through the gas stream show an excess below about 5 keV over that expected from photo-electric absorption. This low-energy excess shows no pulsations and can be explained by X-rays scattered around the gas stream by the less absorbing ambient stellar wind. The spectra can be modeled by two powerlaw components with the same energy index with different absorbing column density, the strong component being the highly absorbed direct beam from the neutron star and the second low-absorption component of the order of 5% of the strong component representing the scattered X-rays (Haberl and White 1990; Haberl 1990).

Spectra from SXRB's show a narrow iron emission line near 6.4 keV (Makishima 1986). To determine the intrinsic (unabsorbed) line intensity it is important to use the proper absorption at the line energy. Adding the iron line to the high-absorption component overestimates the intrinsic line strength because of the very high absorption from the gas stream (Haberl and White 1990).

2. Vela X-1

EXOSAT made a total of 17 observations of Vela X-1 at a variety of orbital phases between 1984 February and 1985 May (Haberl and White 1990) with observation durations ranging from 2 to 16 hr. The iron line energy was determined from fits to the spectra obtained by the GSPC and was between 6.37 ± 0.10 (1985 April 30) and 6.51 ± 0.12 keV (1985 May 9). The iron line intensity and equivalent width, EW, depend on which of the two absorption components in the spectra the line is attached to. The results of both combinations, first with the line attached to the low N_H component (panels a,b and c) and second to the high N_H component (panels d,e and f) are shown in Figure 1. For the case with the line added to the low N_H component the EW (relative to the fainter low N_H component) is constant except for 3 high points. The considerable scatter in the EW around the average is consistent with the uncertainties. The unabsorbed line intensity shows much more variations but these are matched by changes in the total 1 - 20 keV continuum. In the right part of Figure 1 the continuum flux (d), the line flux (e) and the line EW (f) are shown for the model where the line is associated with the high N_H component. The higher absorption (at phase 0.3 and 0.5) causes the unabsorbed line intensity to be more than two orders of magnitude higher with the EW (relative to the bright high N_H component) a factor of ~ 20 lower. The EW increases during periods of high absorption.

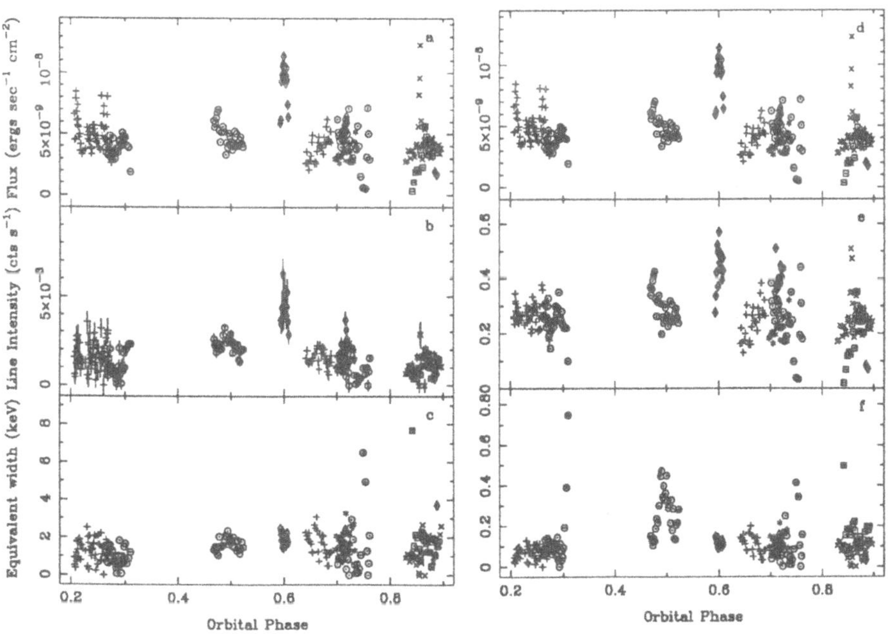

Fig. 1: Spectral properties for the Vela X-1 spectra with an iron fluorescence line added to the low N_H component (left 3 panels) or added to the high N_H component (right 3 panels) shown versus orbital phase (from Haberl and White 1990).

3. GX301-2

EXOSAT made a total of 13 observations of GX301-2 each lasting between 3 hr and 2 days between April 1984 and March 1985 (Haberl 1990). The energy of the line determined from GSPC spectra was between 6.38 ± 0.07 and 6.54 ± 0.13 keV and is consistent within the errors with that of neutral iron, i.e. 6.4 keV. In Figure 2 the total unabsorbed flux in the 1 - 20 keV band is shown in the top panel. The outburst peaks near phase 0.95 with a factor of ~ 100 higher intensity compared to the minimal intensities near phases 0.15 and 0.7. Between the two observations with lowest intensity the intensity is about a factor of 5 higher. The middle panel shows the unabsorbed line intensity as derived adding the line to the low-absorption component. The line equivalent width using the total continuum (low- plus high-absorption component) is shown in the lower panel. The line intensity follows the variations of the total flux and the resulting equivalent width is more or less constant although there is considerable scatter for each spectrum. Overall the equivalent width does not exceed 200 eV at any orbital phase. The unabsorbed line intensity as a function of the flux in the 7-20 keV band is shown in Figure 3a with the line added to the low-absorption component. By integrating over the Fe K-edge above 7.1 keV with a fluorescence yield of 0.34 the expected fluorescent line intensity can be compared with the observations for different depths of the edge, i.e. different iron column densities. The two solid lines in Figure 3a show the expected line intensities for two values of iron column density which correspond to an equivalent hydrogen column density of $2{\cdot}10^{22}$ H cm^{-2} (lower line) and $2{\cdot}10^{23}$ H cm^{-2} (upper line) for cosmic element abundance. The neutral H equivalent column density derived from the spectral fits for the low-absorption component of the ME spectra lies with some exceptions during outburst between $2 \cdot 10^{22}$ H cm^{-2} and $2 \cdot 10^{23}$ H cm^{-2} and is consistent with the corresponding iron column density. In Figure 3b the same is shown for the line added to the high-absorption component. The line intensity shows more scatter as in Figure 3a. The two solid lines in Figure 3b show the expected line intensities for iron column densities corresponding to an equivalent hydrogen column density of $1{\cdot}10^{23}$ H cm^{-2} (lower line) and $1{\cdot}10^{24}$ H cm^{-2} (upper line). The first value for the hydrogen column density is the lower limit of the column density derived from the spectral fits for the line added to the high-absorption component. More points in Figure 3b lie below the lower line as compared to Figure 3a which can be explained either by a lower Fe abundance in the material responsible for the fluorescence or that the H equivalent column density of the fluorescing material is overestimated.

4. Conclusion

Recent observations of SXRB's with *EXOSAT* have shown the presence of a dense gas stream in the wind of the primary in at least three systems. This region of high wind density causes high photo-electric absorption at orbital phases when the line of sight passes through the gas stream. Due to the high low-energy absorption of the direct X-rays from the neutron star a second less absorbed component becomes visible in the X-ray spectra, which can be modeled as X-rays scattered around the gas stream by the ambient wind. Fluorescent iron line emission from different regions in the wind will be absorbed by different column densities. Using the high column density caused by the gas stream to determine the intrinsic iron line intensity overestimates the line

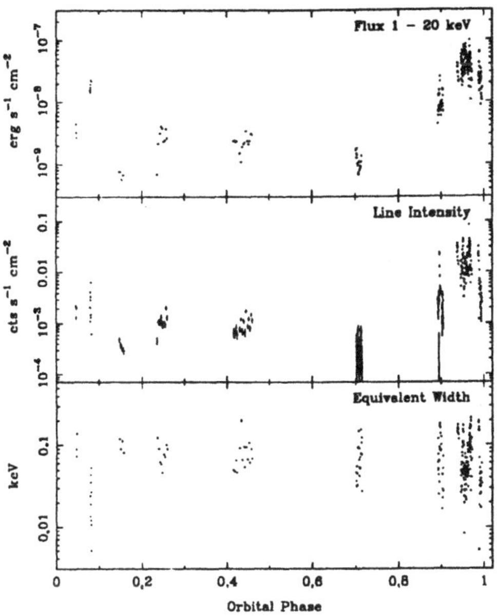

Fig. 2: The unabsorbed flux, the iron fluorescence line intensity and the line equivalent width from GX301-2 as function of orbital phase. The iron line is added to the low N_H component of the parallel absorption model (from Haberl 1990).

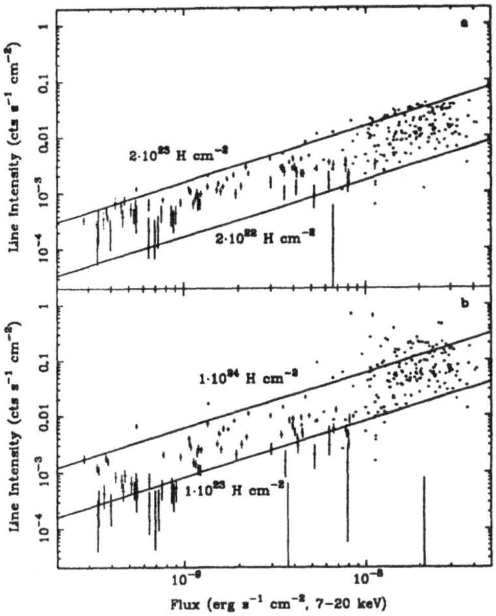

Fig. 3: The unabsorbed Iron line intensity as a function of the total continuum flux in the 7–20 keV band. In (a) the line is added to the low absorption component and in (b) to the high absorption component (from Haberl 1990).

intensity. On the other hand the column density of the low-absorption component is lower than the average absorption. The real intrinsic line intensity will be between these two extreme values and will depend on the contribution of the gas stream to the total mass loss rate of the primary star.

References

Branduardi, G., Mason, K.O., and Sanford, P.W. 1978, *M.N.R.A.S.*, **185**, 137.

Haberl, F., White, N.E., and Kallman, T. 1989, *Ap.J.*, **343**, 409

Haberl, F., and White, N.E. 1990, *Ap.J.*, **361**, 225.

Haberl, F. 1990, submitted to *Ap.J.*.

Makishima K. 1986, in *Proceedings of a Workshop on The Physics of Accretion onto Compact Objects*, ed. K.O. Mason, M.G. Watson, and N.E. White (Springer), p. 249.

Swank, J.H., Becker, R.H., Boldt, E.A., Holt, S.S., Pravdo, S.H., Rothschild, R.E., and Serlemitsos, P.J. 1976, *Ap.J.(Letters)*, **209**, L57.

White, N.E. 1985, in *Interacting Binaries*, ed. M.L. Eggleton and J.E. Pringle (Dordrecht: Reidel), p. 249.

PROPERTIES OF IRON LINE EMISSION IN THE X-RAY BINARY GX301-2

D. A. Leahy
Max Planck Institut fur Physik and Astrophysik,
Institut fur extraterrestrische Physik
8046 Garching bei Munchen, FDR
(permanent address: Dept. of Physics and Astronomy
University of Calgary, Canada T2N 1N4)

ABSTRACT

The X-ray binary GX301-2 was observed extensively by the Tenma satellite. GX301-2 exhibits strong fluorescent iron line emission and a strong K-edge absorption feature. Time variations in the X-ray luminosity, absorption column density, and the iron emission line are used to study the properties of the circumstellar matter in this system.

1. INTRODUCTION

GX301-2 is an X-ray pulsar with a 701 second pulse period and a 41.5 day orbital period. The light curve and its relation to the properties of the binary system are described by White and Swank (1984). The binary companion is the supergiant (B2Iae) star WRAY 977 which has mass loss rate greater than 3×10^{-6} solar masses per year (Parkes et al., 1980). The high mass loss rate results in large amounts of circumstellar matter and also feeds the accretion onto the neutron star which powers the X-ray emission. GX301-2 has strong iron line emission in its spectrum and also shows the effects of large hydrogen absorption column densities (2×10^{23} cm^{-2} to 1.8×10^{24} cm^{-2}).

High resolution spectral data were obtained using the Tenma satellite gas scintillation proportional counters (Tanaka et al., 1984, Koyama et al., 1984). Aspects of the Tenma data on GX301-2 which have already been reported are: an observation of the spectrum by Makino et al.(1985); timing results by Sato et al.(1986); an unusual eclipse-like feature by Leahy et al.(1988); and results of iron-line emission studies by Leahy et al.(1989a,b).

2. VARIATIONS OF SPECTRUM AND IRON LINE

Background-subtracted spectra were fitted by the standard X-ray pulsar spectral model (White et al., 1983) consisting of a power law spectrum with exponential cutoff. Absorption effects due to intervening matter, consisting of light elements and iron, were separately included ($\exp(-\sigma N_H)$ and $\exp(-\sigma_{Fe} r_{Fe} N_{Fe})$). N_H is the hydrogen column density and $r_{Fe} N_{Fe}$ is the iron column density. N_{Fe} is the iron equivalent column density; the standard cosmic abundance of iron is $r_{Fe} = 4.0 \times 10^{-5}$. The best fit parameters for the law parameters K-edge energy, iron line energy, E_{Fe}, iron line intensity, I_{Fe}, 7-30 keV continuum intensity, and $\log(N_{Fe})$, have been presented by Leahy et al. (1989a) as a function of time for the 6 days of observations.

A plot of I_{Fe} against continuum energy shows a clear correlation and strong evidence for the fluorescent production mechanism for the iron line. However the relation between iron line strength and continuum intensity should be linear if the the fluorescing matter has constant properties. We find an observed power law dependence of $I_{Fe} \propto I_{cont}^{0.67}$ by a least squares fit (with linear correlation coefficient of 0.93). The linear function is a much poorer fit to the data (for the 45 data points, the chi-squared is 707 for the power-law fit and 1965 for the linear fit). The daily average values of iron line intensity are also strongly correlated with daily average continuum intensity. A power law fit gives $I_{Fe} \propto I_{cont}^{0.75}$ with a correlation coefficient of 0.98. This indicates a change in the fluorescing matter dependent on source intensity.

The iron line energy has been found to be a function of column density. A decreasing iron ionization and cooling of the fluorescent gas with increase in observed column density is evident. A difference in temperature between absorbing gas and emitting gas is seen in comparing the Fe emission line energy with the K-edge absorption energy (Leahy et al, 1989a). The absorption samples the gas along the line-of-sight, while the fluorescent line emission comes from all gas irradiated by X-rays. When the Fe line energy is high, the irradiated gas mean temperature rises (to 2×10^5 K) while the gas along the line-of-sight remains cool. A difference in $\log(\xi)$ of 0.5 is required between the line-of-sight gas and the irradiated gas. This can be accomplished by the pulsar if it has a beaming factor of 3; alternately the density of the gas along the line-of-sight is a factor of 3 higher than the average for the irradiated gas; or some combination of the two effects.

3. THE EFFECTS OF CIRCUMSTELLAR MATERIAL

Scattering, photoelectric absorption, and fluorescent iron line production are all important in determining the observed shape of the spectrum. To study these effects by material around the neutron star on the observed continuum spectrum, Monte Carlo simulations were performed (Leahy et al, 1989b). Two matter geometries were considered: a thin spherical shell ($\Delta r/r = 0.1$) and a uniform sphere, each around a central isotropic X-ray continuum source. The emergent spectrum was calculated.

The observed equivalent width vs. column density relation is consistent with the calculated relation curves for either sphere or shell case, but has significant fluctuations. These are likely caused by inhomogeneities which cause the measured column density to be different than the column density averaged over 4π steradians around the source. The low energy excess seen in a number of the spectra (Leahy et al.,1989a) provide independent evidence for inhomogeneities. Another cause of deviation from the theoretical curve is expected: the change of ionization of the fluorescing matter caused by source luminosity variations. However, any dependence of column density on 7-30 keV continuum intensity is masked by other variations in the column density. There is evidence for the effect of luminosity variations from the non-linear dependence of iron intensity on continuum intensity. A possible cause for this dependence is a gradual change from a spherical accretion geometry for low accretion rate to disk-like geometry for high accretion rate. There is other evidence for large inhomogeneities in the circumstellar matter in GX301-2: An intensity dip (described in detail in Leahy et al., 1988) occured for about 5 hr. on April 27, 1984.

4. PULSE PHASE DEPENDENCE OF THE SPECTRUM AND IRON LINE

The spectral parameters have been studied for dependence on pulse phase for each day of observation and for the Apr.24 flare. A detailed discussion of the variability of the parameters and their statistical significance is given by Leahy and Matsuoka (1990).

The phase dependence of column density is in agreement with expected different column densities for the direct and scattered components and estimates of the varying contributions of direct and scattered radiation as a function of pulse phase. The variations of power law index, e-folding energy and cutoff energy are primarily due to intrinsic variations caused by varying line of sight to the emission region on the surface of the neutron star.

A real variation in iron line intensity with pulse phase is seen. The observed variation is consistent with fluorescence in the atmosphere of the companion to the neutron star. The time delay between the main continuum pulse peak and the iron line peak due to companion illumination for each of the six days was calculated, and has 2 contributions: time travel delay and sweeping delay. The observed daily delay times agree with the calculated delay times.

5. SUMMARY

Many important and interesting effects have been deduced from iron line studies in X-ray binaries, as exemplified by GX301-2 here. More detailed and extensive observations of iron line emission will allow further understanding of these systems.

REFERENCES

Koyama, K., Ikegami, T., Inoue, H., Kawai, N., Makishima, K., Matsuoka, M., Mitsuda, K., Murakami, T., Ogawara, Y., Ohashi, T., Suzuke, K., Tanaka, Y., Waki, I., & Fenimore, E.E. 1984, Publ. Astron. Soc. Japan, **36**, 659.

Leahy, D. and Matsuoka, M. 1990 Astrophys.J., **355**, 627.

Leahy, D. and Matsuoka, M. in Proceeding of COSPAR Symposium No.14-Advances and Perspectives in X-ray and Gamma-ray Astronomy, Helsinki, 1988.

Leahy, D., Matsuoka, M., Kawai, N. and Makino, F. 1989a, Mon. Not. Roy. Astr. Soc., **237**, 269.

Leahy, D., Matsuoka, M., Kawai, N. and Makino, F. 1989b, Mon. Not. Roy. Astr. Soc., **236**, 603.

Leahy, D., Nakajo, M., Matsuoka, M., Kawai, N., Koyama, K., & Makino, F. 1988, Publ. Astron. Soc. Japan, **40**, 197.

Makino, F.,Leahy, D., & Kawai, N. 1985, Space Sci. Rev., **40**, 421.

Ohashi, T., Inoue, H., Koyama, K. Makino, F., Matsuoka, M., Suzuki, K., Tanaka, Y., Hayakawa, S., Tsunemi, H., & Yamashita, K. 1984, Publ. Astron. Soc. Japan, **36**, 699.

Parkes, G., Mason, K., Murdin, P. & Culhane, J. 1980, Mon. Not. R. astr. Soc., **191**, 547.

Sato, N., Nagase, F., Kawai, N., Kelley, R., Rappaport, S., & White, N. 1986, Astrophys. J., **304**, 241.

Tanaka, Y., Fujii, M., Inoue, H., Kawai, N., Koyama, K., Maejima, Y., Makino, F., Makishima, K., Matsuoka, M., Mitsuda, K., Murakami, T., Nishimura, J., Oda, M., Ogawara, Y., Ohashi, T., Shibazaki, N., Suzuki, K., Waki, I., Yamagani, T., Kondo, I., Murakami, H., Hayakawa, S., Hirano, T., Kunieda, H., Masai, K., Nagase, F., Sato, N., Tawara, Y., Kitamoto, S., Miyamoto, S., Tsunemi, H., Yamashita, K., & Nakagawa, M. 1984, Publ.Astron. Soc. Japan, **36**, 641.

White, N., Swank, J., & Holt, S 1983, Astrophys.J., **270**, 711.

White, N., & Swank, J. 1984, Astrophys.J., **287**, 856.

Behaviour of the iron line in Cygnus X-1 during dips and persistent emission

Monika Bałucińska-Church[1] & Paul Barr[2]

[1] School of Physics and Space Research, University of Birmingham, U.K.
[2] Exosat Observatory, Astrophysics Division, ESTEC, Holland.

0. Abstract

The Exosat GSPC observations of Cygnus X-1 have been analysed to study the behaviour of the iron line during dips and persistent emission. The energy of the emission line is approximately 6.3 keV for the persistent emission. The absolute intensity of the line and its equivalent width increase during dips. For one dip a broad emission line (FWHM $\sim$ 2.6 keV) at 5.8 keV can be seen. This behaviour is discussed in terms of Comptonisation of the incident X-rays in relatively cold material, possibly associated with the stellar wind of the secondary.

1. Introduction

Cygnus X-1 is a well known black hole candidate and an X-ray binary source with a 5.6 day orbital period. The emission line at an energy of around 6.2 keV with an equivalent width $\sim$ 120 eV was discovered by Barr, White and Page (1985, hereafter BWP) using the Exosat GSPC spectra who identified it as an iron emission K-line which is Compton-redshifted in an optically thick ($\tau \sim 5$) photoionized corona above and below the accretion disc. The intensity of the emission line as well as the FWHM and the equivalent width did not show clear evidence of being a function of the 5.6 day period.

The 1.5 - 20 keV spectra collected by the Tenma satellite during dips reveal an absorption edge at 7.18 keV (Kitamoto et al. 1984). These authors do not report an emission line during dips. Recently Kitamoto et al. (1990) report an iron emission line at 6.5 keV with an equivalent width of around 60 - 80 eV found in the spectra collected by the Tenma satellite during the persistent emission. They also claim that for the spectrum obtained close to zero orbital phase (which consists of two narrow dips) the iron line is either weak (EW less than 38 eV) or broadened.

The results reported from Tenma and Exosat are inconsistent. The emission line at 6.2 keV is also visible in the Exosat ME data (Bałucińska and Hasinger 1990). Its intensity is constant out of the dips whereas going into the dips it increases by more than a factor of four. Simultaneously the equivalent width changes from 40 eV to 190 eV. It should be noticed that the estimation of the parameters of the line like the intensity and the equivalent width were made with the energy of the line and the FWHM fixed at 6.2 keV and 1.2 keV, respectively.

In this paper we analyse all Exosat GSPC spectra which are now available. They contain the spectra previously analysed by BWP and also other spectra collected later by the Exosat GSPC. We show that the iron line during the dips is located at around 5.8 ± 0.4 keV with an equivalent width $\sim$ 180 eV, an intensity of $\sim 0.033^{+0.030}_{-0.019}$ ph/cm^2/s and FWHM $\sim 2.6 \pm 1.2$ keV. In the persistent emission the spectra exhibit the emission line at $6.3^{+0.3}_{-0.2}$ keV with an equivalent width $\sim$ 70 eV which is less intense than during the dips with the absolute intensity of $0.005^{+0.003}_{-0.002}$ ph/cm^2/s. In addition to the emission line, an absorption edge at $\sim$ 7.5 keV is seen in some spectra during the persistent emission.

2. Results of the Spectral Analysis

All twelve observations of Cygnus X-1 made using the Exosat GSPC and available now from the Exosat database have been re-analysed. Five of them have previously been analysed and published by BWP. Two of the observations include five dips: two dips on July 8, 1984 and three on October 15, 1985. The dips are also visible in the Exosat ME spectra (Bałucińska and Hasinger 1990). The spectra collected during the dips have been analysed separately. The other data were collected during persistent emission.

The best fit parameters of **the spectra during the dips** were obtained using a power law with a gaussian emission line and a correction for cold matter absorption (Morrison and McCammon 1983). They are shown below.

observation	line energy [keV]	FWHM [keV]	line intensity [ph/cm^2/s]	EW [eV]	χ^2/d.o.f.
1984, July 8 (1st dip)	6.0 ± 0.3	1.9 ± 1.2	$0.028^{+0.020}_{-0.013}$	129	185/173
1984, July 8 (2nd dip)	5.7 ± 0.5	3.5 ± 1.4	$0.06^{+0.05}_{-0.03}$	235	186/173
1985, October 15	5.8 ± 0.2	2.6 ± 0.7	$0.010^{+0.005}_{-0.003}$	176	211/183

The mean values of the parameters during the dips are: the line energy $E_{el} = 5.8 \pm 0.4$ keV, FWHM=2.6 ± 1.2 keV, the line intensity $I_{el} = 0.033^{+0.030}_{-0.019}$ ph/cm^2/s, equivalent width EW=180 eV.

Twelve **spectra during persistent emission** were analysed. Two of them (the observations from July 8, 1984 and October 15, 1985) were extracted outside of the dips. To obtain the parameters of the emission line, a power law with a gaussian emission line was fitted in the energy range 3 - 15 keV. The spectrum of Cygnus X-1 is multicomponent (Barr and van der Woerd 1990, Bałucińska and Hasinger 1990) in the soft energy range but in the range examined by us, the main contribution to the continuum comes from the power law component. The model of power law with gaussian emission line adequately described five of the spectra and gave a reduced χ^2 not greater than 1.06. The other spectra did not give acceptable fits and showed a residual absorption edge near 7.5 keV. Fitting the power law with the gaussian emission line and the absorption edge improved the reduced χ^2 substantially ($\chi^2 \sim 0.85$ - 1.05) and did not change the centroid energy of the emission line. The parameters of the line and the edge change slightly from spectrum to spectrum. They give mean values of: energy of the emission line $E_{el} = 6.3^{+0.3}_{-0.2}$keV, FWHM=$1.4^{+1.2}_{-0.7}$keV, intensity of the emission line $I_{el} = 0.005^{+0.003}_{-0.002}$ ph/cm^2/s, energy of the absorption edge $E_{abs} = 7.5^{+0.3}_{-0.4}$keV, optical depth $\tau = 0.06 \pm 0.02$, equivalent width of the emission line EW = 70 eV.

3. Discussion

The Exosat GSPC spectra show the emission line to have different absolute line intensity as well as equivalent width during the dips and the persistent emission which confirms the results obtained with the Exosat ME spectra (Bałucińska and Hasinger 1990). In the persistent emission some of the spectra show an absorption edge at an energy of around 7.1 - 7.8 keV.

The energy of **the emission line in the persistent emission** is typically $6.3^{+0.3}_{-0.2}$ keV which is in agreement with the value 6.2 keV reported by BWP. The line energy estimated for each spectrum separately is widely spread down to 5.9 keV. Only one spectrum shows the emission line at 6.5 keV.

Recently Kitamoto et al. (1990) reported an emission line at 6.5 ±0.1 keV which is higher than the line energy obtained with the Exosat spectra. However, they fitted a single power law to the complete spectrum from 1.5 - 20 keV, and it is known that there is a soft excess below 3 keV (Priedhorsky et al. 1979, Barr and van der Woerd 1989, Bałucińska and Hasinger 1990). This will lead to overestimating the steepness of the spectrum at 6 keV. We have performed simple simulations which show that the energy of the line and also the line width depend on the assumed spectral index. For example, an increase in the photon index from 1.4 to 1.6 will change the line energy from 6.0 to 6.5 keV. In the present work, the slope of the continuum has been obtained in the energy range above 3 keV to avoid the soft component, making the spectrum at 6 keV slightly flatter and giving a line at 6.4 keV, or less.

Fabian et al. (1989) suggest that the iron line at 6.4 keV can be produced from hard X-ray irradiation of cool gas around a black hole. The broadening and redshift of the line are caused by the Doppler effect and gravitational redshift in the the disc which is formed by the cool matter. This model is able to explain a redshift down to 6.2 keV, but not lower. The best fit parameters of the emission line in the persistent emission that we derived are not formally inconsistent with this model; however during dips we obtain a line energy significantly lower than 6.2 keV, i.e. inconsistent with the model.

The emission line during the dips is clearly stronger. The absolute line intensity is very significantly greater by a factor of 6 in the dips compared with the persistent emission. Its equivalent width is larger by a factor of 2.5. The errors associated with the mean energy of the line from several dips do not allow us to make any positive statement that a redshift is observed. On the other hand looking at the dip from October 15, 1985 one can easily see that the centroid energy is redshifted from $6.43^{+0.18}_{-0.11}$ keV in the persistent emission to 5.8 ±0.2 keV during the dips. It suggests that such an effect must originate in the matter which is responsible for the dips. However, the other two dips do not show clear evidence of the redshift.

The sample of the spectra during the dips is not very rich but the stronger emission line during the dips was also reported from the Exosat ME data collected at the same time (Bałucińska and Hasinger 1990). The equivalent width of the line varied from 90 eV in the persistent emission up to 190 eV in the dip. This behaviour is in agreement with the results reported in this paper. Kitamoto et al. (1984) analysed the Tenma spectra collected during dips but they reported a strong absorption edge at 7.18 keV instead of an emission line.

A broad line (FWHM $\sim$ 2.7 keV) at 5.9 keV was observed in 4U1543-47 (van der Woerd, White and Kahn 1989). The properties of the line are well described by Comptonization of K_α photons (6.41 keV) in a cool, finite plasma cloud. An optical depth of $\sim$ 4.5 was obtained in the case of 4U 1543-47 from both the redshift and the broadening of the line. In our case the width of the line is about 2.6 keV and the mean redshift and the broadening in dips gives an optical depth $\sim$ 4.6 - 5.

Intensity dips are believed to be related to the clumpiness of the cool matter situated most probably in the outer regions of the stellar wind of the companion star having a temperature of $\sim$ 30,000 K (Kitamoto et al. 1984). The X-ray source influences the companion by reducing gravity towards the Lagrangian point by X-ray heating and by the radiation pressure. These effects destroy the spherical symmetry of the stellar wind, and may slow down the wind and so increase its density (e.g. Haberl, White and Kallman 1989). This suggests that one can expect a dense clumpiness in the stellar wind with increased optical depth.

The absorption edge in the persistent emission we see at an energy of around 7.1 - 7.8 keV is a significant feature in six spectra. The edge can be produced by photoionization of iron (Fe 1s) in different ionization states from neutral to Fe^{+18} (Reilman and Manson 1978). This implies that the temperature of the region where the absorption edge is generated might be as high as $1.6 \cdot 10^7$ K. Kitamoto et al. (1984) reported an absorption edge at 7.18 keV implying an ionization state not higher than Fe^{+5}. Such an absorption feature could be generated in the stellar wind of the companion star. Only in two cases reported in this paper are the results consistent with the value of Kitamoto et al. (1984). In the other four cases the energy of the absorption edge is higher, but significantely lower than 9 keV, thus definitely excluding K-shell absorption from Hydrogen- or Helium-like iron.

Acknowledgment: MB-C thanks the Exosat Observatory, Estec for financial support during her stay at Estec.

References

Bałucińska, M., and Hasinger, G., 1990, Astron. Astrophys. *in press.*
Barr, P., and van der Woerd, H., 1990, Ap. J. (Letters), 352, L41.
Barr, P., White, N. E., and Page, C. G., 1985, M.N.R.A.S., 216, 65p.
Fabian, A. C., et al., 1989, M.N.R.A.S., 238, 729.
Haberl, F., White, N. E., and Kallman, T., 1989, Ap. J., 343, 409.
Kitamoto, S., et al., 1984, P.A.S.J., 36, 731.
Kitamoto, S., et al., 1990, P.A.S.J., 42, 85.
Morrison, R., and McCammon, D., 1983, Ap. J., 270, 119.
Priedhorsky, W. C. Terrel, J., and Holt, S., 1983, Ap. J., 270, 233.
Reilman, R.F. and Manson, S. T., 1978, Phys. Rev. A, 18 (5), 2124.
van der Woerd, H., White, N. E., and Kahn, S. M., 1989, Ap. J., 344, 320.

The EXOSAT GSPC Iron Line Catalogue

M. Gottwald[1,2], N.E. White[2,3]

[1] Max-Planck-Institut f. extraterrestrische Physik, D-8046 Garching, FRG
[2] EXOSAT Observatory, Astrophysics Division, SSD, ESA, ESTEC, Postbus 299, 2200-AG Noordwijk, The Netherlands
[3] NASA, Goddard Space Flight Center, Greenbelt, MD20771, USA

1. Introduction

Between May 1983 and April 1986 the European X-ray astronomy satellite EXOSAT made about 1800 pointed observations of several hundreds of X-ray sources. One of its instruments onboard, the gas scintillation proportional counter (GSPC), was capable to measure X-ray spectra with an energy resolution which exceeded that of conventional proportional counters by a factor of 2. With a peak sensitive area of about 100 cm^2 and a field of view of 45′ FWHM the GSPC was well suited to study in detail the spectra of bright X-ray emitters.

2. Data Analysis

In the course of preparing the GSPC part of the EXOSAT database all EXOSAT observations of sufficiently bright sources (>5 cnts s^{-1} in the Medium Energy Experiment) were found useful to extract the corresponding X-ray spectra. In total 609 GSPC database entries of 240 individual sources were obtained. This sample of X-ray spectra can be regarded as a rich repository for a search for iron line features in X-ray sources. Most of the 609 spectra proved to satisfy quality criteria such as good background subtraction and minimum acceptable GSPC counting rate in order to allow detailed spectral analysis.

Continuum fits in the energy range 3–10 keV were restricted to a set of relatively simple one- and two-component models. In very bright sources the contiuum was modelled from 3–15 keV. Fits without a line were tried for a thermal bremsstrahlung, a powerlaw, a cutoffpowerlaw and a cutoffpowerlaw plus a blackbody model. The latter was only applied to low-mass X-ray binaries (LMXB) and bright unidentified sources (e.g. White et al. 1985, 1986). All four models included an absorption term. Addition of a gaussian line profile, described by the free parameters centroid line energy, line FWHM width and line flux, to all of the four above mentioned models represented the fits to the spectra under the assumption of the presence on an iron emission line. The best fit model was determined from the χ^2 statistics of the individual fits. Only line detections with $>3\sigma$ significance were considered as positive results. As long as the line significance was between 3–5σ the line FWHM width was fixed at 0.25 keV.

3. Results

205 spectra out of the total sample of 609 showed an emission line with $>3\sigma$ significance. This number comprises 54 individual sources. Most of them belong to the classes of LMXB (19) and high–mass X–ray binaries (HMXB) (11). The remaining 24 objects consist of bright unidentified objects (6), supernova remnants (4), cataclysmic variables (3), cluster of galaxies (8), AGN (1) and others (2). The line energies found in our analysis are shown in fig. 1. Here filled and open symbols represent line detections with $>5\sigma$ and 3–5σ significance, respectively. For sources where there was only a single line detection the error bar indicates the true statistical error (90% confidence) while for multiple line detections the extent of the rectangular bar indicates the range of line energies found in the fits for a particular source.

We found evidence that the iron lines in LMXB have a higher average line energy (6.67 keV) than the high–mass systems (6.49 keV) supporting the results published on smaller samples of iron line detections (e.g. Hirano et al. 1987 and references therein). However in each of the two object classes there are exceptions. The line energy in Ser X–1 was found to lie below 6.5 keV and in contrast the transient EXO2030+375 exhibited a line predominantly at 6.63 keV. A very wide range of values, running from <0.1 keV (NGC1275 and W49B) to the other extreme case of 2.5 keV (1543-475), was found for the line FWHM width. The minimum detectable line flux reached about 3×10^{-4} cnts cm^{-2} s^{-1} while the source with the strongest iron line, Sco X–1, had an average line flux of 2.8×10^{-1} cnts cm^{-2} s^{-1}. Fig. 2 shows the X–ray sky in the light of the iron emission line with the size of the symbols dependent on the line flux. Open circles stand for 3–5σ and filled circles for $>5\sigma$ line detcetions. Since most of the line emitters detectable with the EXOSAT GSPC belong to object classes having a small scale height perpendicular to the galactic plane the iron line sources can be mainly found near galactic latitudes b – 0°.

The quivalent width as a function of the source X–ray brightness is given in fig. 3. At the lefthand side of this figure one finds the faint sources where only lines with high equivalent widths are detectable, as e.g. in W49B (equivalent widths >7.5 keV). The brightest X–ray source in the GSPC, Sco X–1, appears at the righthand side with counting rates exceeding 300 cnts s^{-1} and equivalent widths of <100 eV.

References:

Hirano, T., et al.: 1987, Publ. Astron. Soc. Japan **39**, 619

White, N.E., Peacock, A., and Taylor, B.G.: 1985, Astrophys. J. **296**, 475

White, N.E., et al.: 1986, Monthly Notices Roy. Astron. Soc. **218**, 129

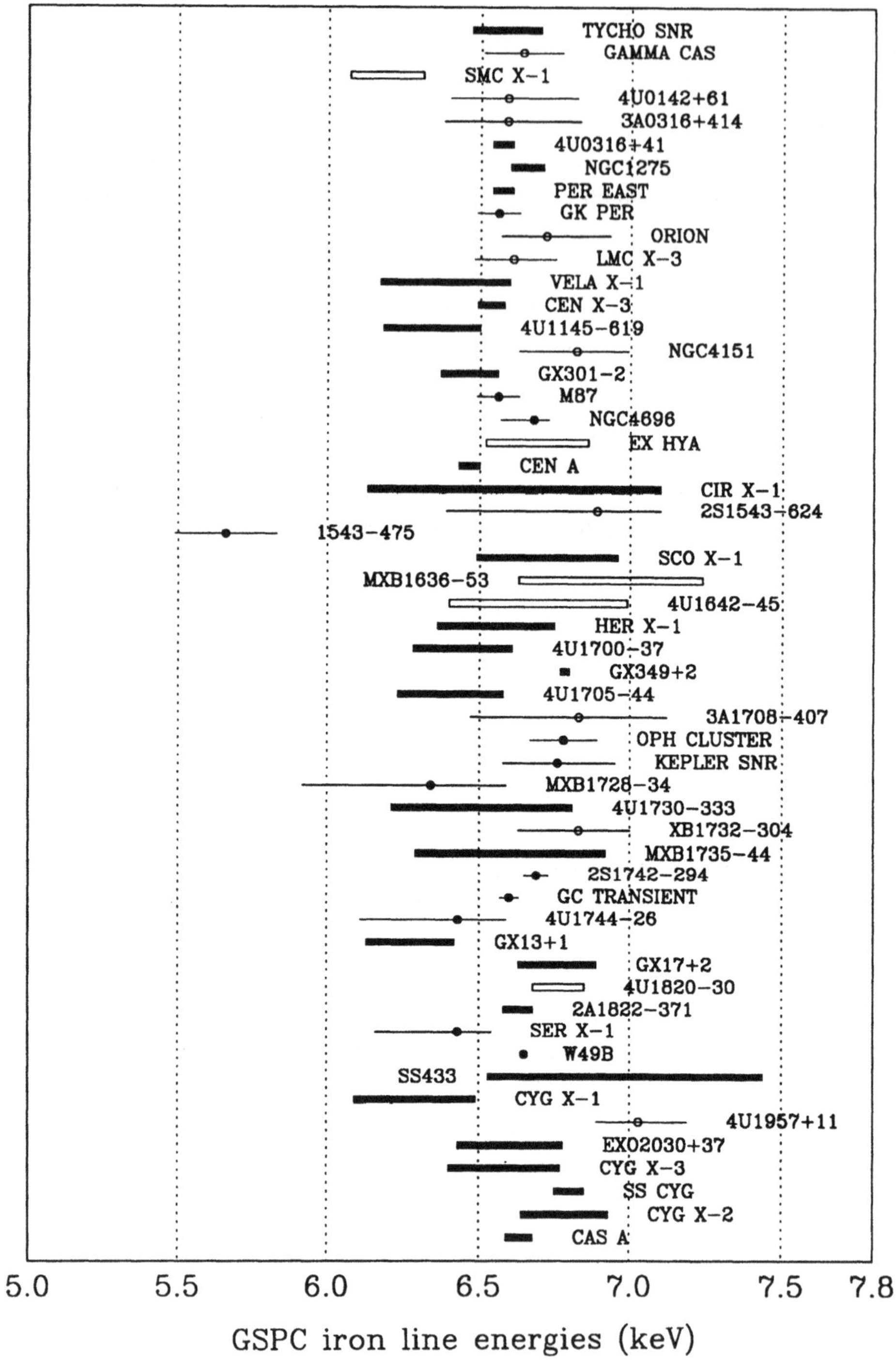

Fig. 1: The line energies for all sources with significant iron emission lines. For the meaning of the symbols see text.

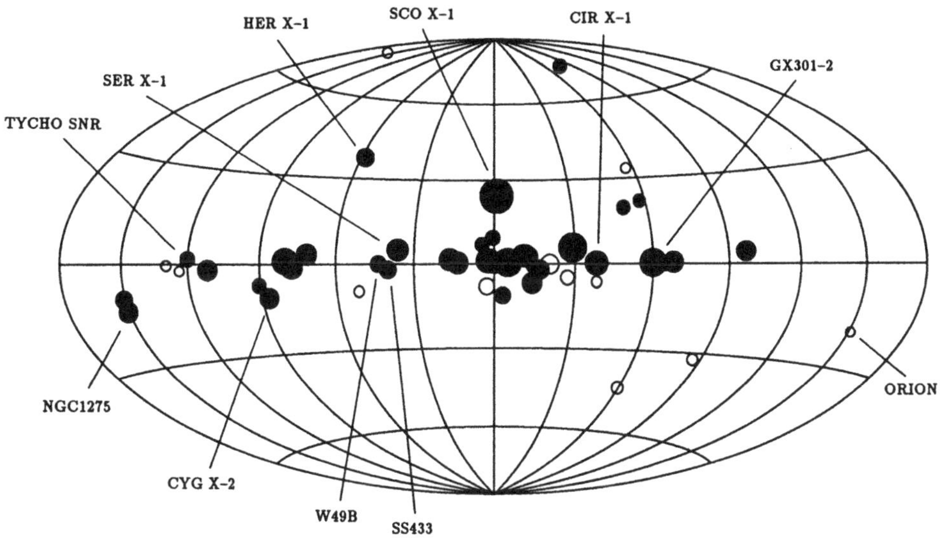

Fig. 2: The X-ray sky in the light of the iron emission line in galactic coordinates.

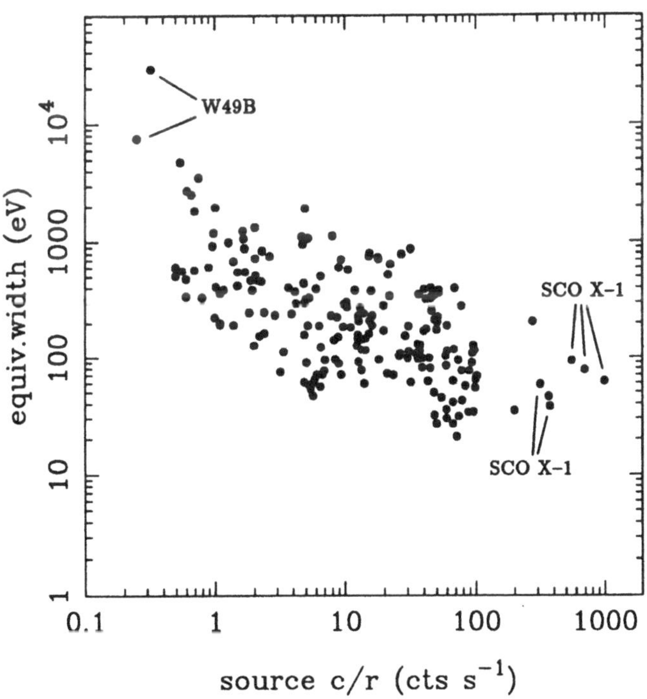

Fig. 3: The measured line equivalent width as a function of the source X-ray brightness (GSPC counting rate).

Narrow Iron K Line and Warm Iron K Edge in Pulse Phase Spectra of Hercules X-1

Peter Kahabka

Max-Planck-Institut für Extraterrestrische Physik
D-8046 Garching, Germany

Abstract

Spectra of the accreting neutron star Hercules X-1 observed with the EXOSAT GSPC detector on days 61,65,92 and 96 1984 were analysed for iron line features. A broad gaussian emission line with a FWHM of 1.32 ± 0.13 keV was found during *on*, *turn-off* and *short-on* states in the 35-day-cycle at an energy of $\sim$ 6.55keV in the pulse phase averaged spectra. In an observation on day 160 1984 a pulse phase dependent spectral analysis was performed. A warm K_α iron edge was introduced in the fit. A narrow line (FWHM $\sim$ 0.4keV) and an edge ($E_K = 8.43 \pm 0.13$keV) of warm iron FeXIX to FeXXII within the 99% confidence level were found. This is expected if the iron line flux originates in the magnetospause 1.8 10^8 cm distant from the neutron star. An electron temperature $T_e = 4.10^6$K, an electron density $n_e = 7.3\ 10^{18}$ cm^{-3}, a shell thickness $\delta = 1.2\ 10^5$ cm and a source luminosity L $\sim$ 10^{37} erg sec^{-1} (4.5 kpc) are in agreement with the observed iron line and soft X-ray flux.

1. Introduction

With *OSO-8* and *HEAO-A1* $K\alpha$ iron line emission of Her X-1 was discovered (Pravdo et al. 1977a,b, 1978a). The line center was $\sim$ 6.8keV, the line was either broad with a FWHM of 1.5 keV or split up into two components separated by $\sim$ 1keV. The broadening of the line was explained by Comptonization in a plasma with an optical depth of $\tau \sim 6$, the line splitting by a plasma corotating at a distance of $\sim 8.10^8$cm from the neutron star. Both scenaria are higly problematic and the situation appears to be much simpler. Kahabka et al. (1985) and Kahabka (1987) have analysed the spectra of Her X-1 during one 35-day-cycle in 1984 with respect to the iron line flux. Broadened lines (FWHM $\sim$ 1.3keV) and no splitting into two lines was found. The line was stated to be asymmetric. The line intensity was following the 35-day-lightcurve. Energy and FWHM of the line appeared not to vary with 35-day-phase. Neither corotation in magnetospheric plasma 1.810^8cm distant from the neutron star nor Comptonization of iron line photons in plasma in the inner edge of the accretion disk could cause such a broadening. A pulse phase selected analysis of the iron line flux was performed to find a hint for the broadening mechanism.

2. Observations

EXOSAT observed Her X-1 32 times starting from 28th June 1983 up to 13th June 1984. One observation lasted from 30 minutes up to 40 hours. Two 35-day-cycles of Her X-1 were scanned in about 4 days intervals. The first cycle was observed from 1st March 1984 to 5th April 1984, the second cycle from 30th May 1985 to 16th June 1985.

In the time span from 28th June 1983 to 26th August 1983 Her X-1 was in a prolongued low-state. Parmar et al. (1984) reported about the observation of the iron line during this phase. Her X-1 was seen again in a normal 35-day-cycle on 1st March 1984. The observations of the iron line flux during this cycle have been reported by Kahabka et al. (1985). The observation from 11th to 13th June 1984 covering one full 1.7 days binary orbit was used to perform a pulse phase selected analysis of the iron line flux. 11.3 hours of the data, when the source was outside of eclipse and absorption dips, were analysed. The data analysis has been performed at the *MPE* in Garching. Only *DIRECT* mode data have been used. The photon arrival times have been corrected to the solar barycenter. Arrival time corrections for the binary orbit of Her X-1 have been applied using the ephemeris of Deeter, Boynton and Pravdo (1981). The pulse period used was 1.237792 sec.

3. Results

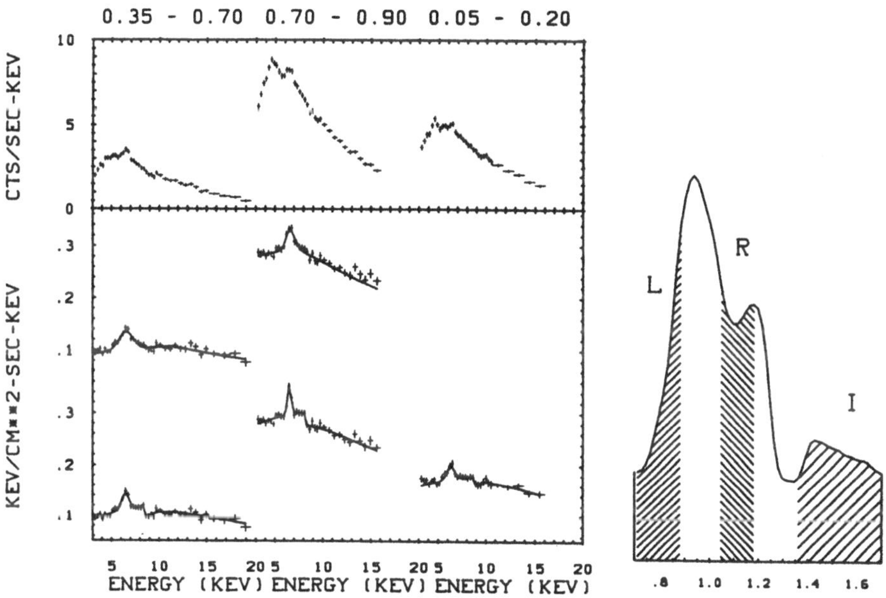

Figure 1 : *Counts spectra (upper panel) and energy flux spectra (lower panel) of Her X-1 for three different pulse phase intervals on* 11th *June 1984 measured with the EXOSAT GSPC detector for three different pulse phase regions: a) "off"-pulse* ($\Phi_{35} = 0.35 - 0.70$), *b) left pulse flank* ($\Phi_{35} = 0.70 - 0.90$) *and c) right pulse flank* ($\Phi_{35} = 0.05 - 0.20$). *In the right figure the pulse profile of Her X-1 is shown for the energy range 3.9 - 6.0 keV. The spectra are shown in the hatched phases of the pulse. The iron line profile was fitted by a gaussian function and a gaussian function plus a warm iron edge, Ref [5].*

First, the model of a powerlaw continuum plus cold gas absorption and a gaussian emission line was fitted to the pulse phase averaged observed spectra in the energy range 1 to 15 keV. The powelaw index was ~ 0.9. The following parameters of the iron line were deduced: The line centroid is $E_l = 6.55 \pm 0.02$ keV, the FWHM is $F_l = 1.32 \pm 0.13$ keV, the line flux is $F_l = 7.5 \pm 0.8\ 10^{-3}$ photons cm^{-2} sec^{-1} and the equivalent width is $EW_l = 345 \pm 31$eV. These are the mean parameters deduced from observations during the 35-day phases $\phi_{35} \sim 0.23$, 0.33, 0.67, 0.13 and 0.23 on days 61, 65, 76, 91 and 96 1984 including one turn-off and one short-on observation. This confirms the broad iron line picture. Second spectra were fitted in selected phases of the pulse. This was done at an observation performed on 11th June 1984 at the 35-day phase $\phi_{35} \sim 0.16$. The pulse was split into 20 phase bins. The model of a powerlaw continuum plus cold gas absorption and a gaussian emission line was fitted to the spectra. The iron line flux was vanishing at the phase of the center of the hard (> 10.keV) pulse. But otherwise the iron line flux was following the continuum flux at energies below ~ 6 keV. This confirms the finding of Pravdo (1977, 1978a), that there is missing iron line flux at the phase region of spectral hardening. Third, spectra were fitted in 3 selected phases of the pulse. If phase 1.0 refers to the intensity maximum of the hard (> 10. keV) pulse, then spectra were fitted to phases 0.35 to 0.70 (the interpulse), 0.70 to 0.90 (the rising flank of the main pulse) and 0.05 to 0.20 (the declining flank of the main pulse). The model of a powerlaw continuum plus cold gas absorption, a gaussian emission line and a warm iron edge was applied to the data. There was the new finding, that the spectrum at the leading flank of the main pulse could be fitted with a narrow iron line FWHM = 0.38(−0.04, +0.15) keV at an energy of $E_l = 6.52(-0.04, +0.04)$ keV and with a warm iron edge at an energy of $E_K = 8.36(-0.25, +0.11)$ keV and a column density of absorbing iron atoms of $N_{Fe} = 0.7(-0.2, +0.3)\ 10^{19}\ cm^{-2}$. The spectrum in the interpulse region could also be fitted with this model, but the gaussian line is broader, FWHM = 0.94(−0.13, +0.25) keV, $E_l = 6.45(-0.07, +0.08)$ keV, $E_K = 8.43(-0.13, +0.13)$ keV and $N_{Fe} = 2.2(-0.5, +0.6)\ 10^{19}\ cm^{-2}$. The spectrum in the declining flank of the main pulse was fitted with the parameters FWHM = 1.0(−0.17, +0.5) keV, $E_l = 6.3(-0.13, +0.1)$ keV, $E_K = 8.8(-0.27, +0.5)$ keV and $N_{Fe} = 1.1(-0.4, +0.3)\ 10^{19}\ cm^{-2}$.

4. Discussion

The energy of the iron K_α-edge, E_{K_α} within the 99% confidence level,

$$E_{K_\alpha} = 8.43(\pm 0.4)\ \text{keV}$$

determines the mean state of ionization of iron. It is FeXIX to FeXXII. For the column density of absorbing iron atoms one gets

$$N_{Fe} = 1.1(\pm 0.8)\ 10^{19}\ cm^{-2}.$$

The optical depth τ_T of the absorber can be determined from the relation

$$\tau_T = N_{Fe}\ \sigma_T\ Y_{Fe}^{-1}.$$

With the Thomson cross section $\sigma_T = 6.65\ 10^{-25}$ atoms cm^{-2} and the relative iron abundance $Y_{Fe} = 3.10^{-5}$, Basko (1980) one gets

$$\tau_T = 0.25(-0.18, +0.19).$$

Iron has a relative cosmic abundance of $\chi_{Fe} \approx 3.10^{-5}$ and a fluorescence yield of $\omega_K \approx 0.347$. For an optically thin plasma the flux in the iron line is

$$F_l = \frac{\omega_K}{4\pi D^2} \int_{\mathrm{Target}} \chi_{Fe} n r^2 dr d\Omega \int_{E_K}^{\infty} \sigma_{abs}^{Fe} \frac{S(E)}{r^2} dE,$$

with the distance D to the source, the gas density n, the volume element $r^2 dr d\Omega$, the energy of the K-edge E_K, the photoionization cross section of iron σ_{abs}^{Fe} and the source spectrum S(E). The energy of the fluorescence line, of the edge and the absorption cross section are dependent on the state of ionization of the plasma. The photoionization cross section above the K-edge of iron can be approximated as

$$\sigma_{abs}^{Fe}(E) = C \frac{6.3\ 10^{-18}}{Z_{eff}^2} \left(\frac{E_K^{Fe}}{E}\right)^{2.8} \mathrm{cm}^2 ,$$

with $Z_{eff} = 26$ for hydrogen like iron FeXXVI, $Z_{eff} = 25.73$ for iron FeXXV and $Z_{eff} = 25.47$ for FeXXIV. The corresponding edge energies are 9.28, 8.83 and 8.63 keV. $C = 1$ for FeXXVI and $C = 2$ otherwise (Bai 1980). An optically thick medium modifies the flux in the iron line in comparison to the optically thin case. A shift in the line centroid $h\nu_0$ of the line and a line broadening $\delta h\nu$, with

$$\delta h\nu \ / \ h\nu_0 = 2\sqrt{\ln 2}\ \sqrt{\tau\ kT_e/mc^2}$$

occur, with the optical depth τ, the electron temperature T_e and $mc^2 = 511$ keV (Pozdnyakov et al. 1979). With the integral for the observed flux in the iron line F_l the distribution and physical condition of the fluorescencing matter can be deduced. In case of a shell geometry

$$\int_{\mathrm{Target}} dr\ d\Omega = \Omega\delta ,$$

with the thickness of the shell δ and the opening angle of the shell Ω, the integral is then

$$\int_{E_K}^{20} \sigma_{abs}^{Fe}\ S/D^2\ dE = 6.82\ 10^{-21} ,$$

with the observed source flux $S/D^2 = 0.1\ E^{-1}$ photons $\mathrm{cm}^{-2}\ \mathrm{sec}^{-1}\ \mathrm{keV}^{-1}$. If one sets $\omega_K = 0.347$ and $\chi_{Fe} = 3.\ 10^{-5}$, then one obtains

$$F_l = 2.82\ 10^{-27}\ n\Omega\ \delta .$$

With the observed values of the energy of the line centroid $E_l = 6.55 \pm 0.06$ keV and of the iron K-edge $E_K = 8.4 \pm 0.4$ keV one can determine the state of ionization of iron to FeXIX-FeXXII. This leads to an ionization parameter of

$$\xi = 57 \pm 4,$$

if one applies the models of Kallman and Mc Cray (1982) and Ross (1979) to the case of a $\tau_T = 0.4$ shell for a source luminosity of 10^{37} erg sec^{-1}. From the observed width of the iron line information can be deduced about the temperature, the optical depth and the velocity dispersion of the fluorescing gas. A plasma with a mean velocity v causes a line broadening of $h\delta\nu = h\nu_0$ (v/c). The velocity of a corotating neutron star with angular frequency Ω is composed of the free-fall velocity v_{ff} and the Alvén velocity v_a. They amount for a neutron star with a mass of 1.3 $M_\odot$, a radius of the magnetopause of 1.8 10^8 cm and with $\Omega = 5.08$ sec^{-1}, $v_{ff} \approx 1.4$ 10^9 cm sec^{-1} and $v_a \approx 0.9$ 10^9 cm sec^{-1}. For the line broadening from both components one gets $h\delta\nu \leq h\nu_0 \frac{0.3 v_{ff}}{c} \approx 90$ eV (Basko 1980) and $h\delta\nu \leq h\nu_0 \frac{v_a}{c} \approx 200$ eV. The Doppler broadening by electron scattering contributes for $kT_e \leq 0.4$ keV and $\tau \leq 0.44$ with $h\delta\nu \leq 170$ eV. The maximum line width then is $h\delta\nu \leq 280$ eV. This explains the observed line width at the leading flank of the main pulse. In the interpulse region a line with a FWHM ≥ 550 eV is observed. An extra line broadening effect is necessary to explain that width. The parameters of the magnetopause of the neutron star could be further restricted with the observed soft X-ray flux $L^{th} \approx 0.34$ 10^{37} erg sec^{-1} as demonstrated in Kahabka (1987). The electron density is $n_e = 7.3$ 10^{18} cm^{-3}, the shell thickness is $\delta = 1.2$ 10^5 cm and the opening angle is $\Omega = 0.38$ 4π. A temperature of the plasma $T = 4.4$ 10^6 K (Mc Cray et al. (1982)) was assumed and a radius of the magnetopause of 1.8 10^8 cm.

References

[1] Bai, T., Astroph. J. **239**, 328 (1980).

[2] Basko, M.M., Astron. Astroph. **87**, 33 (1980).

[3] Deeter, J., Crosa, L., Gerend, D. and Boynton, P.E., Astroph. J. **206**, 816 (1976).

[4] Kahabka, P., Ögelman, H., Pietsch, W., Trümper, J. and Voges, W., Space Sci. Res., **40**, 355 (1985).

[5] Kahabka, P., "Der 35-Tage-Zyklus von Hercules X-1", Dissertation, Technische Universität München (1987).

[6] Kallman, T.R. and McCray, R., Astroph. J. **219**, 292 (1978).

[7] McCray, A., Shull, P.E., Boynton, P.E., Deeter, J.E., Holt, S.S. and White, N.E., Astrop. J. **262**, 301 (1981).

[8] Parmar, A.N., Pietsch, W., McKechnie, S., White, N.E., Trümper, J., Voges, W. and Barr, P., Nature (1984).

[9] Pozdnyakov, L.A, Sobol, I.M. and Sunyaev, R.A., Astron. Astroph. **75**, 214 (1979).

[10] Pravdo, S.H., Bolt, E.A., Holt, S.S. and Sermelitsos, P.J., Astroph. J. Lett. **216**, L23 (1977a).

[11] Pravdo, S.H., Becker, R.H., BoLt, E.A., Holt, S.S., Serlemitsos, P.J. and Swank, J.H. Astroph. J. **215**, L61-L64 (1977b).

[12] Pravdo, S.H., Bussard, R.W., Becker, R.H., Holt, S.S. and Serlemitsos, P.J. Astroph. J. **225**, 988 (1978a).

[13] Pravdo, S.H., Bolt, E.A., Holt, S.S. and Sermelitsos, P.J., Astroph. J. Lett. **225**, L53 (1978b).

[14] Ross, R.R., Weaver, R. and McCray, R., Astroph. J. **219**, 292 (1978).

GINGA Observation of X1820−303 and Its Iron-line Emission

B. J. Kellett[1], E. N. Ercan[2], A. M. Cruise[1]

[1]Astrophysics Div., Rutherford Appleton Lab., Chilton, Didcot, Oxon. OX11 0QX, U.K.
[2]Dept. of Physics, Boğaziçi University, Istanbul, Turkey.

Abstract

We present spectral observations of the low-mass X-ray binary source X1820−303 obtained with the *GINGA* Large Area Counter. Two two-component spectral models are found to provide good fits to the data; the superposition of a bremsstrahlung component and a blackbody, or a power-law with exponential cut-off and a blackbody. In both cases, inclusion of an iron line at 6.7 KeV improves the fits. After considering the importance of Comptonization for the spectral and temporal behaviour, we selected the power-law and blackbody model. The model requires a scattering cloud surrounding the neutron star, with a Thomson optical depth of ~7.

By studying the spectrum as the source intensity varies, we find that the blackbody and iron-line fluxes both show an increase by a factor of ~3, while the power-law component only shows a ~20% increase. This clearly shows that the iron-line and blackbody emission regions are related, and we suggest that they both originate from the inner parts of the accretion disk.

1 Introduction

X1820−303 lies close to the centre of the globular cluster NGC6624, and exhibits a wide range of X-ray phenomena. It is the shortest-period binary system known (with a period of just 685 seconds). It also shows variations on both longer and shorter timescales; a ~176 day variation between 'high' and 'low' states on the longer side, and QPO and aperiodic variations on shorter timescales. X1820−303 is also a burst source, although bursts are only detected in the 'low' state (see Morgan *et al.*, and references therein). This makes X1820−303 a prime target for detailed spectral and temporal studies.

We present here a study of the X-ray spectral properties of X1820−303 from the *GINGA* Large Area Counter (LAC).

2 Observations

The data discussed here was accumulated in 48 pulse height channels with the MPC2 mode (for details of the *GINGA* LAC see Turner *et al.*, 1989, and Makino, 1987). Data in this mode was

Table 1: Best Fit Spectral Parameters from (PLEC+BB+L) model.

Parameter	Epoch I	Epoch II	Epoch I + II
Γ	$2.21^{+0.12}_{-0.14}$	$2.16^{+0.15}_{-0.16}$	$2.15^{+0.18}_{-0.20}$
$E_c^{\dagger}$	$4.71^{+0.35}_{-0.28}$	$4.69^{+0.38}_{-0.23}$	$4.70^{+0.40}_{-0.29}$
kT_{BB}	2.35 ± 0.12	$2.43^{+0.08}_{-0.03}$	$2.38^{+0.12}_{-0.05}$
N_H *	$9.45^{+2.22}_{-2.29}$	$9.95^{+0.2.32}_{-2.56}$	$9.44^{+2.34}_{-2.26}$
L Energy	6.70	6.70	6.70
L Flux*	$0.012^{+0.008}_{-0.006}$	$0.021^{+0.017}_{-0.011}$	$0.016^{+0.011}_{-0.008}$
L RMS	$1.09^{+0.32}_{-0.39}$	$1.28^{+0.32}_{-0.66}$	$1.18^{+0.34}_{-0.37}$
χ^2_ν	0.26	0.20	0.16

† E_c is the cut-off energy in KeV. * Columns are in unit of 10^{21}cm^{-2}. * L(ine) flux is in units of Photons.cm^{-2}.s^{-1}.

obtained in two separate observations, separated by about 2 days in May, 1987. (For a full account of this work, see Ercan *et al.* (1990)). With *GINGA*, the simultaneous measurements of the source and background are not possible, thus the background must be estimated by introducing a separate exposure on a blank field near the source made close in time to the source observations. Since the source count rate is over ~3000 cts.s^{-1}, the uncertainty in the background estimation is small compared with the Poisson noise.

3 Spectral Analysis

We have investigated the X-ray spectrum of X1820−303 using the data integrated over ~1000–2000 sec time intervals. These spectra were then fitted with the usual range of single-component and two-component spectral models. No single-component model gave a satisfactory fit to the data. Of the models involving two spectral components, only thermal bremsstrahlung+blackbody (TB+BB), and power-law with exponential cut-off+blackbody (PLEC+BB) gave good fits to the data. Inclusion of a ~6.7 KeV emission line to our best fit two-component (TB+BB) and (PLEC+BB) models fits even better. For 'physical' reasons given below, we prefer the latter model, details of which are given in Table 1.

4 Discussion

These *GINGA* observations of X1820−303 show the source having a luminosity of ~5.4×10^{37} erg.s^{-1} in the first epoch (Epoch I) of our observations and ~6.7×10^{37} erg.s^{-1} in Epoch II of our observations, assuming a distance of 6.4 kpc to the source (see eg. Vacca *et al.* (1986), and references therein).

It has been argued that the X-ray spectra of low-mass X-ray binaries do not depend strongly on inclination of the system, suggesting that the central X-ray emission region is relatively spherical (White, 1986). It has also been pointed out that the X-ray spectra of many low-mass X-ray binaries cannot be adequately described by a single spectral function, and that more complex two-component spectral models should be used, generally involving a blackbody component together with another component with an exponential cut-off (e.g. bremsstrahlung, blackbody, or cut-off power-law). Such models have provided acceptable fits in most cases (see e.g. White and Mason, 1985) and these two-components have been identified with two physically distinct sites of X-ray emission; namely the neutron star surface and the inner accretion disk region (see e.g. White *et al.*, 1986; Mitsuda *et al.* 1984). Recently, several models stressing the importance of Comptonization for the spectral and temporal behaviour of low-mass X-ray binaries have been developed by Schultz and Wijers (1989), Lamb (1989), Ponman *et al.* (1990), and Damen *et al.* (1990). All these models try to develop more physically realistic models then the usual "two-component" interpretations.

In the light of these arguments it seems that on physical grounds the (PLEC+BB+L) model provides a more satisfactory explanation of the observed spectrum of X1820−303. This model would then suggest the existence of a scattering cloud surrounding the neutron star with a Thomson optical depth, τ, of ~7, using the relation between τ, Γ, and E_c given by Sunyaev and Titarchuk (1980) for a spherical geometry. (Ponman *et al.* (1990), found τ ~7.6±0.6, when fitting their scattering envelope model to the EXOSAT high-state spectrum of X1820−303, in very good agreement with our high-state result). A schematic picture of what X1820−303 might look like is shown in figure 1.

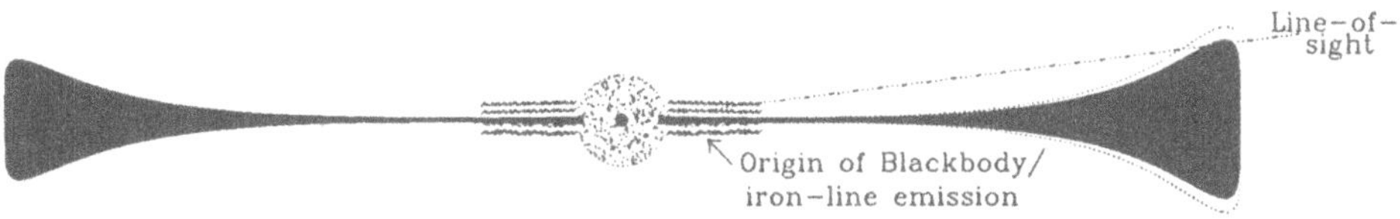

Figure 1: Schematic picture of X1820−303. The neutron star is surrounded by a scattering cloud that produces the Comptonized spectral component, while the blackbody component and iron line originate in the inner region of the accretion disk. Structure in the rim of the disk is thought to cause the dips and sinusoidal variations observed.

By dividing the two epochs of data into shorter sub-sections it is possible to study the correlations between the blackbody flux, the power-law flux, and the iron line flux (F_{BB}, F_{PL}, and F_{LINE} respectively), as the source intensity varies. In fact, from the lowest to highest intensities in these sub-divisions, the source shows a ~50% increase in countrate. The fluxes in the 3 spectral components are shown in figure 2, where the squares (□) show Epoch I and the diamonds (◇) show Epoch II. It is worth noting that BOTH the blackbody and iron line fluxes show a ~3-fold increase, while the power-law component only shows a $\lesssim$20% increase. This is suggestive of an association between the blackbody and line emission regions, confirming the suggestion of Kallman and White (1989). The general correlations shown in figure 2 may also indicate that the BB flux (and hence the line flux) are at least partly due to reprocessing of the Comptonized component by the accretion disk.

5 Conclusion

These observations demonstrate the usefulness, in the study of X-ray spectra, of instruments with large collecting areas and low backgrounds such as the LAC on *GINGA*.

We have found that the *GINGA* LAC data of X1820−303 are best fitted by a two-component spectral model, and on physical grounds we selected the power-law with exponential cut-off and a blackbody (plus a 6.7 KeV iron-line) model. The blackbody component and the iron-line flux both show an increase in importance, rising by around a factor 3 in flux. This strongly suggests that this emission originates in a common component, and we suggest that this is located in the inner parts of the accretion disk.

6 Acknowledgements

We are grateful to all who are involved in *GINGA* (ASTRO-C) mission at ISAS, especially Profs. Makino and Tanaka. This work is partially supported by the Research Foundation of Boğaziçi University and Science and Engineering Research Council (SERC), UK. We also acknowledge the

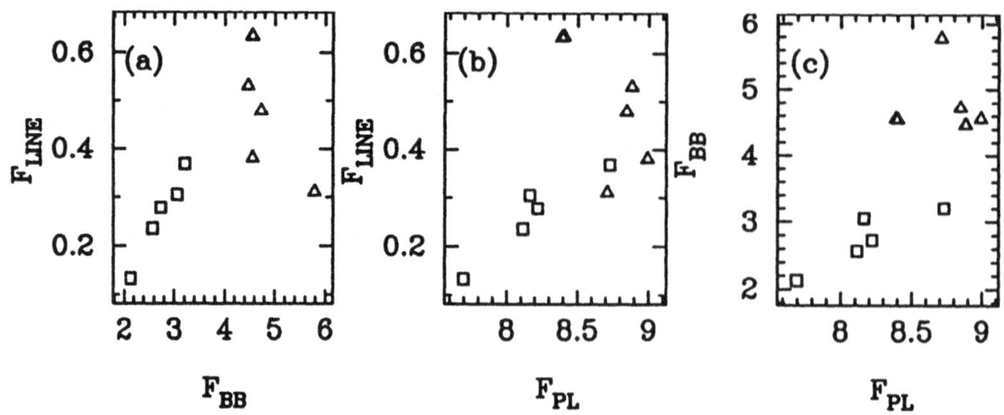

Figure 2: The 1–20 KeV fluxes of the power-law with exponential cut-off (PL), blackbody (BB), and 6.7 KeV Fe-line (LINE) spectral components plotted against each other. The squares (□) are the Epoch I sections, while the triangles (△) show the Epoch II data.

support provided in terms of both software and hardware by the STARLINK Project which is funded by the UK SERC.

References:

Damen, E., Wijers, R.A.M.J., van Paradijs, J., Penninx, W., Oosterbroek, T., Lewin, W.H.G., & Jansen, F.: 1990, *Astron. Astrophys.*, **223**, 121.

Ercan, N.E. *et al.*: 1990, (in preparation).

Kallman, T., & White, N.E.: 1989, *Astrophys. J.*, **341**, 955.

Lamb, F.K.: 1989, *Proc. 23rd ESLAB Symp.*, **ESA SP-296**, 215.

Makino, F., & the ASTRO-C team: 1987, *Astrophys. Letters Commun.*, **25**, 223.

Mitsuda, K., Inoue, H., Koyama, K., Makishima, K., Matsuoka, M., Ogawara, Y., Shibazaki, N., Suzuki, K., Tanaka, Y., & Hirano, T.: 1984, *Publ. astr. Soc. Japan*, **36**, 741.

Morgan, E.H., Remillard, R.A., & Garcia, M.R.: 1988, *Astrophys. J.*, **324**, 851.

Ponman, T.J., Foster, A.J., & Ross, R.R.: 1990, *Mon. Not. R. astr. Soc.*, **246**, 287.

Schultz, N.S., & Wijers, R.A.: 1989, *Proc. 23rd ESLAB Symp.*, **ESA SP-296**, 601.

Sunyaev, R.A., & Titarchuk, L.G.: 1980, *Astron. Astrophys.*, **86**, 121.

Turner, M.J.L., Thomas, H.D., Patchett, B.E., Reading, D.H., Makishima, K., Ohashi, T., Dotani, T., Hayashida, K., Inoue, H., Kondo, H., Koyama, K., Mitsuda, M., Ogawara, Y., Takano, S., Awaki, H., Tawara, Y. & Nakamura, N.: 1989, *Publ. Astron. Soc. Japan* **41**, 345.

Vacca, W.D., Lewin, W.H.G., & van Paradijs, J.: 1986, *Mon. Not. R. astr. Soc.*, **220**, 339.

White, N.E., & Mason, K.O.: 1985, *Spac.Sci.Rev.*, **40**, 167.

White, N.E.: 1986, "The Physics of Accretion onto Compact X-ray Objects", eds. K.O.Mason, M.G.Watson, & N.E.White, p.377.

White, N.E., Peacock, A., Hasinger, G., Mason, K.O., Manzo, G., Taylor, B.G., & Branduardi-Raymont, G.: 1986, *Mon. Not. R. astr. Soc.*, **218**, 129.

The iron lines of SS433

W. Brinkmann
Max–Planck Institut für Physik und Astrophysik
Institut für extraterrestrische Physik
8046 Garching, W.-Germany
and
N. Kawai, M. Matsuoka
Cosmic Radiation Laboratory
The Institute of Physical and Chemical Research
2-1 Hirosawa, Wako, Saitama 351-01, Japan

Introduction

At X-ray energies SS433 had appeared to be a relatively undistinguished, variable source with a 2-6 keV luminosity of the order of $L_x \gtrsim 10^{35}$ erg/s. The strong intrinsic variability of the source and the two observationally rather unpleasantly long periodicities, the binary period of $P_b \sim 13$ days modulated by the much longer precession period of $P_p \sim 164$ days, inhibited any easy modelling of the X-ray emission.
This situation changed considerably after X-ray observations with the EXOSAT satellite revealed the presence of Doppler shifted iron lines in the spectrum of SS433 (Watson *et al.*, 1986). As the centroids of the lines follow the precessional Doppler curve (see e.g. Margon, 1984) the X–ray emission must originate from the jets, emitting thermal radiation at temperatures kT $\gtrsim$ 2 keV. The results of Tenma observations (Matsuoka *et al.*, 1986) were consistent with this picture although a higher temperature for the underlying continuum was obtained.
SS433 has been observed by Ginga in May 1987 for 6 days covering the primary eclipse of the X–ray source around binary phase $\Phi \sim 0.0$, at a precessional phase $\Psi \sim 0.76$. The main results of these observations (Kawai *et al.*, 1989) were the detection of a very broad iron line, centered at about 7 keV on a very hot (kT $\gtrsim$ 30 keV) thermal continuum. The equivalent width of the line was found to be $\gtrsim$ 1.8 keV, considerably higher than expected from a plasma with cosmic abundances at these high temperatures. During the partial eclipse of the X-ray source the temperature of the continuum dropped to about kT $\sim$ 12 keV indicating the obscuration of the hotter, innermost parts of the jet.
A second observation of SS443 was conducted a year later, in May 1988, for 5 days. The observations covered again the primary eclipse of the X–ray source, the precessional phase was $\Psi \sim 0.01$. The temperature of the continuum appeared to be much lower than in 1987, but this can mainly be attributed to the different Doppler boosting factors in the two observations (Brinkmann *et al.*, 1990).

The Line region

The strong iron lines can be fitted by single Gaussian lines, generally with acceptable χ^2's but with centroid energies which are definitely different from the expected Doppler shifted line centers. Figure 1a shows the result of such a single line fit to a data segment taken before eclipse. The remaining residua show a distinct, always similar pattern indicating that the one - line - model is not the physically most realistic choice. A fit of two Gaussian lines to the data results in nearly always vanishing residuae in the line region and gives, in general, a much better χ^2 value, as clearly seen in figure 1b, which shows a two line fit to the same data set.

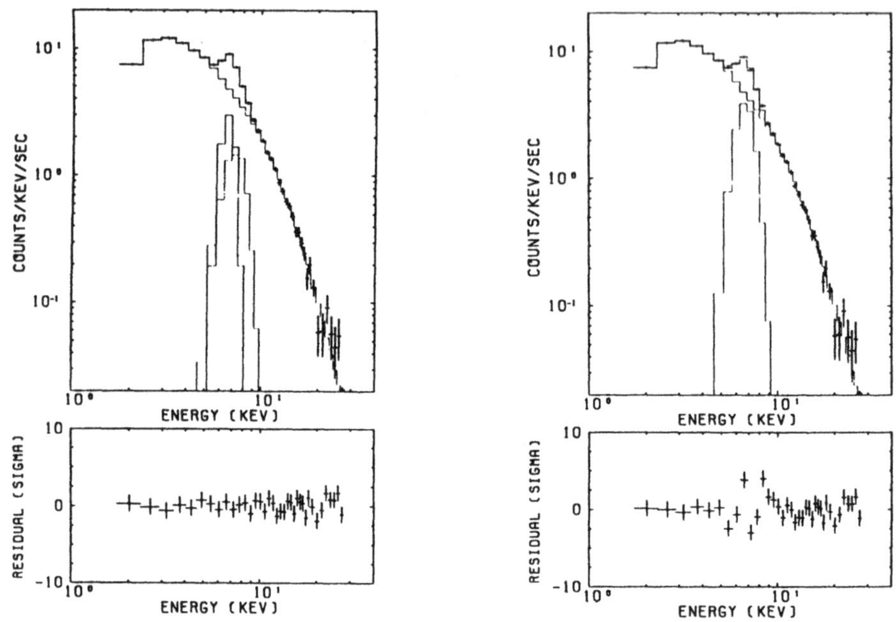

Fig. 1: a) Single Gaussian line fit to data out of eclipse — b) fit of two Gaussian lines to the same data set

This leads to the conclusion that we are actually observing the blend of two lines: a narrow component with an intrinsic width below the instrument's energy resolution with a centroid energy close to, but systematically offset from the exact Doppler shifted position and a second, broad component with its centroid at around 7 keV.

a) <u>The narrow line</u>

A detailed analysis demonstrates that the narrow line is intrinsically located at the 'correct' Doppler shifted position; the apparent centroid shift towards the $\approx$ 7 keV centre of the broad line is artificial, caused by fitting a Gaussian line profile to a component which is not Gaussian in shape, but not distinguishable from it with the LAC's energy resolution. Taking into account this systematical shift the real centroid positions of the narrow lines are consistent with the assumption, that they are always emitted from the 'upper' jet, i.e., the jet emanating from above the system's equatorial plane. This picture eases considerably the visibility problems raised by the necessity to hide the red-shifted jet at early precessional phases (Watson *et al.,* 1986, Brinkmann *et al.,* 1989).

The small variations in intensity and the centroid position of the narrow line show that it is only little affected by the eclipse process. This can clearly be attributed to the fact, that the bulk of this Doppler shifted line emission originates from fairly far out in the slightly cooler regions of the jet and, therefore, only a relatively small fraction of this line flux can actually be eclipsed by the primary. It should further be noted, that the EW of this line is consistent with the assumption of cosmic iron abundance in the outflowing gas.

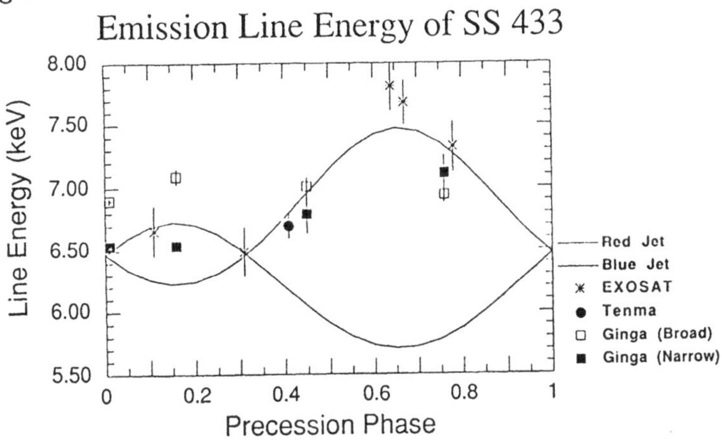

Fig. 2: Line centroid positions of the two fitted Gaussian lines as function of the precessional period

b) <u>The broad line</u>

The centroid of the broad emission line is always consistent with a value of $\approx$ 7 keV irrespectively of the precessional phase of the jets. However, the line width and the line flux are correlated with the overall intensity, i.e., they are affected by the eclipse of the primary. This implies that the emission of the broad line originates from matter in the central parts of the disc, close to the jet.

Attempts to model the observed continuum spectrum in and out of eclipse indicated a deficiency of soft X-ray flux from the jet (Brinkmann *et al.*, 1990). This missing flux could be supplied by an additional component, with roughly thermal bremsstrahlung characteristics at a temperature of about 5 – 6 keV. The total intensity of this component is of the order of $\lesssim 9\ 10^{34}$ erg/s, i.e., it amounts to only less than 8% of the observed X-ray luminosity.

Thermal emission from scattering material at the base of the jet can easily account for this extra component. The fact that the gross geometrical conditions of the jet – scattering region do not change drastically with precessional phase indicates that the reprocessing material on the inner parts of the disk follows the precessional motion of the jets.

The schematic picture of the envisaged scenario in Figure 3 demonstrates that we are always seeing three components of the X-ray emission. One, which originates from distances greater than the height of the scattering 'blob' on the disc is seen directly, i.e., with the appropriate Doppler shift in energy and Doppler boost in intensity. The second component consists of reprocessed radiation from the bottom part of the jets. Both red- and blue shifted radiation can be scattered into the line of sight, the characteristics of this emission are determined by the relative sizes of the emission region of the jets and of the height of the scattering component. And, finally, we see the direct thermal emission from the scattering material, which must be distributed asymmetrically in azimuth around the outflowing jet.

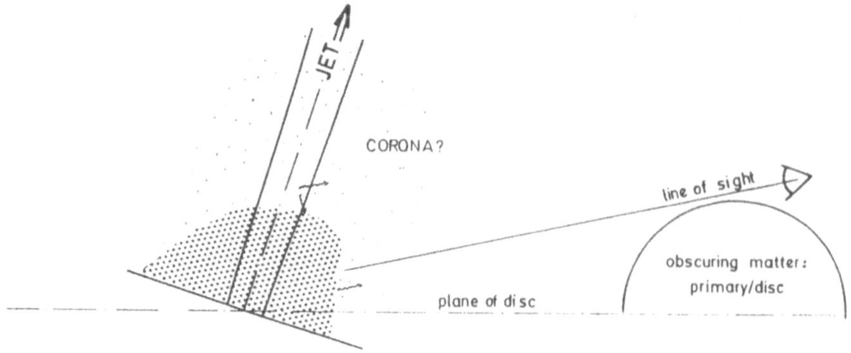

Fig. 3: Sketch of the structure of the X-ray emission region.
Picture is not to scale.

The number of photons in the broad line is generally larger than the number of photons in the narrow line. It should be mentioned that the K β resonance line of FeXXVI at $\sim$ 6.96 keV is exactly at the centroid position of the broad line. One possible explanation for the large line flux is based on the fact that the jets are already very hot when they emanate from the opaque disc. This implies that there is, even further down in the disc, an acceleration and heating region where we will find at least the same high temperatures in optically thick plasma conditions. The photon flux leaking out along the jet can change the ionization structure at the jet's base considerably such that the emission of the Fe Kβ is enhanced.

On the other hand, the outflowing jet is a highly non spherical phenomenon. The optical depth along the jet's axis is much larger than perpendicular to it. Therefore, we can expect conditions as found commonly for resonant line scattering which is known to produce highly anisotropic radiation patterns (Mewe 1990, Raymond 1988).

A deeper insight into the physical conditions of the central parts of the accretion disc requires, however, a detailed numerically modelling.

References

Brinkmann, W., Kawai, N., Matsuoka, M., and Fink, H.H.: 1990, Astron. Astr. in press

Brinkmann, W., Kawai, N., and Matsuoka, M.: 1989, Astron. Astr. **218**, L13

Kawai, N., Matsuoka, M., Pan, H.C., and Stewart, G.C.: 1989, Publ. astr. Soc. Pac. 41, 491

Margon, B.: 1984, Ann. Rev. Astr. Astr. **22**, 507

Matsuoka, M., Takao, S., and Makishima, K.: 1986, Mon. Not. R. Astr. Soc. 222, 605

Mewe, R.: 1990, in "Physical Processes in Hot Cosmic Plasmas", (eds. W. Brinkmann, A. Fabian, F. Giovannelli), Kluwer Acad. Publ., Dordrecht, p.39

Raymond, J.: 1988, in "Hot Thin Plasmas in Astrophysics", (ed. R. Pallavicini), Kluwer Acad. Publ., Dordrecht, p.3

Rothenflug, R. and Arnaud, M.: 1985, Astron. Astrophys. **144**, 431

Watson, M.G., Stewart, G.C., Brinkmann, W., King, A.R.: 1986, Monthly Not. Roy. Astron. Soc., **222**, 261

MODELLING OF X-RAY EMISSION FROM SS433

E.V. Seifina, N.I. Shakura, K.A. Postnov and M.E. Prokhorov
Sternberg Astronomical Institute, Moscow University
SU-119899 Moscow V-234, USSR

Introduction

The results on X-ray emission of the unique object SS 433 obtained from the board of Japanese X-ray Astronomy Satellite "*Ginga*" turned out to be a shock of lightning which stroke imagination of the X-ray astronomers (Kawai et al. 1989). The work of Kawai (1989), who have managed to fit X-ray spectra of SS 433, provided even more impression. It is the results that shed the light of truth on the physics of this object and also make more clear the observations previously obtained both in optical and X-ray band (see e.g. for review Cherepashchuk 1988, Zwitter et al. 1989).

To begin with let's remember the main *Ginga* 1987-1989 results on the X-ray spectra and X-ray light curves of SS 433 (see Kawai 1989, Brinkmann et al. 1990). At first, at the epoque of *Ginga* observations the X-ray Iron emission was decomposed onto broad and narrow components and the X-ray continuum evidenced higher temperature of the hot emitting matter, than that deduced from *Einstein* and *EXOSAT* data obtained earlier (Grindley et al. 1984, Watson et al. 1986); the overall X-ray luminosity was also higher . At second, X-ray eclipses both in the Iron line emission and the continuum were observed to be very different at different precession phases of the source. All these facts allow us to construct a following basical model of the SS 433.

Basical model

We agree with many investigators believing we are dealing with supercritical disk accretion onto a compact object (whose nature is discussed below) in a binary system. Qualitative picture of this process was suggested by Shakura and Sunyaev (1973). Dynamics of this disk accretion regime is governed by some critical mass inflow coming into the outer disk edge, $\dot{M}_{cr} \cong 3\cdot 10^{-8}$ $M/M_{\odot}$ $(M_{\odot}\ yr^{-1})$, and in case of black holes or neutron stars M is its mass. Thermal luminosity at highly supercritical regime ($\dot{M} \gg \dot{M}_{cr}$) approaches the Eddington critical luminosity $L_{cr} \cong 10^{38} M/M_{\odot}$ (erg s^{-1}). Besides, under radiation pressure action, practically all matter infalling into outer edge of the disk, inside a certain radius, $R^*(cm) \cong 10^6 M/M_\odot\ (\dot{M}/\dot{M}_{cr})$, leaves the disk with the velocity close to the parabolic one. The kinetic power of this outflow is of order of $\cong L_{cr}$. The optical observations of SS 433 show that the rate of mass transfer from the secondary component through the inner Lagrangian point is of order of $10^{-4}-10^{-5}\ M_\odot\ yr^{-1}$, i.e. the highly supercritical disc accretion actually takes place in this case. Processes of angular momentum exchange are of primarily importance in creating deep inside the radius R^* the hydrodynamic sheared flows having a poloidal velocity practically perpendicular to the disk plane. It is in these regions that one can expect two nozzles to be formed which could canalize the matter escaping the central source under a certain pressure at its surface. Tensions in the sheared flows smooth the velocity difference, so at the radius $R \geq R^*$ the radial flow is expected to be near uniform with the speed of order of $V_p(R^*) \cong c\ (\dot{M}_{cr}/\dot{M})^{1/2} \cong 3000$ km s^{-1} (here c is the speed of light). Therefore, this outflowing quasi-spherical shell will be looked alike independently on what object - black hole or neutron star - is hidden inside it. But to our mind only a pressure at the surface of neutron star without strong magnetic field is capable of generating the jets along the nozzles.

Broad Fe line emission

Our analysis of *Ginga* X-ray light curves shows that due to a thermal instability, the flow in this shell should be splitted into relatively cool and dense "flowing isles" surrounded by hotter less dense medium. Relative motions of the "isles" and the medium is the source of their heating. We consider broad Fe-line discovered by *Ginga* is primarily due to hydrogen-like Fe ($E \approx 6.95-6.97$ keV) which is in this hot medium. The proposed here mechanism of strong H-like Fe-line formation resemble to that of Fe-line formation proposed by Felten et al (1972) for another X-ray sources. A relatively high ratio of the line intensity to the continuum is caused by their different formation zones. The continuum is formed in the deeper and hotter layers. The line results from photorecombination and is formed in the layers with scattering optical depths

$$\tau_{sc} = -\int_{\infty}^{R_1} n\ \sigma_T\ dR \approx 1-3$$

(here $n \propto R^{-2}$ is for electron density in the hot medium, σ_T is usual Thompson cross-section). The line is broadened in these layers as well. From the other hand, the photons of this line formed deeply inside (at $R \approx R_1$) should also suffer from several scatterings on the cool "isles".

So these photons will escape the shell somewhere higher at some $R_{eff} \gg R_1$, where effective optical depth relative to these "isles" is of order of unity, $\tau_{eff} \approx N D^2 R_{eff} \cong 1$ (here N is the number of the "isles" per unit volume, D their size). The computer simulations of X-ray light curve of SS 433 gives $R_{eff} \approx 0.1\ a$, where a is the binary separation. Note that the photons of X-ray continuum escape from this radius too. The envelope that small ($\approx R_{eff}$) will be completely obscured by the secondary component during the eclipse.

The analysis of *Ginga* May 1987 observations leads to the conclusion that the broad Fe-line is also partially formed in the relativistic blobbed jet, but this line is simply K-line (E≈6.4 keV) of low ionized Iron blue-shifted to ≈7 keV. Broadening of this line occurs on free electrons of the blobbed jet. Due to Compton effect this line should be asymmetric. However, the broad line belonging to the jet, whose energy would be less than ≈6.5 keV, is not seen on the spectra obtained during 1988-1989 observations. The blobbed jet at these moments is likely to be more rarefied and correspondingly hotter. At the moments of observation in 1988 and 1989 the precession phases of the source were such that the outer edge of the accretion disk obscured (partially in 1988 and completely in 1989) the central shell of radius R_{eff}. So at full eclipse in 1988 data and all the way in 1989 data the only broad line emission observed was the residual emission (≈4 counts s^{-1} -- Kawai, 1989) coming from outer halo ($R>R_{op}$) around the binary. Melting of the "isles" in the halo rises the electron density there and we observe the photons scattered by these electrons.

Narrow Fe-line emission

The narrow Fe-line seen in 1987 data is formed in the surface layers of the relativistic jet blobs heated up due to its motion relative to slow moving surrounding plasma (v≅3000 km s^{-1}, see above). This emission could be identified with K_α-line for atoms of Iron with partially populated L-shell (the K_α-line energies for different ionized Iron see in Mewe et al. 1980, Sylwester et al. 1980 and ref. therein). The line seen in 1988 spectra can be identified with Helium-like emission (6.7 keV) from blobbed jet red-shifted due to transverse Doppler effect. It is likely that at this moment the blobs were less denser and had correspondingly higher temperature and we observe the emission from all volume of such blobs. The narrow 1989 line is formed similarly to the 1988 line (i.e., comes from all volume of the blobs), but the temperature of the blobs is close to the surface temperature of 1987 blobs. If in 1987 data we have no seen the narrow line from the red jet due to its being screened by the disk, in 1989 data the red narrow counterpart should be seen, may be slightly dimmer.

The continuum emission

The shape of X-ray eclipse evidences the 1987 continuum is mainly due to emission from the shell with radius R_{eff}. The substantial emission is also provided by higher levels of outflowing shell, because the melting of the "isles" contributes significantly to the bulk emission measure and the latter decreases more slowly than R^{-1}. In 1988-1989 X-ray luminosity of the shell got lower and thus the jet became to influence relatively more on the X-ray eclipse shape. The physical variability does not allow us to separate reliably continual X-ray emission from the outflowing wind and that from the relativistic jet.

Discussion and conclusion

The relative duration of X-ray eclipse allows us to determine the mass ratio in this system $q=M_x/M_{op}$ to be less than 0.25 (see also Antokhina et al. 1990). This estimate permits to consider a neutron star as primary component of the binary. We believe that after taking into account additional optical light carried by the "floating isles" off the Roche lobe of the primary, the estimation of q following from optical light curve analysis ($\approx$1) will be reduced to that resulting from X-ray light curve synthesis.

References

Antokhina E.A., Seifina E.V., and Cherepashchuk A.M.,1991, *IBVS*, in press
Brinkmann W., Kawai N., Matsuoka M., and Fink H.H.,1990, MPE Preprint No 189
Cherepashchuk A.M.,1988, in *Sov.Sci.Rev.E.Astrophys.Space Phys*,7, p.183
Felten J., Rees M.J., and Adams T.F.,1972, *Astron. Astrophys.* 21, p.139
Grindley J.E., Band D., Seward F., Leahy D., Weisskopf M.C., and Marshall F.E.,1984,*Astrophys.J.*, 277, p.286
Kawai N.,1989, in *Proc. 23d ESLAB Symp. on Two-Topics in X-ray Astronomy, Bologna, Italy (Ed. N.E.White)*, p.453
Kawai N., Matsuoka M., Pan H.-C., and Stewart G.C.,1989, *PAS Japan*, 41, p. 491
Mewe R., Schrijver J., and Sylwester J.,1980, *Astron. Astrophys. Suppl.*, 40, p.323
Shakura N.I, and Sunyaev R.A.,1973, *Astron. Astrophys.* 24, p.337
Sylwester J., Mewe R., and Schrijwer J.,1980, *Astron. Astrophys. Suppl.*, 40, p.335
Watson M.G., Stewart G.C., Brinkmann W., and King A.R., 1986, *Mon.Not.Roy.Astron.Soc.*, 222, p.261
Zwitter T., Calvani M., Bodo G., and Massaglia S.,1989, *S.I.S.S.A. 12 A*

Iron Kα Lines and the Absorption Structure of Intermediate Polars

A J Norton[1], M G Watson[2] & A R King[3]

1: High Energy Astrophysics Group, Physics Dept., University of Southampton, Southampton S09 5NH
2: X-Ray Astronomy Group, Dept. of Physics & Astronomy, University of Leicester, Leicester LE1 7RH
3: Astronomy Group, Dept. of Physics & Astronomy, University of Leicester, Leicester LE1 7RH

Abstract : The X-ray spectra of Intermediate Polars typically show a large absorbing column density and a strong iron Kα emission line. Using observations made with *EXOSAT* and *GINGA*, we show that the lines probably have a fluorescent origin but that their equivalent widths are too large to be due to fluorescence solely in the material which gives rise to the measured column density, assuming a single value for N_H in the vicinity of the emission region. To explain this we suggest that the spectra are absorbed by a continuous distribution of column densities. We show that a power law form for this distribution gives rise to line equivalent widths of several hundred eV, as observed, and also provides acceptable fits to the X-ray continua. We further suggest that a distribution of column densities would arise naturally as a result of the "blob-model" for accretion in magnetic Cataclysmic Variables.

1 Introduction & Observations

Intermediate Polars (IPs) are asynchronously rotating magnetic Cataclysmic Variables (CVs), characterized by pulsed X-ray emission from the polecaps of the accreting white dwarf. Their X-ray spectra typically show strong Iron Kα emission lines and a high column density [1]. In this paper we use the line parameters to investigate the absorption structure close to the white dwarf.

Eight confirmed IPs were observed by *EXOSAT* [2,3,4], in a total of 24 observations (see [1] for details). In addition we now have 3 observations made with *GINGA* [5], giving an additional spectrum for each of TV Col, GK Per and FO Aqr. The spectra can be described by power law continua and either homogenous or partial covering absorbers. Models using thermal bremsstrahlung continua generally produced fits in which the temperature was poorly constrained or unbounded, probably implying that the true bremsstrahlung temperature is above the energy detection limit of these observations or that the emission is non-isothermal. The fits to the spectra of all sources (except

Table 1 : Spectral Parameters – Homogenous and Partial Absorber Models

Source	Line Energy (keV)	Equivalent Width (eV) Homogenous Abs.	Equivalent Width (eV) Partial Abs.	Intrinsic Col. Density ($\times 10^{21}$ cm^{-2}) Homogenous Abs.	Intrinsic Col. Density ($\times 10^{21}$ cm^{-2}) Partial Abs.
EXOSAT					
GK Per	$6.49^{+0.10}_{-0.10}$	660^{+90}_{-80}	650^{+160}_{-70}	116^{+5}_{-5}	242^{+194}_{-115}
V1223 Sgr	$6.30^{+0.36}_{-0.35}$	540^{+160}_{-150}	530^{+190}_{-230}	16^{+4}_{-3}	116^{+150}_{-76}
TV Col	$6.50^{+0.60}_{-0.64}$	320^{+330}_{-300}	300^{+380}_{-300}	23^{+13}_{-10}	71^{+50}_{-53}
AO Psc	$6.44^{+0.47}_{-0.42}$	380^{+140}_{-130}	470^{+180}_{-170}	3^{+6}_{-3}	60^{+34}_{-24}
FO Aqr	$6.49^{+0.80}_{-0.72}$	480^{+320}_{-300}	300^{+990}_{-170}	40^{+39}_{-24}	171^{+810}_{-176}
BG CMi	——	< 200	< 200	56^{+22}_{-18}	44^{+48}_{-31}
EX Hya	$6.97^{+0.31}_{-0.31}$	690^{+140}_{-120}	510^{+110}_{-100}	$0.07^{+0.13}_{-0.05}$	$0.6^{+1.8}_{-0.5}$
H0542–407	$6.38^{+0.13}_{-0.90}$	320^{+880}_{-320}	320^{+410}_{-320}	0^{+6}	$1.0^{+8.0}_{-0.5}$
GINGA					
GK Per	$6.59^{+0.37}_{-0.36}$	390^{+140}_{-190}	300^{+170}_{-170}	12^{+5}_{-5}	241^{+616}_{-241}
TV Col	$6.58^{+0.10}_{-0.10}$	700^{+110}_{-110}	380^{+150}_{-130}	32^{+7}_{-5}	277^{+198}_{-149}
FO Aqr	$6.41^{+0.07}_{-0.07}$	780^{+80}_{-80}	580^{+100}_{-120}	45^{+4}_{-4}	211^{+42}_{-38}

BG CMi) were significantly improved by the inclusion of a narrow emission line with an energy in the range 6–7 keV.

Table 1 contains a summary of the results of spectral fitting. Where several *EXOSAT* observations were made of a single source, the best fitting spectral parameters have been averaged over all observations to give the values shown. For GK Per, the *EXOSAT* observations used were made during outburst and the *GINGA* observation was during quiescence. In the partial absorber model, the covered fraction of the source was typically $\approx 80\%$ and the column density to the uncovered fraction was usually $< 10^{21}$ cm^{-2} (see [1] for details).

2 Simple Emission Line Models

Both the *EXOSAT* and *GINGA* data indicate that the spectra of IPs may be characterised by thermal bremsstrahlung continua with $kT > 25$ keV. A plasma at this temperature will give rise to thermal line emission at $\sim$ 6.9 keV with an equivalent width < 100 eV [6]. Such lines are too weak to be detected in these spectra.

The mean line energy measured in these observations is 6.51 keV ($\pm$0.12 keV) and is therefore not consistent with a thermal origin for the emission process. (Nb. The mean line energy *excluding* measurements from EX Hya is 6.47 keV – see Discussion.) Fluorescent Kα lines arising as a result of reprocessing in low ionisation (Fe I – XVII), cool ($T < 10^5$ K) iron ions have an energy in the range 6.4 – 6.5 keV, varying slowly over a large range of ionisation stages. Hence, a fluorescent line is entirely consistent with the line energies measured in these spectra, except probably for EX Hya.

The simplest geometrical pictures of how fluorescent line emission may arise predict certain relationships between the line equivalent width and the hydrogen column density measured in the X-ray spectrum. In Figure 1 we show lines representing two of these. The first (labelled 'a') is for the simplest case where a symmetrical absorbing cloud causes the line of sight absorption, scattering and fluorescence; it may be approximated by the expression : EW (eV) $\sim N_H(\mathrm{cm}^{-2})/10^{21}$. The second line ('b') represents a system in which the direct radiation is completely blocked by opaque material and only scattered and fluorescent emission is seen; such a situation may arise in an eclipsing system, for instance, and give rise to a $\sim$constant equivalent width as a function of column density. These relationships, which assume the cosmic abundance of iron, are adapted from [ref. 6]. The points represent the data from Table 1, for comparison.

Figure 1 : Line equivalent widths and column densities for each object. Circles are *EXOSAT* data, squares are *GINGA* data; open shapes are for homogenous absorber models, filled shapes are for partial absorber models. The lines represent the expected relationship assuming fluorescent emission and (a) the absorption, scattering and fluorescence occur in material distributed symmetrically around the source, (b) the direct radiation is blocked and only scattered and fluorescent emission are seen.

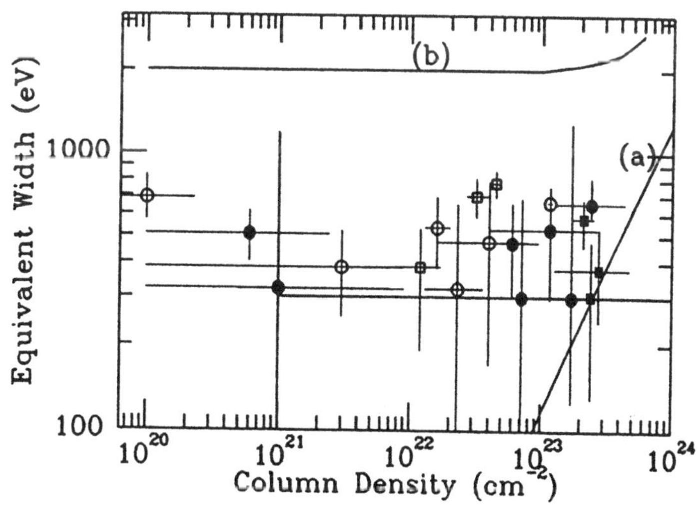

If we assume that the material responsible for the measured photo-electric absorption is also responsible for the fluorescence, the column densities determined from either homogenous *or* partial covering models for the absorption are systematically too small (or the equivalent widths too large) to conform with the description given by model 'a'. Conversely, the equivalent widths are not large enough to be produced by a system in which the direct radiation is completely blocked ('b'). A more complicated explanation is needed. One solution may be to propose a large over-abundance of iron in the system, but it is not apparent why this should be the case. A more acceptable solution is to suppose that the fluorescence takes place in material other than that which gives rise to the measured (single value) absorption. It is this possibility which we investigate here.

3 A Column Density Distribution Model

Accretion in magnetic CVs probably occurs via material which is composed of many 'blobs' with different sizes and densities [7]. The largest, densest blobs penetrate deep into the white dwarf photosphere before being decelerated by a shock. Most of the X-rays released below the shock have to pass through a large column of absorbing material (the white dwarf photosphere) before emerging. Conversely, the smaller, less dense blobs penetrate only slightly (if at all) into the white dwarf photosphere before releasing X-rays. This emission experiences a far lower column density. A continuous range of blob sizes and densities will therefore give rise to a continuous distribution of absorbing column densities through which the X-ray emission emerges.

We investigated the behaviour of several models for the column density distribution. Those which are able to reproduce the observed spectral shapes most closely *and* predict line equivalent widths of the magnitude which we see are power law distributions of column densities. To calculate the predicted line equivalent width from these spectral models, we considered the contributions to the observed spectrum made by an incident flux distribution, F, passing through discrete column densities in the range 10^{17} – 10^{24} cm^{-2}. Each of the components was weighted by a factor proportional to $(N_H)^n$ where n is the power law index of the column density distribution. We further assumed that the contribution to the equivalent width produced by each discrete column density in the summation is given by : EW (eV) $\sim N_H(\mathrm{cm}^{-2})/10^{21}$. The total line equivalent width as a function of n was then calculated as the sum of the line fluxes from each component divided by the sum of the continua at 6.4 keV :

$$\mathrm{EW}(n) = \frac{\sum_{N_H=10^{17}}^{10^{24}} (N_H)^n F(6.4)\exp\left(-\sigma(6.4)N_H\right)(N_H/10^{21})}{\sum_{N_H=10^{17}}^{10^{24}} (N_H)^n F(6.4)\exp\left(-\sigma(6.4)N_H\right)}$$

This relationship between the line equivalent width and the power law index of the column density distribution is independent of the form taken by the incident flux distribution, F, and is shown in Figure 2.

Figure 2 : The line equivalent widths and column density distribution power law index for each object. Circles represent *EXOSAT* data and squares represent *GINGA* data. The line represents the expected relationship between equivalent width and the power law index of the column density distribution.

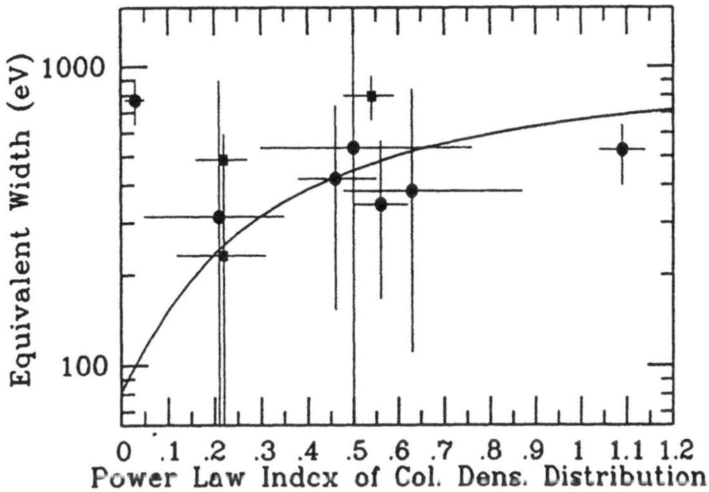

We also fitted this spectral model to each *EXOSAT* and *GINGA* spectrum, with 'n' as a free model parameter. This spectral model is not required by the continua, but *can* fit them just as well as simpler models do : the χ_r^2 values for these fits were similar to those resulting from a partial covering absorber model. The best fit parameters for each object, averaged over all *EXOSAT* observations of a given source, are listed in Table 2 and plotted in Figure 2 for comparison with the expected relationship. We see that the fitted and predicted line equivalent widths are in general agreement, although the fitted parameters are poorly constrained owing to the complexity of the model used.

4 Discussion

We conclude that a spectral model containing a power law distribution of column densities can describe both the continuum and line features of the X-ray spectra of IPs. Furthermore, this spectral form may arise naturally in the 'blob model' for accretion in magnetic CVs.

Although the IPs have been considered as a homogenous set of objects for this analysis, there are significant differences between the individual sources [1]. The most important difference here concerns EX Hya. The energy of the emission line in its spectrum is well constrained and strongly indicates a *thermal* origin. Furthermore, the bremsstrahlung temperature fit to the spectrum is < 10 keV and a plasma at this temperature will produce an emission line with an equivalent width of several hundred eV as observed. Hence, a thermal emission line is a more likely explanation for the feature seen here.

Table 2 : Spectral Parameters – Power Law Column Density Distribution Model

Source	Equivalent Width (eV)	Col. Dens. Dist. Power Law Index	Equivalent Width (eV)	Col. Dens. Dist. Power Law Index
	EXOSAT		GINGA	
GK Per	520^{+110}_{-120}	$1.09^{+0.05}_{-0.05}$	230^{+200}_{-220}	$0.22^{+0.09}_{-0.10}$
V1223 Sgr	420^{+320}_{-270}	$0.46^{+0.09}_{-0.08}$		
TV Col	530^{+750}_{-530}	$0.50^{+0.26}_{-0.20}$	490^{+100}_{-100}	$0.22^{+0.05}_{-0.06}$
AO Psc	350^{+220}_{-180}	$0.56^{+0.06}_{-0.06}$		
FO Aqr	380^{+460}_{-270}	$0.63^{+0.24}_{-0.15}$	800^{+130}_{-140}	$0.54^{+0.05}_{-0.06}$
BG CMi	< 400	$0.58^{+0.20}_{-0.15}$		
EX Hya	770^{+140}_{-130}	$0.03^{+0.02}_{-0.02}$		
H0542–407	310^{+580}_{-310}	$0.21^{+0.14}_{-0.16}$		

This and the very small intrinsic column density provide further evidence for the unique nature of EX Hya amongst the IPs [8,9].

The prototype polar system, AM Her, also shows a strong iron Kα line in its spectrum which has been interpreted as arising from a combination of thermal *and* fluorescent emission [10]. The line energy measured in AM Her lies between that expected for the two processes and so cannot be used to distinguish between emission mechanisms [11]. Explanations of the line solely in terms of fluorescence or solely in terms of thermal emission, require model continua which are respectively either too absorbed or too steep for the continuum data. A mix of temperature components ($kT \sim 10$ keV – 30 keV) with both thermal and fluorescent contributions to the line emission was found to be one way of producing the required spectrum [10]. In IPs , similar conditions may exist, but we conclude that in most of these systems the contribution of any thermal emission lines is small and the measured equivalent widths are a result of fluorescence only.

The approach adopted here is one possible explanation for the X-ray spectra of IPs. A column density distribution model is unlikely to be a unique description of the data and models with multiple emission components may be equally successful. However, any scenario for the accretion process which gives rise to a structured, inhomogenous flow will produce a distribution of column densities to absorb the X-ray emission. Other, more complex, forms of the column density distribution may therefore be more appropriate to the 'real' situation than the simple power law used here.

Acknowledgements

Some of the data analysed for this work were obtained from the *EXOSAT* data archive and the *GINGA* data for TV Col and GK Per were obtained from performance verification observations. The analysis was carried out using the STARLINK computer systems at Southampton and Leicester Universities.

References

[1] Norton, A.J. & Watson, M.G., 1989a. *Mon. Not. R. astr. Soc.,* **237,** 853.

[2] de Korte, P.A.J., Bleeker, J.A.M., den Boggende, A.J.F., Branduardi-Raymont, G., Brinkmann, A.C., Culhane, J.L., Groenenschild, E.H.B.M., Mason, K.O. & McKechnie, S.P., 1981. *Space Science Reviews,* **30,** 495.

[3] Turner, M.J.L., Smith, A. & Zimmerman, H.U., 1981. *Space Science Review,* **30,** 513.

[4] White, N.E. & Peacock, A., 1988. in *X-ray Astronomy with EXOSAT*, Journal of the Italian Astronomical Society, eds. Pallavicini, R. & White, N.E.

[5] Turner, M.J.L., Thomas, H.D., Patchett, B.E., Reading, D.H., Makishima, K., Ohashi, T., Dotani, T., Hayashida, K., Inoue, H., Kondo, H., Koyama, K., Mitsuda, K., Ogawara, Y., Takano, S., Awaki, H., Tawara, Y. & Nakamura, N., 1989. *Publ. Astron. Soc. Japan,* **41,** 345.

[6] Makishima, K., 1986. In *The Physics of Accretion onto Compact Objects*, eds. Mason, K.O., Watson, M.G. & White, N.E.; Springer-Verlag.

[7] Hameury, J-M., King, A.R. & Lasota, J-P., 1986. *Mon. Not. R. astr. Soc.,* **218,** 695.

[8] Rosen, S.R., Mason, K.O. & Córdova, 1988. *Mon. Not. R. astr. Soc.,* **231,** 549.

[9] Norton, A.J. & Watson, M.G., 1989b, *Mon. Not. R. astr. Soc.,* **237,** 715.

[10] Swank, J.H, Fabian, A.C. & Ross, R.R., 1984. *Astrophys. J.,* **280,** 734.

[11] Rothschild, R.E., Gruber, D.E., Knight, F.K., Matteson, J.L., Nolan, P.L., Swank, J.H., Holt, S.S., Serlemitsos, P.J., Mason, K.O. & Tuohy, I.R., 1981. *Astrophys. J.,* **250,** 723.

Simulation of the Boundary Layer in White Dwarf with the SPH Method

D. Molteni, G. Giannone

Istituto di Fisica dell'Università, via Archirafi 36
90123 Palermo, Italy

ABSTRACT
The boundary layer between an accretion disk and the surface of a compact object is simulated using the Smoothed Particle Hydrodynamics (SPH) method.The equations include radiation, which is treated in the flux limited diffusion approximation.
The integrated equations and some numerical simulations of the 3D motion of the gas are presented.

Keywords : accretion disk, boundary layer, numerical methods.

1. INTRODUCTION

The boundary layer between the accretion disk formed by gas falling onto a compact star seems to be responsible of a great fraction of the total energy emitted; it may be also the common birthplace of relevant phenomena as jet, winds, QPO .However it is very difficult to describe analytically the structure of the boundary layer since it is a region of transition from highly supersonic motion to subsonic motion where radiation pressure and turbulent shocks should be important .Numerical solutions represent a unique opportunity to understand the physical processes operating in this region.

2. THE SPH METHOD

The Smoothed Particles Hydrodynamic method (SPH) is essentially a lagrangean scheme where particles represent parcel of fluid in motion according pressure and body forces. A detailed treatment of the method can be found in Monaghan (Ref. 1).
Summarizing we can say that according the method every bounded function ρ(r) can be approximated by the interpolating formula:

$$\rho(\mathrm{r}) = \int W(\mathrm{r}-\mathrm{r}', h)\rho(\mathrm{r}')d\mathrm{r}'$$

where W , the interpolating Kernel, is a continuous function , and h is the smoothing length and in the limit for h->0 W is the Delta distribution function. Using this formalism, we now write the tipical fluid equation in the form
Typical fluid equations are

$$\frac{Dv}{Dt} = -\frac{1}{\rho}\nabla P_g + g$$

$$\frac{D\varepsilon_g}{Dt} = -\frac{P_g}{\rho}\nabla \cdot v$$

they became in the SPH formalism:

$$\frac{D v_i}{Dt} = -m\sum_j\left(\frac{p_j}{\rho^2{}_j} + \frac{p_i}{\rho^2{}_i}\right)\nabla_i W_{ij} + g_i$$

$$\frac{D\varepsilon_i}{Dt} = \frac{m}{2}\sum_j\left(\frac{p_j}{\rho^2_j} + \frac{p_i}{\rho^2_i}\right) \mathbf{v}_{ij} \cdot \nabla_i W_{ij}$$

where

$$g_i = -\nabla_i \phi_i$$

the operator D_/Dt stands for the usual substantial derivative, and the index i is referred to the i-th particle..In this scheme the continuity equation is identically satisfied, and energy and angular momentum are very well conserved. An artificial viscosity is needed to handle shocks. We used the artificial viscosity proposed by Lattanzio et al. (Ref. 2) with an artificial thermal diffusion term that reduces shock smearing.

2. DIFFUSION OF RADIATION WITH THE SPH METHOD

To handle the radiation we adopted the flux limited Levermore-Pomraning diffusion approximation (Refs. 3, 4) that avoids the infinite velocity escape of the radiation at low optical depth. This approximation has been used by Eggum to simulate inner regions of accreting disks around black holes (Ref. 5),and by Kley to simulate the boundary layer on a white dwarf (Ref. 6).

The treatment of diffusion problems with the SPH method is relatively recent (Ref. 7) and exhibits more difficulties than the integration of the convective motion .
A general diffusion equation

$$\frac{dT}{dt} = \frac{d}{dx}(Q\frac{dT}{dx})$$

in the SPH scheme appears as

$$(\frac{dT}{dx})_a = \sum_b \frac{(Q_a + Q_b)(T_a - T_b)}{\rho_b x_{ab}} \frac{\partial W_{ab}}{\partial x_b}$$

(the (a,b) indices refer to particles). Since the (pseudo)particles interact by pairs, the system is thermally insulated. To avoid this problem we can add a term equivalent to cooling into vacuum when the opacity at the particle's position is sufficiently low(Ref. 7). However this is valid for star geometries where the density is function only of one variable (r) , but not for geometries such as disks where the opacity is function of two variables (r, z) and it can be low not only at the surface and we can have a fictious cooling for inner disk regions.
We therefore adopted a procedure that singles out the particles on the boundaries of the system and apply only to them the cooling.

3. THE BASIC EQUATIONS

We used the classical fluid - radiation equations as in the work by Kley, but they are written in lagrangean form, (see also Mihalas and Mihalas (Ref.8).
The momentum equation is

$$\frac{Dv}{Dt} = -\frac{1}{\rho}(\nabla P_g + \nabla \cdot \overline{\overline{S}}) - \nabla \phi$$

the equation for the total specific energy is given by

$$\frac{D\varepsilon_t}{Dt} = -\frac{1}{\rho}(P_g \nabla \cdot v + (\overline{\overline{S}} \cdot \nabla) \cdot v + \nabla \cdot \mathbf{F})$$

the equation for the density of radiation energy E_r is given by

$$\frac{DE_r}{Dt} = -E_r \nabla \cdot v - \nabla \cdot (\overline{\overline{S}} \cdot v) - \nabla \cdot \mathbf{F} - \frac{k_{ass} + k_s}{c} v \cdot \mathbf{F} + \rho c k_a (B - E_r)$$

the flux of radiation in the flux limited diffusion approximation is

$$\mathbf{F} = -\frac{cD_f}{\rho (k_{ass} + k_s)} \nabla E_r$$

where

$\overline{\overline{S}}$ is the Eddington flux tensor

4.RESULTS

Various tests have been done to verify if conventional physical situation are reproduced and some simulations have been performed to study the role of the inner boundary conditions at the star surface. Although the presented simulation is axis simmetric we used a full 3D code in view of future study of non axis simmetric configurations.
The adimensionalization parameters for accretion on white dwarf are :

x_0=20000 Km

$\rho_0 = 4.7\ 10^{-6}$ gr/cm3

$V_0 = (GM/x_0)^{1/2}$

$T_0 = (\gamma - 1)\mu m_a V_0^2/k_b$

(the width of the kernel was h=.05),. in this units the star radius is 0.8 and the time for one keplerian orbit at x_0 is 2π .

The tests made with full radiation treatment show that it can be obtained a classical thin disk similar to the Shakura Sunyaev αP disks (Ref. 9), the differences are due the numerical viscosity coefficient we used: $\eta = \alpha C_{sound} \rho h$. (ref. 10)
To simulate the boundary layer physics a crucial point is the boundary condition at the star surface: the preliminary results presented are obtained imposing a pure reflection of the particles at distance less then R_{star}, so that the inner disk particles continue their rotation at the keplerian velocity , but their accumulation produces a high density ring interacting with the incoming gas of the disk.

The time reached in the simulation here exihbited corresponds to about 1 keplerian period at the x_o reference distance.

Fig. 1 shows the distribution of the particles, and their velocity (crosses represent the starting points of the vectors) seen edge on. It is clearly apparent the slow azimutal motion of the fluid vith $v_z = 2\ 10^{-2}\ v_\phi$.

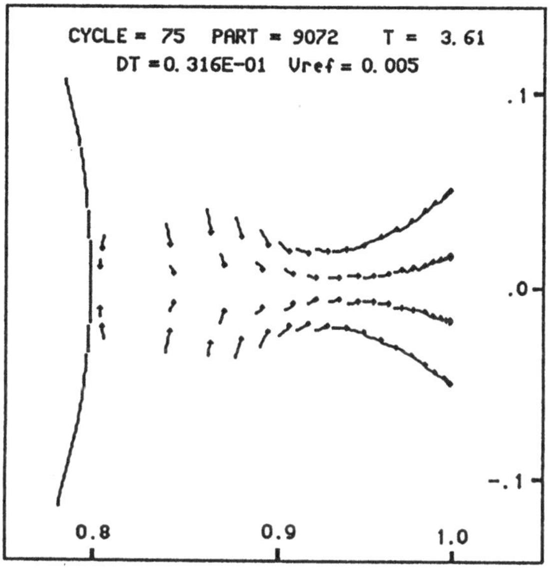

Fig.1 : Particles at the surface of the WD forming the boundary layer.

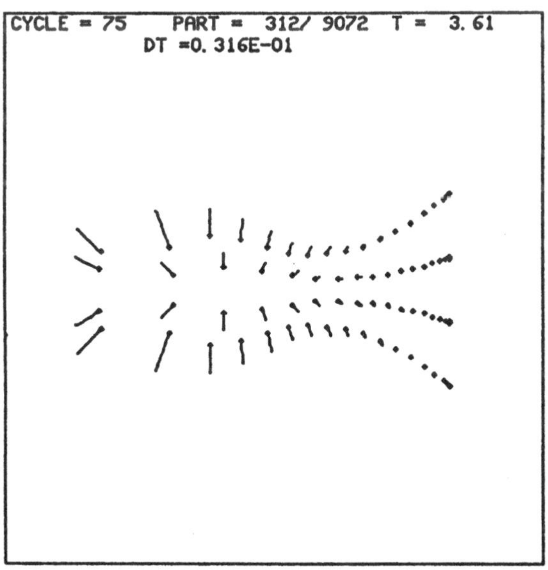

Fig. 2 : Radiaton field of the particles in the boundary layer seen edge on.

Fig. 2 shows the radiation flux distribution for the same particles of fig.1 (crosses represent the starting points of the vectors). Despite the fact that at this time the fluid has just reached the star surface, it appears clearly that the radiation tends to spread out the matter and to slow down the further accreting fluid.

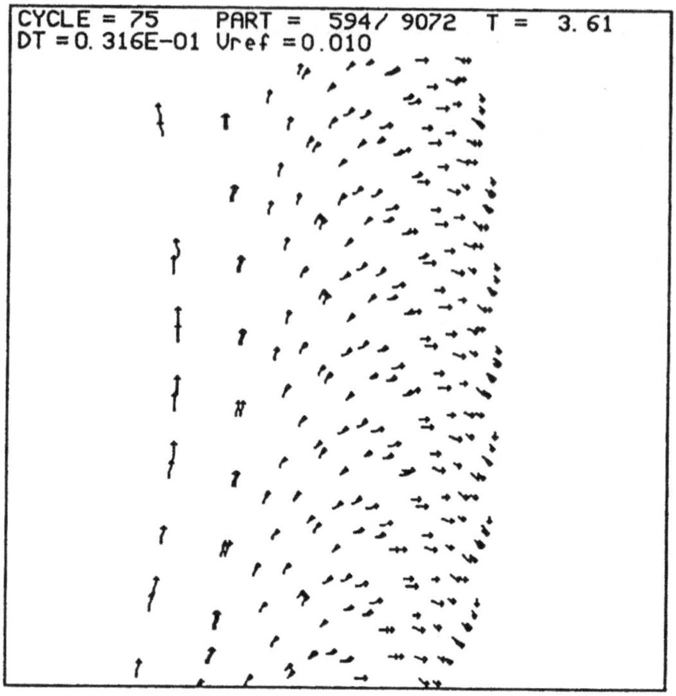

Fig. 3 : Difference between the particle velocities and the keplerian velocities in the boundary layer projected on the xy plane.

Fig. 3 shows - projected on the xy plane - the difference between the particles velocity and the keplerian velocity at the particle position. That is zero vectors correspond to particles moving in keplerian orbits.

These results are preliminar and other tests are in progress, neverthless the tecnique seems quite promising and efficient. We are grateful indebted to prof. J.J. Monaghan for his kind supplying the core of his SPH code and to prof. L Scarsi, director of the CNR IFCAI Institute, for the computer facilities provided.

References

1. Monaghan J. J. , *Comp. Phys. Repts.*,**3**, 71, 1985
2. Lattanzio J. C., Monaghan J. J., Pongracic H., Schwarz M. P., *S.I.A.M., Sci. Stat. Comp.*, 1985
3. Levermore C. D., Pomraning G. C., *Astrophys. J.*, **248**, 321, 1981
4. Levermore C. D., *J. Quant. Spectrosc. Radiat. Transfer* ,**31**, 149, 1984
5. Eggum G. E., Coroniti F. V., Katz J. I., *Astrophys. J.*, **323**, 634, 1987
6. Kley W., *Astrnon. Astrophys.* , **208**, 98, 1989
7. Brookshaw L., *Proc. ASA* , **6**, (2), 207, 1985
8. Mihalas D., Mihalas B.W., *Foundations of Radiation Hydrodynamics*, Oxford University Press, New York, 1984
9. Shakura N. I., *Astrnon. Astrophys.* , **24**, 337, 1973
10. Molteni D., Giannone G.,*Accretion Disks and Magnetic Fields in Astrophysics*, 145, Kluwer Academic Publishers, 1989

III. Active Galactic Nuclei

IRON LINES IN AGN AND BLACK HOLE CANDIDATES

A.C. Fabian & I.M. George
Institute of Astronomy, Madingley Road, Cambridge CB3 OHA, UK.

1. INTRODUCTION

Fluorescent iron lines are now a common feature of the observed spectra of AGN and Galactic Black Hole candidates. This is principally due to the large collecting area provided by the *Ginga LAC*, which can easily detect emission lines of only 100 – 200 eV equivalent width. Rapid variability of the iron line shows that it originates from the inner regions of the accretion flow and provides us with a feature of known rest wavelength produced very deep in the potential well. There is good reason to hope that these lines will be the tool with which the existence and true nature of black holes will be established.

To put the discovery of the fluorescent iron lines in some perspective, we first give a brief, historical review of the observations and interpretation of the X-ray spectra of AGN and black hole candidates. Needless to say, this review is biased to our point of view. We then describe how recent observations have added to our knowledge of the innermost regions of AGN.

2. AN HISTORICAL PERSPECTIVE

It was noted early in the 1980s that the 2–20 keV spectrum of AGN is well approximated by a power–law of energy index $\alpha \sim 0.7$ (Mushotzky 1982; Rothschild *et al.* 1983). This result was consolidated by *EXOSAT* studies (*e.g.* Turner & Pounds 1989), which showed the distribution of power-law indices sharply peaked around $\alpha = 0.7 \pm 0.1$, although a number of objects have indices which lie a small, but significant distance from this value. Higher energy observations demonstrated that AGN spectra steepen between $\sim$ 500 keV and 2 MeV. The Galactic Black Hole candidate Cyg X–1 is often represented by a thermal spectrum (*e.g.* Sunyaev & Trumper 1979), but a power–law of index $\alpha = 0.6$ is a good fit to the 2 – 20 keV band in the low state (White *et al.* 1984).

Next came the discovery of a quasi-blackbody thermal bump in the EUV and soft X-ray spectra of some AGN, which carries a significant fraction of the total luminosity (Malkan 1983; Arnaud *et al.* 1985; Turner & Pounds 1989). The 'soft X-ray excesses' are observed to be variable, sometimes on short ($t \lesssim 1000$ s) timescales (Lawrence *et al.* 1985; Kaastra & Barr 1990), implying they originate from very close to the central engine. The temperature ($T \sim 2 \times 10^5$ K) and luminosity of this emission provided the first indication of large quantities of relatively cold gas within the innermost regions of AGN. A low–energy X-ray excess is also observed in Cyg X-1, although in this case the temperature is a few times 10^6 K, as is expected from a simple blackbody scaling argument.

The original explanation for the soft X-ray emission in AGN was in terms of the high energy tail of the thermal emission from the putative accretion disc. Guilbert & Rees (1988) have suggested an alternative, however, in which the soft X-rays arise as the result of the thermal re-emission of hard non-thermal X-rays absorbed by a mixture of cold, dense gas clouds in and around the central engine. Provided the clouds are dense enough to remain cold (*i.e.* at a few 10^5 K; see also Ferland & Rees 1988), they absorb much of the nonthermal power and reradiate it as the observed quasi-blackbody bump. A direct prediction of the model is the occurrence of features (absorption edges, fluorescent lines and a Compton reflection hump[1]) in the X-ray

[1] Although the 'reflected' photons do not obey the laws of reflection, the name has stuck. By the 'reflection spectrum' we refer to the back-scattered spectrum formed as a result of electron scattering, absorption and fluorescence, whilst by 'reflection hump' we refer to the broad peak in this spectrum between $\sim$ 10 and $\sim$ 100 keV.

spectrum. Lightman & White (1988) extended the idea by modelling the shape of the reflected continuum in some detail.

Although the Guilbert & Rees model has the quasi-blackbody produced as a by-product of reprocessing the nonthermal power, it should be realized that the joint occurrence of cold gas (by cold we mean effectively neutral so far as X-rays are concerned) and hard X-rays *must* create X-ray spectral features. Even if the hard X-ray source is in front of the cold gas, provided that some of the hard X-rays are incident on the cold gas, the observed spectrum will then contain features due to the reflection component. This is clearly also true in the case of galactic sources, and the broad, iron emission-line observed with *EXOSAT* in the spectrum of Cyg X-1 (Barr *et al.* 1985) was so interpreted by Fabian *et al.* (1989) as the fluorescent iron line in the reflection spectrum. It was suggested that the hard X-ray source is located above the accretion disc, and the breadth of the line is then due to Doppler and gravitational frequency shifts from different parts of the disc. As pointed out by Fabian *et al.*, such an interpretation opens up the exciting possibility of mapping the inner disc around a black hole using a spectral line of known rest wavelength and thereby of measuring the mass of the black hole. This has since been explored further by Stella (1990) and Laor (1990).

Following the launch of *Ginga* in 1987, a number of observational discoveries quickly followed confirming the widespread existence of spectral features in AGN. Firstly, fluorescent iron lines were found to be common in AGN with equivalent widths typically $\sim$ 150 eV (Nandra *et al.* 1989; Pounds *et al.* 1989; Matsuoka *et al.* 1989/90; Pounds *et al.* 1990; George *et al.* 1990; Singh *et al.* 1990; Kunieda *et al.* 1990; Piro *et al.* 1990; Yamauchi *et al.* 1990; Turner *et al.* 1990; Day *et al.* 1990). Iron lines were, of course, already well-known in a number of AGN which showed strong low–energy absorption, such as Cen A (Mushotzky *et al.* 1978) and NGC 4151 (Warwick *et al.* 1989 and references within) and could be attributed there to fluorescence in the matter causing the absorption. The importance of the recent discoveries is that in these AGN there is no (or little) evidence for any absorption beyond that in our own interstellar medium.

Secondly, a 'hard tail' was evident in *Ginga* spectra of several AGN above $\sim$ 10 keV (Matsuoka *et al.* 1989/90; Pounds *et al.* 1990). This was interpreted by the first authors as due to partial covering by very optically thick material in their original preprint (Matsuoka *et al.* 1989) and as a reflection hump by Pounds *et al.* (1990) – an interpretation supported by Matsuoka *et al.* (1990) in the published version of their paper. Evidence for similar reflection spectra in the Galactic black hole candidates has also been found and discussed by Tanaka (1989, 1990). As with most discoveries, there were earlier detectious of these features without any clear interpretation, such as the report of iron lines in the spectra of quasars by Ohashi *et al.* (1988) and the report of a hard tail in the *Tenma* spectrum of IC4329A (Miyoshi *et al.* 1988).

Finally, rapid variability of the iron line was observed in NGC 6814 by Kunieda *et al.* (1990). In this remarkable observation, to which we shall refer again later, the X-ray source varied by a factor of 2 or more on a timescale of $\lesssim 50\,$s. Furthermore, the iron line intensity was observed to followed the continuum variations with a lag of less than 256 s (see also Kunieda 1990).

In the light of these observations there was a general convergence during and since the 23rd ESLAB Meeting held in Bologna in 1989 towards a view that at least half of the hard X-ray luminosity of AGN is reprocessed in cold matter below the X-ray source, so giving the iron line and reflection hump (see the many papers in those Proceedings, ed Hunt & Battrick 1989, and subsequent papers). The cold gas also probably causes the soft X-ray excess, perhaps boosted in luminosity by its own gravitational energy release.

3. THE ORIGIN OF THE IRON LINE

It is clear that the X-ray spectra of AGN (and Galactic black hole sources) are much more complicated in shape than a simple power-law. There is still an underlying continuum, which we assume has a power-law form, due to some (currently uncertain) hard X-ray emission mechanism. (It should be noted that even this continuum may be the result of reprocessing the gamma-ray spectrum due to photon–photon collisions.) In some sources the directly visible component of this continuum is supplemented by emission reprocessed by the cold gas which gives substantial quasi–blackbody emission at soft X-ray energies (*i.e.* dominating the flux below say 200 eV), a fluorescent iron line at 6.4–6.9 keV and a Compton reflection hump above $\sim$ 10 keV (Fig. 1) In addition, a broad annihilation line around 500 keV is predicted and there may be some absorption features due to cold and/or ionized iron in the 7 to 9 keV band.

The observed energy of the iron line, $\sim$ 6.4 keV (Nandra, 1990), is strong evidence that we are dealing with 'cold' iron, *i.e.* iron ionized to a state less than FeXVII (although the possibility that we are observing higher ionization states cannot be excluded if we allow for some redshift). The emission must be due to fluorescence if the line varies rapidly since 2-body collisional processes are relatively slow. This effectively rules out a purely collisional thermal origin for the line in NGC 6814 on the same grounds that thermal bremsstrahlung cannot generate the underlying continuum. The gas clouds would have to be so dense to make the cooling time short enough, and so large to produce the luminosity, that Comptonization would dominate the energy transfer process (Lightman *et al.* 1978).

The variable iron line in NGC 6814 must originate in the central engine, very close to the source of the hard X-rays. Since most other Seyfert-1 AGN have similar spectra (actually NGC 6814 has a flatter power-law and a stronger line than most), then it is usually assumed that the iron line in most Seyfert-1s originates from the central regions. It should be remembered, however, that there will be additional fluorescent iron emission from any other gas in and around the nucleus, from the broad–line region (*e.g.* from the BLR clouds which are generally assumed to have a covering fraction of less than 10 per cent), from any molecular torus such as present in Seyfert-2 galaxies (Krolik & Kallman 1987) and from the disc of the host galaxy itself (Fabian 1977). Nevertheless in the following we concentrate on the iron line from the central engine since having a line of known rest energy should enable us to eventually map this region in detail.

4. CLUES TO THE GEOMETRY OF THE INNER REGION

The equivalent width of about 150 eV observed in many AGN and the strength of the accompanying reflection hump strongly suggest that we are observing cold gas of approximately cosmic abundance in iron which subtends about half the sky as seen by the hard X-ray source. This is generally taken to mean that the hard X-ray source is outside the cold gas; *i.e.* above the gas as viewed by us.

The broad iron line argues in favour of a disc in Cyg X-1 (Fabian *et al.* 1989) although further observations are required to confirm this, particularly since *Tenma* data do not require the line be broad (Kitamoto *et al.* 1990). It is usually assumed that the cold gas is in the form of a disc in the case of AGN. A range of possibilities is shown in Fig. 2.

We have now (after presenting this talk) further investigated models in which the hard X-ray source is totally surrounded by the cold gas, and the composite spectrum is produced primarily by transmission through clouds partially covering the hard source. The number of fluorescent photons produced per unit depth is clearly a decreasing function of depth into the cold gas due to the Compton down-scattering and absorption of continuum photons. The strong iron line produced within the first $10^{24}\,\mathrm{cm}^{-2}$ (*i.e.* $\tau_{es} \sim 1$) into the cold gas will therefore be reduced by absorption if it has to propagate through much more cold gas prior to escape. (This is precisely why reflection models are the most effective way to produce a strong iron line.) In transmission models, the lack of any spectral flattening below $\sim$ 10 keV requires partial covering by cold gas of thickness $\tau_{es} \gtrsim$ several. Consequently the intensity of the iron line will be much

Figure 1. The view of the incident and reflected spectra, (above) from the Villa Monastero, Varenna (courtesy of the conference poster) and (below) based on Monte Carlo simulations in Cambridge (from George & Fabian 1990).

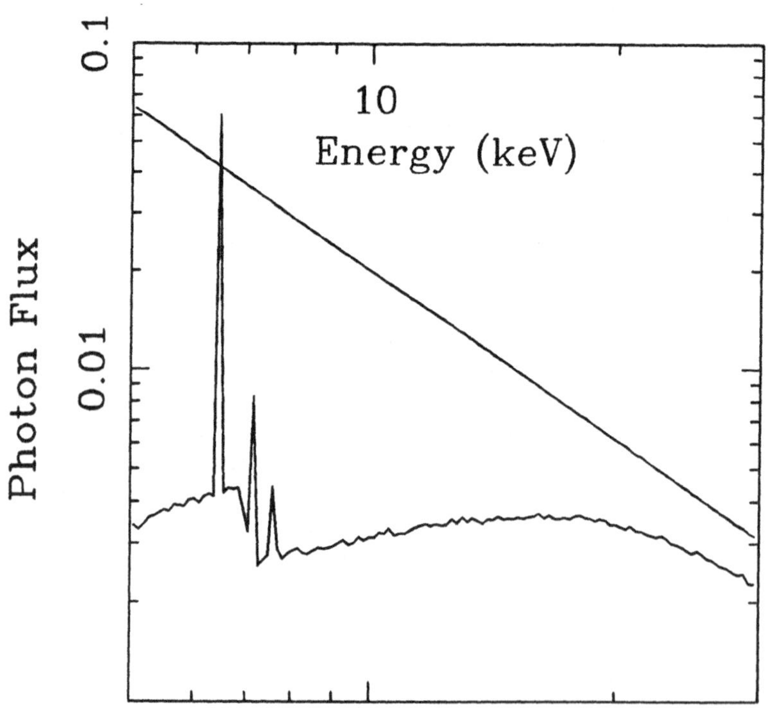

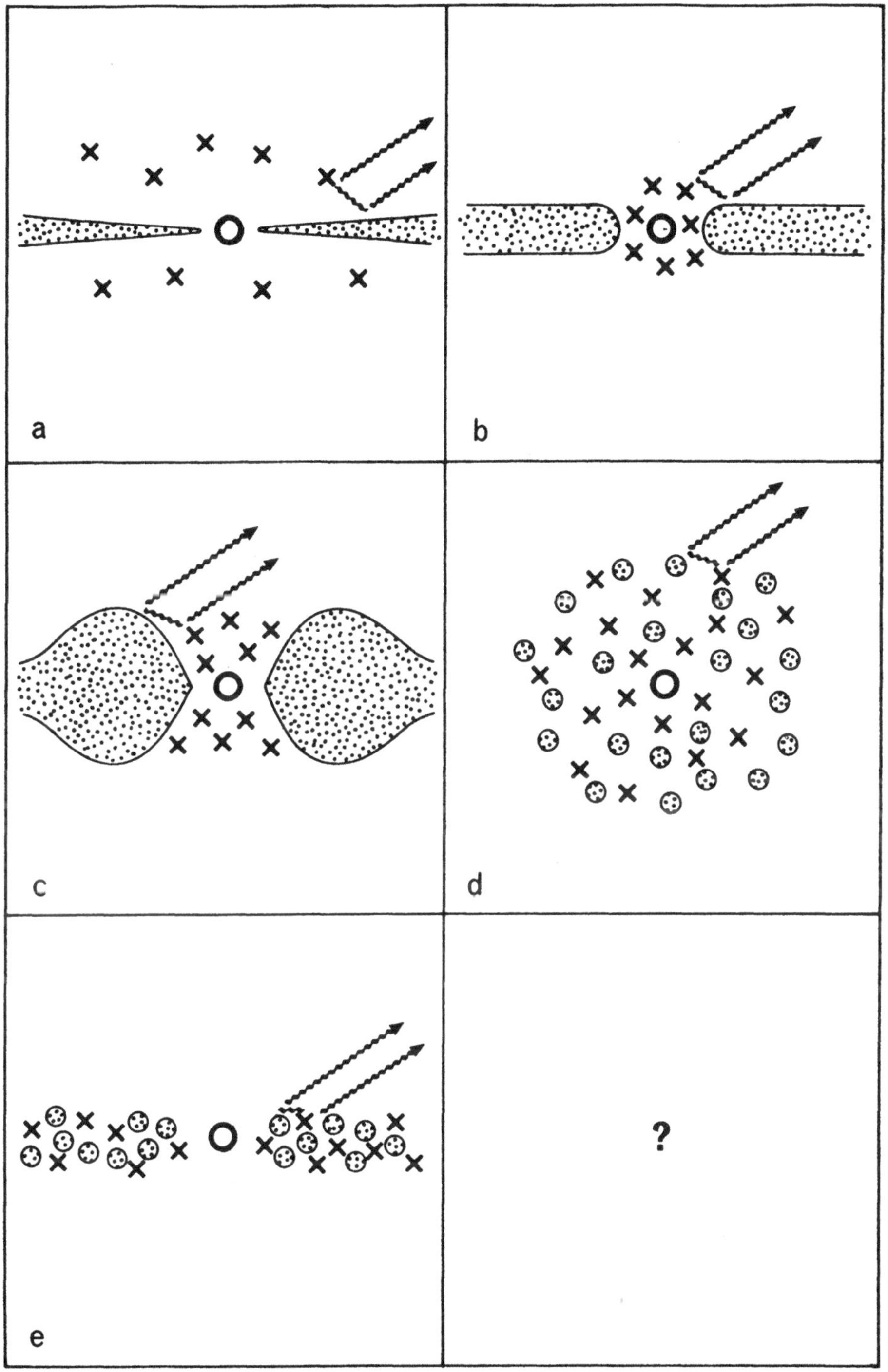

Figure 2. Some possible geometries for the hard X-ray source and cold matter.

reduced to the extent where the observed equivalent width can only be achieved if the covering factor is very large (though $\neq 1$). However, in such cases the predicted strength of the hard tail is many times that observed.

The above does not rule out an embedded source geometry in which the hard X-ray source lies partly inside the cold gas. Indeed, such models seem to be required in the cases of NGC 6814 (Kunieda *et al.* 1990) and Mkn 841 (Day *et al.* 1990) which have iron lines of equivalent widths > 150 eV. These sources also have $\alpha \lesssim 0.6$ and for a stronger reflection component in those sources (as Pounds *et al.* 1990 have demonstrated, the reflection spectrum flattens the apparent 2–20 keV composite spectrum). In an embedded source model, the cold gas could easily subtend more than 2π sr as seen by the hard X-ray source.

The rapid variability and lack of any observable lag between the line and continuum in NGC 6814 suggests to us a clumpy disc. 50 light seconds is smaller than a plausible emission region that is symmetric around the central black hole – unless relativistic beaming is invoked. Consequently it seems that the emission region is just a small asymmetric part of the central engine[2]. The hard X-rays may originate within and from the surface of this clumpy disc.

An alternative explanation for a large equivalent width to the line is that the primary source emits anisotropically, with the radiation incident onto the cold gas being more intense than that escaping to the observer. This situation is expected from optically–thin inverse Compton scattering where the soft photon originate in the cold gas and are then upscattered by energetic electrons above the gas (Ghisellini *et al.* 1990). Head–on scattering leads to more energetic final photons than glancing scatterings.

Finally, in passing we note that as stated above, the Guilbert & Rees model envisages cold gas clouds surrounding the non-thermal source absorbing and re-emitting much of the hard X-ray luminosity. The non-thermal spectrum in their model extends to energies much less than the quasi-blackbody bump. This cannot hold in the case of NGC 4051, in which the lack of simultaneous optical variability during a period of large amplitude, rapid X-ray variability (Done *et al.* 1990) rules out any such continuous spectrum.

5. PROPERTIES OF THE COLD GAS

If the quasi-blackbody luminosity dominates the hard X-ray energy density incident on the cold gas, then the Compton temperature of that gas is less than 10^6 K. Photoionization is however likely to be the most important heating process of the surface gas. For iron to be essentially neutral so far as hard X-rays are concerned, then we need

$$\frac{L_x}{nR^2} \lesssim 600$$

(George, Fabian & Ross 1990; see also Hatchett, Buff & McCray 1976; Begelman & DeKool 1990). Thus if, say, the hard X-ray luminosity $L_x \approx 10^{43}\,\mathrm{erg\,s^{-1}}$ originates about a distance $R \approx 10^{15}$ cm above the disc, then the gas density $n \gtrsim 10^{10}\,\mathrm{cm^{-3}}$. This density is greater than the mean density in a standard 'alpha' disc and so may be consistent with such a disc.

Problems again arise in the case of NGC 6814, where the lack of any observed time lag means that $R < 3 \times 10^{12}$ cm such that $n > 2 \times 10^{15}\,\mathrm{cm^{-3}}$. This is too high (see Begelman & De Kool 1990) for even the mean density in a standard 'alpha' disc in which the viscosity is proportional to the total pressure (radiation + gas pressure). It may still be consistent with the density of magnetic discs in which the viscosity is proportional to just the gas pressure (see *e.g.*

[2]In a pictorial sense, the hard X-ray emission regions may act like lightning bolts do in relation to the thunder cloud, which is the accretion disc in this analogy. They are usually much smaller than the cloud and they may not always occupy the same place. Our observations, especially spectra, of AGN are then temporal and spatial averages of a complex process. Where there are many 'lightning bolts', we may see little overall variability.

Sakamoto & Coroniti 1981). The inner region of a standard disc is usually found to be unstable and so it does not appear unreasonable to us that much of the disc gas is in dense clumps. The hard X-ray emission may then sometimes originate in the spaces between the clumps.

A clumpy disc is attractive for several reasons; the disc is expected to be inhomogeneous if unstable, or if much of the viscosity is magnetic in origin (the Parker instability would then cause clumping in the disc); the chaotic X-ray light curves of many sources emphasise that we are not dealing with a smooth accretion flow; differences in the relative scale heights of the hard X-ray emitting intrablob medium and the cold dense clouds can explain some of the differences in the observed equivalent widths (Day *et al.* 1990); and an intimate mixture of the intrablob medium and the cold clouds can explain the small lag between the continuum and iron line changes in NGC 6814. Unfortunately for simple studies of the inner disc and measurements of the black hole mass (Fabian *et al.* 1989; Stella 1990), the velocities of the cold gas may then be chaotic, especially if the hard X-rays are produced by explosive releases of energy close to the disc, and not simply rotational. If the hard X-rays appear to originate from the centres of 'craters' in the disc, then from special relativistic considerations we expect to see more of the blueshifted wing than of the redshifted side.

6. SOME FURTHER CONSEQUENCES

We must understand the shape of the reflection component before we can unravel the true 'primary' spectrum of the hard X-ray source and the process(es) that give rise to it. Further studies and modelling of the spectra of variable sources are particularly important in this respect.

One first result from decomposing the X-ray spectrum into reflection and 'primary' components is the realization that the 'primary' AGN spectrum has an energy index closer to $\alpha \sim 0.9$ than to the observed value of 0.7. This is then in agreement with the $e^{\pm}$ models, which are thereby revived and appear to be give a viable explanation for the hard X-ray spectra in X-ray compact AGN (Zdziarski *et al.* 1990).

Sources at redshifts of 1 - 2 dominated by the reflection hump may explain the X-ray Background (Fabian *et al.* 1990). The high energy turnover in the Compton hump redshifts to give the 40 keV break in the Background spectrum.

3C 273 has an iron line of about 50 eV equivalent width (Turner *et al.* 1989), which argues (Ghisellini, private communication) that X-ray beaming does not completely dominate the X-ray spectrum.

The agreement with the data of the simple model, in which the hard X-rays originate outside, or on the surface, of the cold gas, is a strong argument for the simple disc geometry. There cannot be any substantial ($\tau_T \gtrsim 1$) highly ionized corona on the disc, or the equivalent width of the cold fluorescent line feature would be much reduced. Of course, there may be much partial ionization, so that some of the oxygen, for example, is ionized at the surface. This would increase the strength of the reflected continuum below the iron absorption edge and may be required by some of the data. Further modelling is required here.

REFERENCES

Arnaud, K.A., *et al.*, 1985. *Mon. Not. R. astr. Soc.*, **217**, 105.
Barr, P., White, N.E. & Page, C.G., 1985. *Mon. Not. R. astr. Soc.*, **216**, 65P.
Begelman, M. & DeKool, M., 1990. Preprint
Day, C.S.R., Fabian, A.C., George, I.M. & Kunieda, H., 1990. *Mon. Not. R. astr. Soc.*, **247**, 15P.
Done, C. *et al.*, 1990. *Mon. Not. R. astr. Soc.*, **243**, 713.
Fabian, A.C., 1977. *Nature*, **269**, 672.
Fabian, A.C., Rees, M.J., Stella, L. & White, N.E., 1989. *Mon. Not. R. astr. Soc.*, **238**, 729.
Fabian. A.C., George, I.M., Miyoshi, S. & Rees, M.J., 1990. *Mon. Not. R. astr. Soc.*, **242**, 14P.
Ferland, G.J. & Rees, M.J., 1988. *Astrophys. J.*, **332**, 141.
Ghisellini, G., George, I.M., Fabian, A.C. & Done, C., 1990. *Mon. Not. R. astr. Soc*, in press.
George, I.M. & Fabian, A.C., 1990. *Mon. Not. R. astr. Soc.*, in press.
George, I.M., Nandra, K., & Fabian, A.C., 1990. *Mon. Not. R. astr. Soc.*, **242**, 28P.
George, I.M., Fabian, A.C. & Ross, R.R., 1990. This Proceedings.
Guilbert, P.W. & Rees, M.J., 1988. *Mon. Not. R. astr. Soc.*, **233**, 475.
Hatchett, S., Buff, J. & McCray, R., 1976. *Astrophys. J.*, **206**, 847.
Kaastra, J.S. & Barr, P., 1990. *Astr. Astrophys.*, **226**, 59.
Kitamoto, S., *et al.*, 1990. *Pasj*, **42**, 85.
Krolik, J.H. & Kallman, T.R., 1987. *Astrophys. J.*, **320**, L5.
Kunieda, H., 1990. These proceedings.
Kunieda, H., *et al.*, 1990. *Nature*, **345**, 786.
Laor, A., 1990. These Proceedings.
Lawrence, A. *et al.*, 1985. *Mon. Not. R. astr. Soc.*, **217**, 685.
Lightman, A.P., Giacconi, R. & Tananbaum, H., 1978. *Astrophys. J.*, **224**, 375.
Lightman, A.P. & White, T.R., 1988. *Astrophys. J.*, **335**, 57.
Malkan, M.A., 1983. *Astrophys. J.*, **268**, 582.
Matsuoka, M., Yamauchi, M., Piro, L. & Murakami, T., 1990. *Astrophys J.*, **361**, 440.
Matsuoka, M., Yamauchi, M., Piro, L. & Murakami, T., 1989. Preprint.
Miyoshi, S. *et al.*, 1988. *Publ. Astron. Soc. Japan*, **40**, 127.
Mushotzky, R.F., 1982. *Astrophys. J.*, **256**, 92.
Mushotzky, R.F. *et al.*, 1978. *Astrophys. J.*, **220**, 790.
Nandra, K., 1990. These Proceedings.
Nandra, K. *et al.*, 1989. *Mon. Not. R. astr. Soc.*, **236**, 39P.
Ohashi, T. 1988 In 'Physics of Neutron Stars and Black Holes', ed Y. Tanaka. U. A. P., 301.
Piro, L., Yamauchi, M. & Matsuoka, M., 1990. *Astrophys. J.*, **360**, L35.
Pounds, K.A. *et al.*, 1989. *Mon. Not. R. astr. Soc.*, **240**, 769.
Pounds, K.A. *et al.*, 1990. *Nature*, **344**, 132.
Rothschild, R.E. *et al.*, 1983. *Astrophys. J.*, **233**, L47.
Sakamoto, P.J. & Coroniti, F.V., 1981. *Astrophys. J.*, **247**. 19.
Singh, K.P *et al.* 1990. Preprint.
Stella, L., 1990. *Nature*, **344**, 747.
Sunyaev R.A. & Trumper, J., 1979. *Nature*, **279**, 506.
Tanaka, Y., 1989. In ESA SP-296, 3.
Tanaka, Y., 1990. These Proceedings.
Turner, M.J.L. *et al.*, 1990. *Mon. Not. R. astr. Soc.*, **244**, 310.
Turner, T.J. & Pounds, K.A., 1989. *Mon. Not. R. astr. Soc.*, **240**, 833.
Warwick, R.S., Yaqoob, T. & Pounds, K.A., 1989. *Publ. Astron. Soc. Japan*, **41**, 721.
White, N.E., Fabian, A.C. & Mushotzky, R.F., 1982. *Astr. Astrophys.*, **133**, L9.
Zdziarski, A.A. *et al.*, 1990. *Astrophys. J.*, **363**, L1 .

IRON LINE EMISSION AND THE DISTRIBUTION OF COLD MATTER IN SEYFERT GALAXIES

K. Nandra, K.A. Pounds & G.C. Stewart

Department of Physics and Astronomy, University of Leicester,
University Road, Leicester LE1 7RH, United Kingdom.

1. INTRODUCTION

The recent discovery of iron lines in several unobscured Seyfert galaxies (*e.g.* Nandra *et al.*, 1989; Pounds *et al.*, 1989; Matsuoka *et al.*, 1990) has prompted an extensive study with *Ginga*. In most cases, the cold line of sight N_H is insufficient to explain the large equivalent widths of the lines (typically $\sim$ 150 eV). Evidence for a spectral flattening at high energies has also been presented in some cases (*e.g.* Pounds *et al.*, 1990; Matsuoka *et al.*, 1990). X–ray reflection from cold material is the most common explanation for these spectral features, whereby the 'hard tail' arises from continuum Compton reflection (Lightman & White, 1988) and fluorescence in cold iron accounts for the emission line. In at least two cases, NGC 6814 (Kunieda *et al.*, 1990) and NGC 5548 (Nandra *et al.*, 1990), variability arguments constrain the reflector to lie close to the X–ray source. The detailed geometry is unclear but an accretion disc (*e.g.* Fabian *et al.*, 1989) or scattering mist of dense clouds (Guilbert & Rees, 1988) have been considered most plausible. Here we present the results of an analysis of 29 *Ginga* Seyfert spectra. In particular we address the question of how common the features reported for individual objects are, in the context of a large sample.

2. THE GINGA SAMPLE

The results presented here consist of an independent analysis of 29 observations of 20 objects. The sample includes a range of Seyfert classifications; however, the majority of the sources included are Seyfert I galaxies and are generally bright (> 1 mCrab), hard X–ray selected sources. Most are low luminosity AGN ($L_x(2-10\text{keV}) < 10^{45}$ erg s^{-1}) and are also relatively unobscured. For a review of the properties of Seyfert galaxies with high intrinsic cold N_H see Warwick (1990). The sample spectra were analyzed using the standard techniques, with background subtraction performed according to the prescription of Hayashida *et al.* (1989). Data selection criteria were applied uniformly.

3. SPECTRAL INDEX

Initially, simple power law plus absorption fits were tried in the 2–18 keV range. Such fits were generally very poor, with an excess at $\sim$ 6.5 keV dominating the residuals, as for the individual sources noted above. Fig. 1 shows the values of the the reduced χ^2 for this model plotted against those obtained with a thermal bremsstrahlung model. The solid line, indicating a 1:1 relationship shows that the power law models generally provide better fits than the thermal model, as found previously. The dotted line shows the value of χ^2_ν at which the model can be rejected at $\sim$ 90 per cent confidence. Only 13 such fits are statistically acceptable at this level suggesting a more complex spectral form. However, we use this model as a simple parameterization of the spectrum in order to allow comparison with previous spectral surveys and to test the validity of the 'canonical' photon index over the full *Ginga* energy range, in the light of recent evidence for a hard tail.

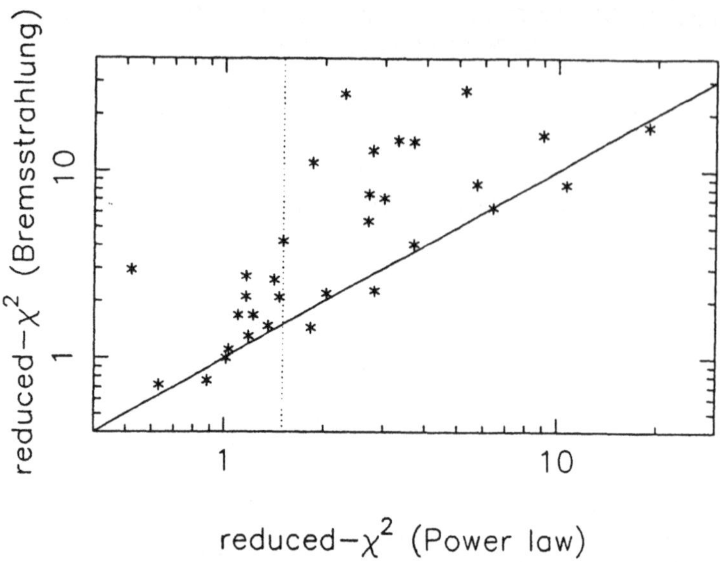

Figure 1: Goodness of fit, as measured by the reduced χ^2 for power law plus absorption and bremsstrahlung plus absorption fits. The solid line represents a 1:1 relationship. The vertical dotted line indicates the level at which the power law fits are acceptable at $\sim$ 90 per cent confidence.

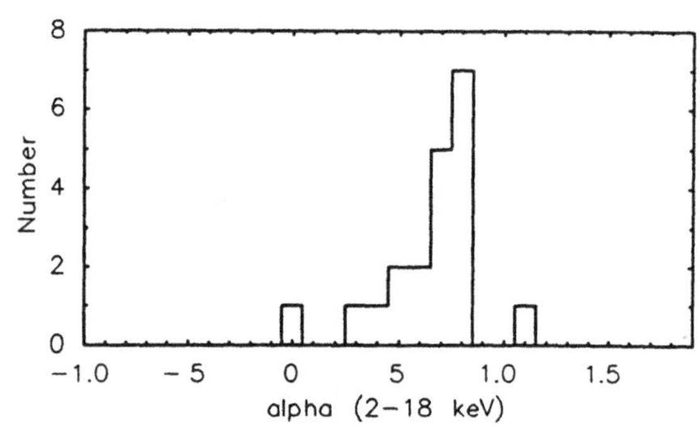

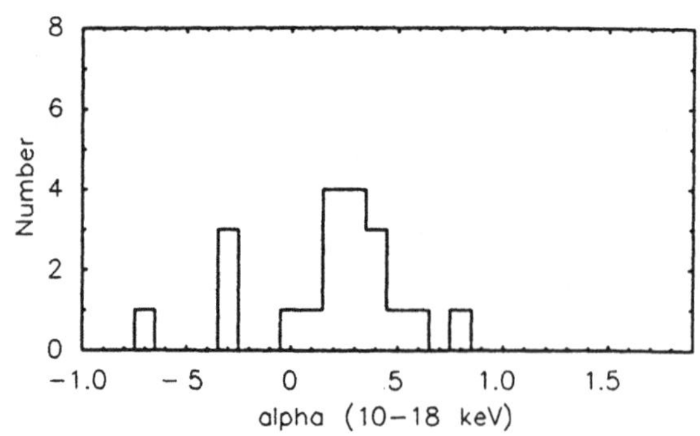

Figure 2: Histograms of the energy indices obtained in a) the 2–18 keV and b) the 10–18 keV energy ranges. The spectral slope is flatter above $\sim$ 10 keV.

2–18 keV index

A histogram of the spectral indices, obtained in the 2–18 keV range, of the *Ginga* sample is shown in Fig. 2(a). The most common value is apparently 0.7, but with individual departures from this. The weighted mean index, $\alpha_{2-18} = 0.722 \pm 0.004$, is consistent with both the *EXOSAT* (Turner & Pounds, 1989)and *HEAO-1* (Mushotzky, 1984) medium energy (2–10 keV) spectral surveys. The standard deviation (which measures both the statistical spread *and* any intrinsic deviation) is 0.25.

b. 10–18 keV: The 'hard tail'

Fig. 2(b) shows the spectral indices derived from a higher energy band, 10–18 keV. Absorption, negligible in this energy range, was excluded from these fits, which gave a flatter slope, on average $\alpha_{10-18} = 0.37 \pm 0.03$ ($\sigma = 0.38$). A KS-test implies that the two distributions are significantly different at > 99 per cent confidence. In addition, a χ^2 test comparing the two distribution also shows a significant difference at > 98 per cent level. The flattening at high energies indicates an additional component to the spectrum, possibly from Compton scattering in cold material.

4. IRON LINE EMISSION

As most of the above fits gave unacceptable χ^2_ν a more complex model must be sought. The most prominent deviation from this simple model appears as an excess close to 6.5 keV. Fits were therefore conducted including a narrow (0.1 keV) gaussian emission, as in the earlier analyses. Inclusion of such a line gave substantially improved fits to the data, although only 19/29 are formally acceptable. The F–test was used to quantify the validity of introducing the new parameter into the spectrum. The derived F–statistics imply 26 detections at > 90 per cent confidence (most at > 99 per cent) in 29 separate spectra. Lines were detected for 18 out of the 20 sources implying that iron K–emission is a very common property of Seyferts.

Line Energy

The average (redshift–corrected) line energy is 6.43 ± 0.07 keV, close to the value for cold iron, with standard deviation $\sigma = 0.38$. Fig. 3 shows a plot of the line energy against source luminosity. A test against a constant line energy hypothesis gives $\chi^2_r = 0.6$, indicating that *all* of the emission lines are consistent with the mean energy. Such a line energy is consistent with fluorescence from cold iron. Thermal line emission from hot plasma or fluorescence in highly ionized iron would be expected to occur at a higher energy, 6.7–6.9 keV and is excluded as a unique value by these data ($\chi^2_r = 2.7$ for a 6.7 keV line). No significant red–shift or blue–shift, (other than the cosmological one) of the mean line energy is observed which has important implications for modelling of the line emission. Processes close to the black hole (*e.g.* gravitation, Doppler motions) would be expected to have a substantial effect on the mean energy, as well as the profile, of the emission line.

Equivalent width

The average narrow line equivalent width is 140 ± 10 eV, with a standard deviation, $\sigma = 80$ eV. The histogram of values from our sample is shown in Fig. 4. This average equivalent width implies columns of $> 10^{23}$ cm^{-2} for spherical covering, whereas most of the sources are relatively unobscured (see Section 6 for a full discussion of this point).

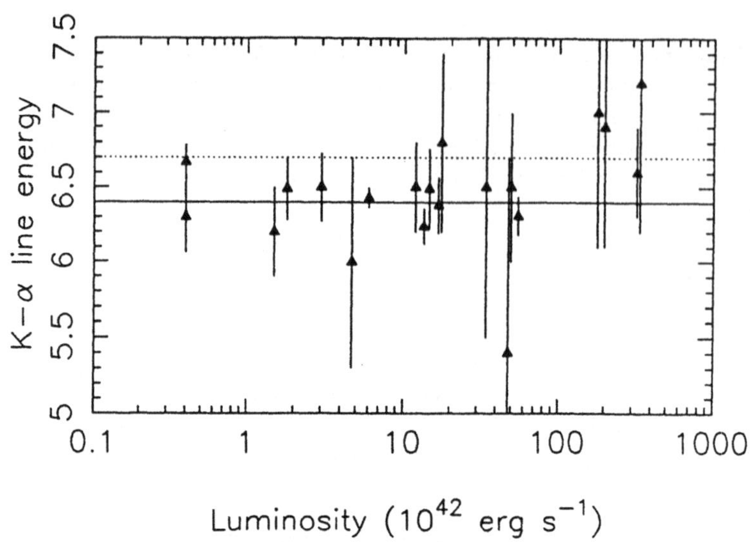

Figure 3: K-α emission line energy versus luminosity. The mean value 6.43 ± 0.07 (solid line) is consistent with fluorescence from iron in a low state of ionization. Thermal emission at ~ 6.7 keV (dotted line) is generally excluded.

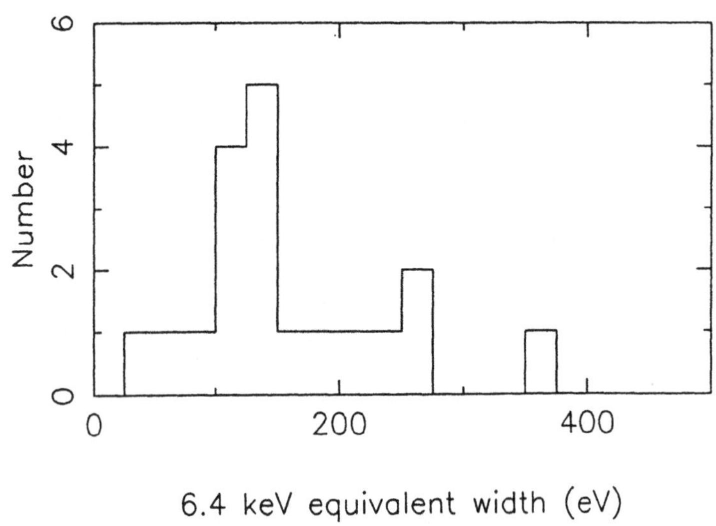

Figure 4: Histogram of 6.4 keV line equivalent widths. The mean is ~ 140 eV.

Line broadening

6 out of the 29 spectra show a significant improvement (at > 95 per cent level) when the line width is left as a free parameter. Typical widths are of order $\sigma \sim 0.7$ keV (corresponding to Doppler broadening at $\sim 10R_s$ if Keplerian motions apply). None of the observations exclude a line width of this order, even in the best determined cases, due to the finite resolution of the LAC detector. The broadening could arise from Doppler motions, gravity, electron scattering or ionization blends. However these may also shift the peak energy and such effects must be reconciled with the mean energy described in Section 4. Note also that the equivalent widths implied in these broad–line fits are systematically larger than when a narrow line is assumed.

5. ABSORPTION FEATURES

Absorption edges, sometimes at a higher energy than the 7.1 keV expected for neutral iron, have been reported in some cases (Warwick, 1990; Nandra *et al.*, 1990a; Pounds *et al.*, 1990). To test for the presence of such a feature, power law plus absorption edge fits were carried out for the full sample. These gave a significant improvement over the simple power law in 20 out of the 29 spectra with a mean edge energy $\sim$ 8 keV. 15 such fits are statistically acceptable. However this may merely represent an artefact if the line is not also included in the model. With a narrow emission line (fixed at 6.4 keV) included 15/29 spectra give a significant improvement over the power law plus line model when the edge is introduced. 12/20 objects show evidence for an absorption edge. Typically column densities $N_{HFe} > 10^{23}$ cm^{-2} (equivalent Hydrogen density) are detectable. Note that these values are much larger than those obtainable via X–ray reflection (George & Fabian, 1990) and favour a substantial line–of–sight column. The average energy is 8.0 ± 0.3 keV implying highly ionized material. A histogram of the edge energies, derived from these fits, is shown in Fig. 5. The mean ionization state implies an ionization state $\sim$ Fe XX and temperature $\sim 10^7$ K if collisionally ionized. A large line–of–sight column of warm material therefore appears to be another common property of Seyfert spectra.

6. SIMPLE MODELS

Monte Carlo calculations have been carried out to determine the line equivalent width and continuum for three different geometries. Here we consider which, if any, of these simple geometries can account for the observed features.

Spherical cloud

Two cases were considered

1. Cold material. All elements were considered neutral and with solar abundances (Morrison & McCammon, 1983).
2. Highly ionized gas. An approximation to highly ionized gas was used in which only the Fe K electrons remain bound.

In Fig. 6(a), the observed equivalent widths are plotted against the covering column densities implied by the soft X–ray absorption. The solid line is the Monte Carlo prediction. A spherical cloud of cold gas with N_H as implied by this low–energy cut–off is clearly insufficient to explain the large equivalent width. However, if the gas is highly ionized it would be largely transparent at low X–ray energies. In this case we expect an imprint at the iron K–edge, for which there is good evidence (Section 5). The iron K–edge column density is plotted against equivalent width in Fig. 6(b), with the model prediction shown as the solid line. Although the warm column

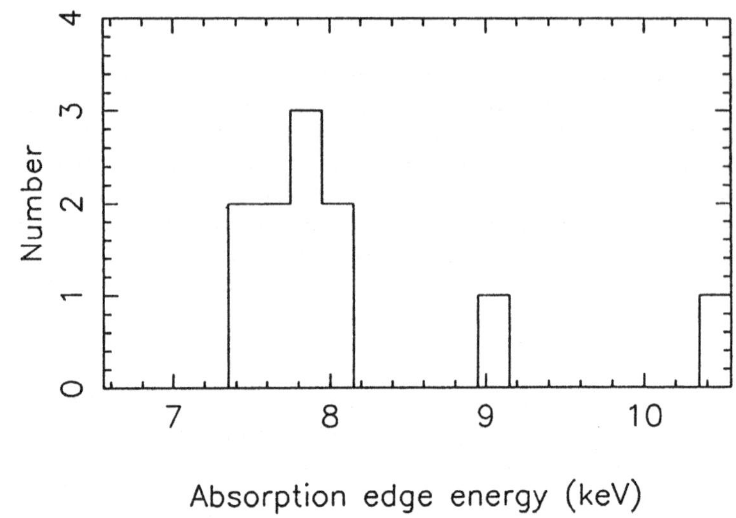

Figure 5: Histogram of the derived edge energies. Absorption by highly ionized iron is indicated as another common feature.

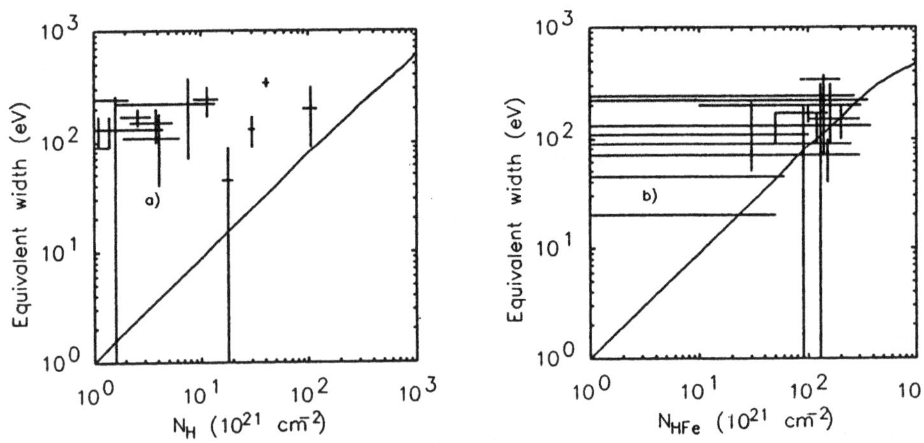

Figure 6: Observed equivalent width versus column density. a) shows the N_H values derived from the low energy cut–off. b) shows the values of N_{HFe} from the iron edge measurements. The Monte carlo prediction (solid line) falls short of the observed equivalent width in both cases, even for 4π coverage.

is still insufficient to provide the whole of the observed line flux, its contribution may well be significant. In addition, the hard tail is not accounted for by either of these 'transmission' models.

Reflection models

Models which produce features via fluorescence and Compton scattering in cold, optically thick material out of the line-of-sight have been applied previously to several individual sources. In most cases the equivalent width suggests that the material subtends a large solid angle at the X-ray source. Here we consider two possible geometries for the reflector:

1. A reflective slab

   Described in detail by George & Fabian (1990). A cold, non-rotating slab is illuminated from above with a power-law incident spectrum. The line equivalent width depends on the inclination of the normal to the slab to the line-of-sight. The maximum value is ~ 150 eV, depending on the inclination. However, the shape of the reflected spectrum is largely independent of inclination and therefore fixed, with the relative normalization of the reflected component allowed to vary.

2. An accretion disc

   As for the slab but the spectrum (and most particularly, the line profile) is modified by Doppler motions and gravitational redshift (in the weak field approximation - Chen, Halpern & Filippenko, 1989; but see also Fabian *et al.*, 1989; Stella, 1990; Laor, 1990). The disc is assumed to extend from $3R_s$ to $1000R_s$ and to remain cold throughout. Here the line profile also depends strongly on the inclination of the disc. Maximum E.W. 150 eV for a face-on disc. Inclinations were fixed at 30° and 60°.

Evidence for the reflected continuum

A fit with a narrow Gaussian line including a reflected continuum component, derived from the 'slab' model described above, as an additional free parameter, was then tried. The inclusion of the reflected component improved the fit significantly in 23/29 spectra. Indeed the Compton reflection component is an excellent fit to the 'hard tail' described previously and 24/29 fits are acceptable with this model. The normalization ratio is compatible with a face-on slab subtending 2π at the X-ray source. This is also implied by the average equivalent width.
If the bulk of the line arises from fluorescence in optically thick material we would expect an anti-correlation between α for a simple power law fit and the observed equivalent width, as for larger equivalent widths the accompanying (flat) reflected continuum is greater. Fig. 7 shows the relationship between α and the equivalent width. A Spearman rank correlation indicates a significant trend at > 98 per cent confidence. The data are compatible with the theoretical prediction of the slab reflection model, shown by the solid lines in Fig. 7.

Results of reflection fitting

- Slab model

  These fits provide a significant improvement to the power law plus line model. Acceptable fits are obtained for 23 spectra. The parameters of the fits indicate that the geometry is compatible, on average, with a face-on slab.

- i=30 disc

  23 fits acceptable. These provide generally worse fits than the slab model, as the line is significantly gravitationally redshifted (to ~ 6.0 keV). The mean reflected/direct normalization ratio is ~ 3.0 (i.e. the reflected component is enhanced by a factor 3 compared to a disc inclined at 30°).

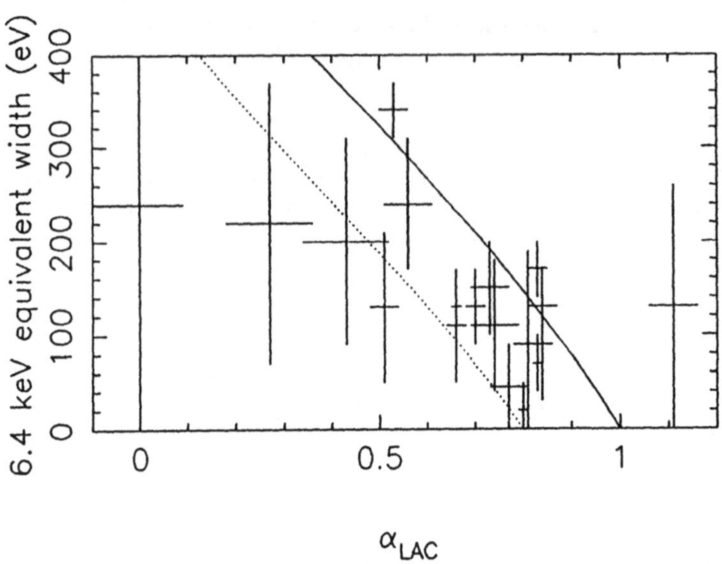

Figure 7: 6.4 keV equivalent width versus energy index. An anti-correlation is indicated at $\sim$ 98 per cent confidence. The curves represent the prediction for the slab model with *intrinsic* indices $\alpha = 1.0$ (solid line) and $\alpha = 0.8$ (dotted line).

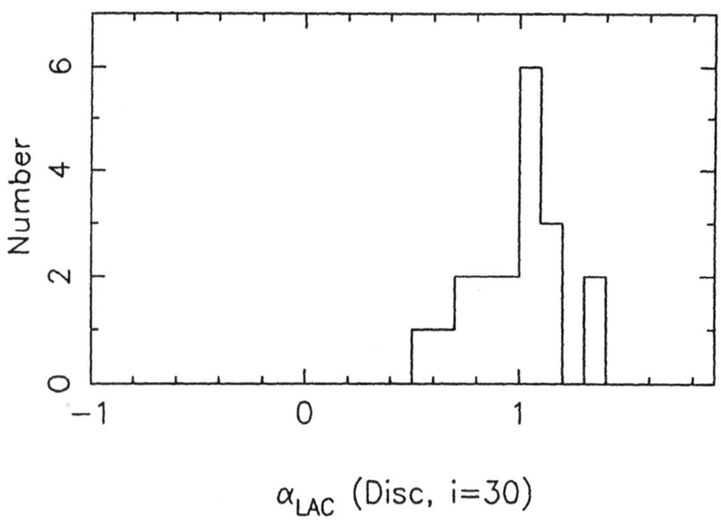

Figure 8: Histogram of spectral indices derived from accretion disc reflection fits, with i=30. The mean index $\alpha \sim 1.0$ is significantly steeper than the 'canonical' $\alpha = 0.7$.

- i=60 disc

  24 fits acceptable. χ^2 statistic generally lower than for the i=30 disc, as the Doppler broadening implies a mean energy closer to 6.4 keV. However a large normalization ratio is implied for these fits, ~ 8.0.

The reflection fits imply an *intrinsic* spectral index significantly steeper than the previous 'canonical' value of $\alpha = 0.7$. In Fig. 8 we show a histogram of the intrinsic indices derived from the i=30 accretion disc reflection fits. The mean index is $\alpha = 1.0$.

7. EMISSION LINE ORIGIN

Although the reflection models have reasonable success in explaining the observed features, certain problems remain. Firstly, the equivalent widths (and also the strength of the reflected continuum) are too large to be explained by simple reflection from an accretion disc subtending 2π at the X-ray source, unless sources with face-on accretion discs are selected preferentially. The emission line energy is $\sim$ 6.4 keV implying cold material but *no* significant gravitational redshift. However, the observed variability in NGC 6814 and NGC 5548 suggest that the line emission arises close to the black hole. The fact that some observations show, and the others are compatible with, a broad line also suggests this. We note here that the warm line-of-sight material may contribute an additional component to the emission line. If so, we expect the energy of the line component from the 'warm absorber' to occur at $\sim$ 6.5 keV, which could account for the difference in the observed mean line energy and that implied by reflection from a face-on disc.

8. CONCLUSIONS

- Iron line emission is a common property of low-luminosity Seyferts.
- The mean line equivalent width is $\sim$ 140 eV at an energy of 6.4 keV, with width $\sigma \sim$ 0.7 keV.
- A spectral flattening is observed above $\sim$ 10 keV.
- Iron K-shell absorption edges are detected in many sources and may be another common feature. Column densities are typically $> 10^{23}$ cm^{-2} at an energy $\sim$ 8 keV, implying substantial highly ionized material in the line-of-sight.

Implications

Continuum:

- The hard tail implies Compton reflection from cold material.
- The *intrinsic power* law index is therefore $\alpha \sim 0.9 - 1.0$.
- $\alpha \sim 0.9$ revives some pair models of the emission mechanism (Zdziarski *et al.*, 1990).
- Soft excesses need to be re-examined, as the implied luminosity and temperature will be reduced in the light of the steeper underlying power law.

- The *Einstein* IPC found $\alpha = 1.0$ in 0.5–3.5 keV band for similar radio–quiet AGN (Wilkes & Elvis, 1987). The reflection models resolve the discrepancy between the IPC indices and those found by *HEAO-1* and *EXOSAT* in the 2–10 keV band, as reflection is negligible in the softer energy range.

Spectral Features:

- The bulk of the line emission arises via fluorescence in optically thick material, close to the nucleus. The mean line energy and equivalent width and the strength of the reflected continuum suggest a geometry more complex than the simple discs considered here, unless face–on accretion discs are preferentially selected. Plausible configurations include a flared disc, optically thick blobs (Guilbert & Rees, 1988), or a 'clumpy' accretion disc (Fabian, 1990).
- A large column density ($> 10^{23}$ cm^{-2}) of highly ionized gas lies in the line of sight and may also contribute to the line emission, resolving the discrepancy between the requirement for the material to lie close to the black hole, and the line energy.

Acknowledgements
The authors thank Ian George and Andy Fabian at the IoA, Cambridge for providing the disc reflection models and for many helpful discussions.

REFERENCES

Chen, K., Halpern, J.P & Filippenko, A.V., 1989. *Astrophys. J.*, **339**, 742.
Fabian, A.C., 1990. This volume.
Fabian, A.C. *et al.*, 1989. *Mon. Not. R. astr. Soc.*, **238**, 729.
George, I.M., Nandra, K. & Fabian, A.C., 1990. *Mon. Not. R. astr. Soc.*, **242**, 28P.
George, I.M. & Fabian, A.C., 1990. *Mon. Not. R. astr. Soc.*, in press.
Guilbert, P.W. & Rees, M.J., 1988. *Mon. Not. R. astr. Soc.*, **233**, 475.
Hayashida, K. *et al.*, 1989. *Publ. astron. Soc. Japan*, **41**, 373.
Kunieda, H. *et al.*, 1990. Nature, **345**, 786.
Laor, A., 1990. This volume.
Lightman, A.P. & White, T.R., 1988. *Astrophys. J.*, **335**, 57.
Matsuoka, M. *et al.*, 1990. *Astrophys. J.*, **361**, 440.
Morrison, R. & McCammon, D., 1983. *Astrophys. J.*, **270**, 119.
Mushotzky, R.F., 1984. *Adv. Space. Res*, **3**, 157.
Nandra, K. *et al.*, 1989. *Mon. Not. R. astr. Soc.*, **236**, 39P.
Nandra, K., Pounds, K.A. & Stewart, G.C., 1990a. *Mon. Not. R. astr. Soc.*, **242**, 660.
Nandra, K. *et al.*, 1990b. *Mon. Not. R. astr. Soc.*, in press.
Pounds, K.A. *et al.*, 1989. *Mon. Not. R. astr. Soc*, **240**, 769.
Pounds, K.A. *et al.*, 1990. *Nature*, **344**, 132.
Stella, L., 1990. *Nature*, **344**, 747.
Turner, T.J. & Pounds, K.A., 1989. *Mon. Not. R. astr. Soc.*, **240**,
Warwick, R.S., 1990. This volume.
Wilkes, B.J. & Elvis, M., 1987. *Astrophys. J.*, **323**, 243.
Zdziarski, A.A. *et al.*, 1990. *Astrophys. J.*, **363**, L1.

SPECTRAL FEATURES IN AGN: PROBLEMS AND QUESTIONS

L. Piro

Istituto di Astrofisica Spaziale, C.N.R., Frascati, Italy

Abstract

GINGA observations of several bright Seyfert 1 galaxies have revealed the existence of a new hard X-ray component above 8 keV, the "high energy bump". This component is probably produced by reprocessing of the direct continuum (through absorption and Compton scattering) on to very thick ($> 10^{24} cm^{-2}$) cold matter surrounding the central source, possibly in an accretion disk. Detection of iron fluorescent lines with equivalent width of 100-200 eV and energy of 6.4 keV in all these objects supports this interpretation. Some relevant predictions of a detailed accretion disk model (iron line parameters, high energy bump level) are compared with GINGA results, including the detection of iron lines with equivalent width $\simeq$ 300-400 eV observed in two other Seyfert galaxies. A partial covering and reprocessing by thick clouds offers also a good explanation to the phenomenology observed by GINGA, and is supported by the latest results on the Broad Line Region obtained by line reverberation measurements in the optical - ultraviolet bands.

Introduction

Before the launch of the GINGA satellite, the X-ray spectrum of Seyfert galaxies in the energy range 1-30 keV was described by a power law with a spectral photon index of 1.7 (Mushotzky 1984, Turner and Pounds 1989). Low energy absorption by a medium with column density $N_H \simeq 10^{22-23} cm^{-2}$ and solar-like abundances (or nearly so) was usually observed in low luminosity objects ($L(2-10keV) < 10^{43.5}$ $erg\ s^{-1}$, Reichert, Mushotzky and Holt 1986). The presence of such a medium , possibly distributed with some degree of anisotropy around the X-ray source, also accounted for the fluorescent iron line at 6.4 keV observed in a few objects (e.g. Matsuoka et al. 1986, Perola et al. 1986). The characteristics of this medium (size$\simeq$ 0.1 pc, column density) are in good agreement with those of the Broad Line Region (Reichert, Mushotzky and Holt 1986).

Over the last years an increasing evidence for the existence of matter much closer to the central powerhouse, possibly in the form of an accretion disk, has gathered, with the discovery of the O-UV "big bump" (Malkan and Sargent 1982) and the soft X-ray excess (e.g Arnaud et al. 1985, Piro et al. 1988, Turner and Pounds 1989). Furthermore, recent O-UV monitoring of Seyfert galaxies (Clavel et al. 1990, 1991), devoted to measure the lag in the line response to changes of the photoionizing continuum, indicate that there is a significant amount of gas very near the central source (a few light days), much closer than previously predicted by the standard BLR model. In order to produce the observed line intensities, this gas should have a very large column density, $N_H \simeq 10^{24} cm^{-2}$ (Ferland and Persson 1989).

Some authors (Guilbert and Rees 1989, Lightman and White 1989) have pointed out that a cold thick medium near the central engine should produce characteristic X-ray spectral features, by reprocessing the direct X-ray continuum, through Compton scattering, photoelectric absorption and fluorescence. I shall show as the recent findings by GINGA of an high energy bump accompanied by iron line emission even in objects with small low energy absorption (Matsuoka et al. 1990, herafter MPYM; Piro, Yamauchi and Matsuoka 1990, herafter PYM; Pounds et al. 1990), can be actually framed in a scenario based on partial covering and/or reprocessing by a thick cold medium in the environment of the central source.

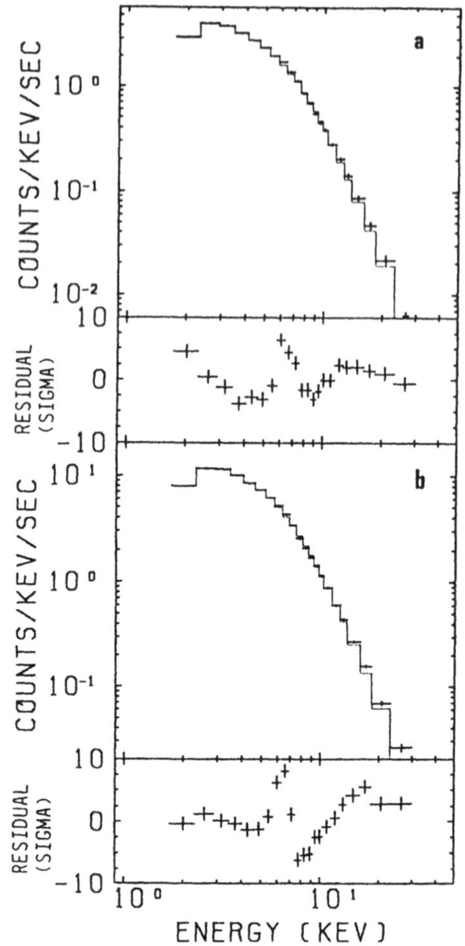

Fig.1a,b *GINGA spectra of NGC 7469 (a) and IC4329A (b) fitted with a simple power law.*

The case of reprocessing from an accretion disk (see also Fabian, this conference) is the most attractive and has been studied in more detail (Matt, Perola and Piro, this conference and 1991, hereafter MPP) yielding a set of observable quantities (e.g. iron line parameters, bump intensity) that can be used to constrain the geometry and the iron abundance, but whose full exploitation should wait for future X-ray detectors with a better energy resolution at 6 keV and/or a larger band and area at higher energies than the proportional counters (LAC) aboard GINGA. Nevertheless, GINGA data already provide significant results and, in particular, show the presence of deviations from model predictions.

GINGA observations of the high energy bump and iron line

A spectral flattening above 8 keV, produced by a new hard component (the high energy bump), along with an iron line at 6.4 keV with an equivalent width $W_\alpha \simeq 100 - 200\ eV$ has been found by GINGA in several bright Seyfert galaxies (MPYM, PYM, Pounds et al. 1990). The deviations from the "standard" model (simple power law with a uniform absorber) produced by these spectral features in two bright Seyfert galaxies are shown in fig.1 (PYM).

The hard component can be detected by the LAC only in relatively bright objects, with a good S/N ratio at high energies. In a small sample of bright Seyfert galaxies studied in collaboration with the japanese colleagues of the RIKEN institute, six galaxies fulfill this requirement. Five objects show the presence of this component (MCG-6-30-15: MPYM; NGC 7469, IC4329A: PYM; NGC 4051, NGC 4593: Yamauchi, Piro and Matsuoka 1990), and one does not (NGC4151: Yamauchi, Piro and Matsuoka 1990; see also Warwick et al. 1989). Those five objects show also an iron line with $W_\alpha \simeq 100 - 200\ eV$, and a small low energy absorption ($N_H < 10^{22} cm^{-2}$). This result indicates that the hard flat component and the iron line are common features in Seyfert galaxies. A similar conclusion has been derived by Pounds et al. by using a composite spectrum, obtained by summing the spectra of 8 Seyfert galaxies to increase the statistics.

Regarding the origin of the hard component, there are some points to be mentioned, that tend to exclude an intrinsic origin. Models where the spectrum is described by a complex intrinsic emission, such as two power laws, or a thermal bremsstrahlung and a power law give a chi square which is worse than that obtained by the partial covering or reflection models described in the following. Furthermore, the power law component that, in the spectral fitting, reproduces the hard spectral flattening, when extrapolated at 50 - 100 keV, would predict a flux much higher than observed (Rotschild et al. 1983). On the other hand, the presence of an iron line with an equivalent width much higher than that expected by the absorbing medium in the line of sight, suggests the presence of an additional cold (because of the line energy at 6.4 keV) inhomogeneous medium around the central source. Such a medium can, in fact, produce both the iron line and the hard component, as shown in the following.

X-ray spectral signatures of cold thick matter

The first model that was employed assumes that an absorbing medium (with solar abundances) is covering only a fraction f_{cov} of the central source with a column density $N_{H cov}$. A good fit is obtained with values $f_{cov} \simeq 0.3 - 0.6$ and very high values of column density, $N_{H cov} \simeq 2 - 6\ 10^{24} cm^{-2}$ (MPYM, PYM). The covered component turns up above 8 keV, producing the observed flattening (fig.2 a,b). Such a medium, if distributed with a spherical geometry, would produce by fluorescence an iron line with $W\alpha \simeq 50 - 100\ eV$. To explain the observed values of 100-200 eV some degree of anisotropy or an iron overabundance (by a factor of 2) would be required.

For such values of the column density the effects of Compton scattering become important. In this case the medium can "reflect" (reprocess) a fraction the central emission into our direction (Guilbert and Rees 1989, Lightman and White 1989). For example, in the scenario suggested by the recent results on the innermost BLR clouds, the clouds in the line of sight will "partially cover" the central source, whereas the clouds around it will "reflect" (reprocess) the central continuum. Since, as shown in the following, the spectral shape of the covered component in the GINGA band is roughly similar to that of the reflected one, it would be rather difficult to disentangle the two spectral components and thus derive information on the distribution of the medium around the central source.

However, if there were no matter along the line of sight or the column density were greater than a few times $10^{25} cm^{-2}$ (then the cut-off of the covered component would fall outside the LAC band), only the reflected component would remain present. I shall discuss in more detail this case.

In fig.3 I show the albedo of a thick ($N_H > 10^{25} cm^{-2}$) slab of cold gas of solar-like abundances, illuminated by a power law spectrum (Lightman and White 1989). Note that the iron line is not shown because the authors did not include it in their formulae, but it is expected to be produced with an equivalent width (relative to the reflected continuum) of about 1 keV (e.g. MPP). The albedo has a broad maximum of 0.7 between 20 and 60 keV. At lower energies it decreases for the effect of photoelectric absorption. At higher energies the reduction is produced by the decreasing of the scattering cross section and the effect of Compton downscattering.

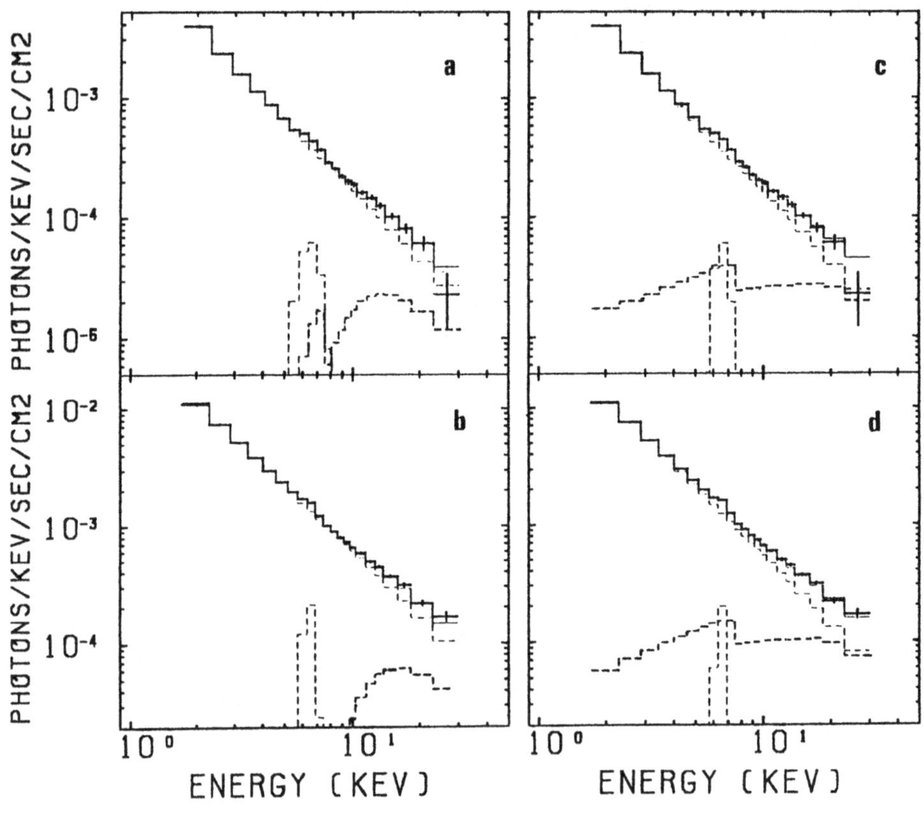

Fig.2 *Photon spectra of NGC 7469 (a,c) and IC4329A (b,d) fitted with the partial covering model with iron line (a,b) and the reflection model with iron line (c,d). The thicker dashed lines represent the covered or the reflected components.*

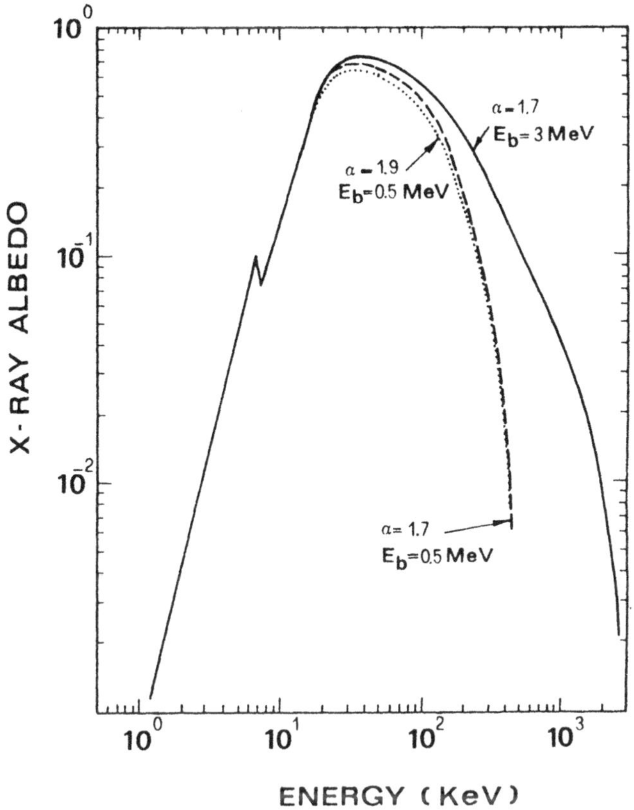

Fig.3 *X-Ray albedo of a slab illuminated by a power law spectrum, for different values of the spectral index α and the high energy cut-off E_b.*

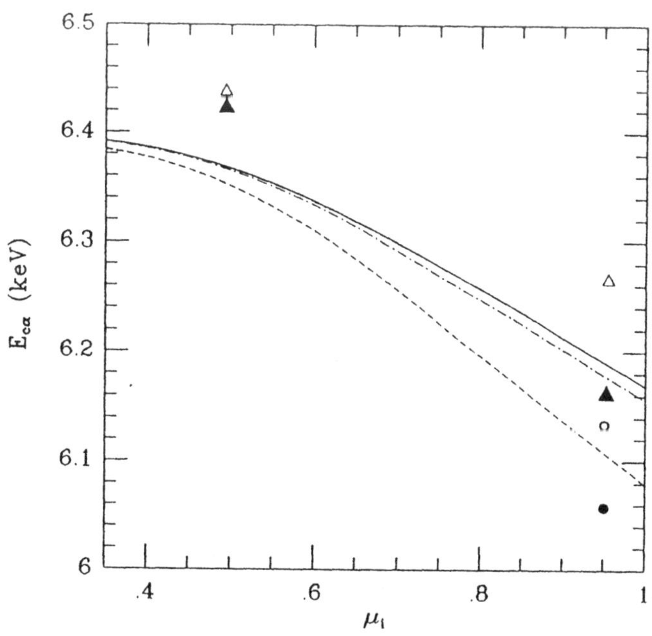

Fig.4 *Comparison of the line energy obtained from a fit with a gaussian profile with the expected energy from the disk model. The line energy of the model is shown as a function of $\mu = \cos i$ for three different values of the disk outer radius: $r_o = 10^4 r_g$ (continuous line), $r_o = 10^3 r_g$ (dash-dot line) and $r_o = 10^2 r_g$ (dash line). The energy of the gaussian is represented by triangles in the case $r_o = 10^4 r_g$ and by circles in the case $r_o = 10^2 r_g$ (filled ones: LAC-GINGA; empty ones: GSPC-SAX)*

TABLE 1.
Reflection model

Source	α	E^a	σ^b	W_α^c	N_{cr}/N_{cd}^d	R^e	$Fe/Fe_\odot$
MCG-6-30-15	1.93±0.05	6.35±0.15	0.6±0.3	200±50	1.1±0.5	0.18±0.10	0.7±0.4
NGC 7469	1.92±0.03	6.35±0.14	0.4±0.2	150±35	0.6±0.2	0.25±0.09	1.0±0.4
IC 4329A	1.97±0.04	6.30±0.08	0.2±0.2	100±20	0.8±0.2	0.13±0.04	0.5±0.2

Note. Errors correspond to 90% confidence level for 1 interesting parameter. a) Line energy in keV corrected for the redshift of the galaxy. b) Line width (σ) in keV. c) Equivalent width in eV. d) Bump intensity relative to the direct continuum at 20 keV. e) $R = W_\alpha N_{cd}/N_{cr}$(=0.25 for a flat disk with solar-like iron abundance).

It is worth noticing that the X-ray albedo is practically independent of the column density (provided that $N_H \geq 10^{25} cm^{-2}$) and of the spectral shape of the incident photons (for $E << E_b$, the break energy of the incident spectrum), as shown in fig.3. Furthermore, the *shape* of the reflected component is only weakly dependent on the inclination angle between the line of sight and the slab (MPP). The "reflection model" composed by a simple power law (representing the intrinsic continuum), and a Compton reflected component with a relative normalization as a free parameter should thus give a good description of the continuum shape in any geometry of the reflecting medium around the central source. The iron line has been described by a gaussian with position, intensity and width as free parameters.

The "reflection" model provides a good fit to the spectra of bright Seyfert galaxies (MPYM, PYM, Pounds et al. 1990), as good as the partial covering model. In fact, the reflected component is roughly similar to the covered one, with a shallower turn up around 6 keV, due to the contribution of photons scattered at small optical depths. (fig.2c,d).

In tab.1 I summarize the relevant parameters of the model obtained for three objects with high S/N ratio, namely MCG-6-30-15 (MPYM), NGC 7469 and IC4329A (PYM). Some general comments can be drawn: a) the spectral index of the intrinsic continuum is significantly steeper that the canonical value 1.7; b) the relative intensity of the reflected component (N_{cr}/N_{cd}, where N_{cd} is the intensity of the direct continuum) and the line equivalent width W_α indicate a fairly large solid angle subtended by the reprocessing medium ($\simeq 2\pi$); d) the line energy is consistent with a cold medium; e) there is some indication of line broadening that can be explained if the medium is in Keplerian motion nearby a massive black hole. More detailed considerations can be derived only by a comparison with specific models.

Reprocessing from an accretion disk

The existence of an accretion disk around the central black hole in AGN was argued on the argument that it provides a simple and efficient way to convert the gravitational energy of infalling matter into thermal energy and radiation (e.g. Shields

1978, Pringle 1981). The discovery of the O-UV "big bump" and the soft X-ray excess provided observational support to the existence of an accretion disk (even though these features cannot be uniquely attributed to an accretion disk, Guilbert and Rees 1989, Lightman and White 1989).

The case of reprocessing by an accretion disk has been studied in detail by MPP, including kinematical and relativistic effects. I shall briefly compare some predictions of the model with the GINGA results summarized in tab.1, referring the reader to MPP for a full description of the model.

The relative intensity of the reflected component for a flat disk depends on the inclination angle and attains a maximum value of 0.6 (at 20 keV) for a face-on disk. The observed values (6th column of tab.1) are consistent with a face-on configuration in all the three objects.

The equivalent width of the iron line, W_α, depends both on the inclination and on the iron abundance. Since the intensity of the iron line and of the reflected bump above 15 keV have almost the same dependence on the inclination angle, their ratio (or the equivalent quantity $R = W_\alpha N_{cd}/N_{cr}$) will depend only on the iron abundance. The ratio of the observed values of R (7th column of tab.1) to the expected one (R=0.25 at 20 keV) thus gives the iron abundance (last column of tab.1), which is consistent with the solar value in all three objects, with a marginal evidence of underabundance in IC4329A.

The profile of the line produced by the disk is highly asymmetric due to the Doppler effect and gravitational redshift and depends on the inclination angle and inner and outer radii. Even though the resolution of proportional counters as those aboard GINGA does not allow to measure the asymmetry, some information on the inclination angle can be derived from the line energy (fig.4).

Given the asimmetry of the line profile, it is important to understand whether a fit with a gaussian profile might give a line energy affected by a sistematic deviation. To this aim, I simulated a spectrum by convolving the model with the LAC response function, for some values of the inclination angle and outer radius. Then I fitted the corresponding PHA spectrum with a gaussian profile. The line energies obtained from the fit do not show any significant difference from the expected values (fig.4), that is explained by the low resolution ($\simeq 1\ keV$) of proportional counters, larger than the line width (in fact, when the resolution is improved by a factor of two, as in the case of a GSPC, the deviations become larger, because the center of the gaussian shifts towards the peak of the line, whose position differs the more from the line (centroid) energy the more the profile is asymmetric).

In conclusion, the line energies listed in tab.1 can be direcly compared with model values (fig.4) to derive the inclination angle of the disk. Yet the result does not agree with the independent estimation of the inclination angle obtained from the relative intensity of the reflected component. In fact the line energy for the face-on configuration indicated by the level of the reflected component is of 6.15 keV, lower than the values observed in all the three objects. The origin of this disagreement in the framework of the accretion disk model remains an open issue. Future missions with solid state detectors aboard, capable to risolve the line profile, should give a definite answer to this point.

Iron lines of large equivalent width

When the disk profile is either conical or parabolic, the intensity of the iron line and the reflected bump does not increase linearly with the solid angle, due to the competing reduction of the projected area of the disk seen by the observer. In particular an iron line with $W_\alpha \simeq 230$ eV can be at most produced for a solar-like iron abundance (MPP). There are at least two objects in which the iron line has an E.W. significantly greater than this limit: NGC 6814 ($W_\alpha = 330 \pm 40\ eV$, Kunieda et al. 1990) and MCG-5-23-16 ($W_\alpha = 350 \pm 60\ eV$, Piro et al. in preparation). Of course this discrepancy could be simply explained in the framework of the accretion disk model by an iron overabundance. Other possibilities include anisotropy of the central continuum of intrinsic origin or produced by an anisotropic distribution of an additional covering medium.

In the case of the partial covering model the equivalent width can attain a value of 1-2 keV, when the central source is fully covered by a medium with $N_H \simeq 10^{24} cm^{-2}$ (Leahy et al. 1989). The observed values of W_α can then be obtained (for solar-like abundance) with an high value of the covering fraction, $f_{cov} \simeq 0.7 - 0.8$, implying a rather enhanced "high energy bump". More information would thus obtain by measuring the level of the high energy bump in these two objects. Unfortunately this task is hampered by the low intensity of MCG-5-23-16 ($\simeq 0.8\ mCrab$) and the low energy absorption of NGC 6814 ($N_H = 6 \times 10^{22} cm^{-2}$), that limits the effective band and increases the uncertainty on the spectral index (which is, in fact, flatter than typical: $\alpha \simeq 1.4$). Only a broad band mission as SAX (Perola this conference) will be able, in the near future, to measure simultaneously and with a good sensitivity, both the iron line and the high energy bump.

Acknowledgments

Thanks are due to G.C. Perola for useful suggestions on this work, to G. Matt for help in producing simulations of disk spectra, and to M. Matsuoka and M. Yamauchi for permission to quote the results on MCG-5-23-16 before publication.

References

Arnaud, K. A., et al. 1985, *M.N.R.A.S.*, **217**, 105

Basko, M. M. 1978, *Ap. J.*, **223**, 268

Clavel, J. et al. 1990, *M.N.R.A.S.*, **246**, 668

Clavel, J. et al. 1991, *Ap. J.*, in press

Ferland, G. J., and Persson, S. E. 1989, *Ap. J.* ,**347**, 656

Guilbert, P. W., and Rees, M. J. 1988, *M.N.R.A.S.*, **233**, 475

Kunieda, H., Turner, T.J., Awaki, H., Koyama, K., Mushotzky, R.F., Tsusaka, Y. 1990, *Nature*, **345**, 786

Leahy, D.A., Matsuoka, M., Kawai, N., Makino, F. 1989 *M.N.R.A.S.*, **236**, 603

Lightman, A. P. and White, T. R. 1988, *Ap. J.*,**335**, 57

Malkan, M. A., and Sargent, W. L. W. 1982, *Ap. J.*, **254**, 22

Matsuoka M., Ikegami, T., Inoue, H., and Koyama, K. 1986, *P.A.S.J.*, **38**, 285

Matsuoka, M., Piro, L., Yamauchi, M., Murakami, T. 1990, *Ap. J.*, **361**, 440 (MPYM)
Matt, G., Perola, G.C., Piro, L. 1991, *Astron. Astrophys.*, in press (MPP).
Mushotzky, R. F. 1984, Adv. Space. Res., Vol.3, No.10-12, p.157.
Perola, G.C. et al. 1986, *Ap. J.*, **306**, 508
Piro, L., Massaro, E., Perola, G.C., Molteni, D. 1988, *Ap. J.(Letters)*, **325**, L25 (PYM)
Piro, L., Yamauchi, M., Matsuoka, M. 1990, *Ap. J. (Letters)*, **360**, L35.
Pounds, K., K.A., Nandra, K., Stewart, G.C., George, I.M., Fabian, A.C. 1990, *Nature*, **344**, 132
Pringle, J.E. 1981, *Ann. Rev. Astron. Ap.*, **19**, 137
Reichert, G. A., Mushotzky, R. F., and Holt, S.S. 1986, *Ap. J.*, **303**, 87
Rotschild, R.E., et al. 1983, *Ap. J.*,**269**, 423
Shields 1978, *Nature*, **272**, 706
Turner, T. J., and Pounds, K. A. 1989, *M.N.R.A.S.*, **240**, 833
Yamauchi, M., Piro, L., and Matsuoka, M. 1990, *"21st Intern. Cosmic Ray Conference"*, in Adelaide, Vol.1 OG 1.1-6 p.19.
Warwick, R.S., Yaqoob, T., Pounds, K.A., Matsuoka, M., Yamauchi, M. 1989 *Publ. Astr. Soc. Jap.*,**41**, 721

THE IONIZATION STRUCTURE OF IRON IN AGN

I.M. George, A.C. Fabian
Institute of Astronomy, Madingley Road, Cambridge CB3 OHA, UK.
R.R. Ross
Physics Dept, College of the Holy Cross, Worcester, MA 01610, USA.

1. INTRODUCTION

Recent *Ginga* observations of AGN have shown iron K-shell features to be common in the spectra of unobscured AGN (Nandra *et al.* 1989, 1990; Pounds *et al.* 1989, 1990; Piro *et al.* 1990; Day *et al.* 1990). Similar features are also often detected in many classes of galactic X-ray binary (*e.g.* Suzuki *et al.* 1984; White *et al.* 1986; Hirano *et al.* 1987 Nagase 1989; Fabian *et al.* 1989). In the case of the galactic sources, the photosphere of the companion, any extended matter surrounding part or the whole system, the dense accreting material, and any hot corona surrounding the accreting matter (disc) or companion, are all sites for the formation of the observed features. The iron emission line, in particular, is a powerful diagnostic — even with the current generation of X-ray spectrometers. Its centroid energy, equivalent width, FWHM, and inclination & phase dependence have been used to place constraints on these reprocessing sites in a number of well studied binaries (*e.g.* Bai 1980; Sato *et al.* 1986; Fabian *et al.* 1989).

The geometry and kinematics of the material responsible for the features in AGN are however currently unclear. Nevertheless ....

- the distribution of centroid energies of emission line from a large sample of objects is sharply peaked around 6.4 keV (Nandra, 1990), indicating that fluorescence from 'cold' iron is responsible.
- a number of sources have a clear indication of a flattening of the X-ray continuum above $\sim$10 keV, as may be expected from a 'Compton reflected' spectral component (Lightman & White 1988; George & Fabian 1990).
- The iron line equivalent widths (100–300 eV) require that the reprocessing region subtends a large solid angle at the X-ray source ($\sim 2\pi$).
- The lack of a lag between continuum and line variations indicate that in at least one AGN (NGC 6814, Kunieda *et al.* 1990) the reprocessor is within 300 light seconds of the continuum source.

Thus in AGN it seems likely that the primary X-ray spectrum is reprocessed by substantial amounts of nearby, cold, optically thick material out of the line of site.

Cold matter can survive the intense radiation field close to the central nucleus (Guilbert & Rees 1988; Ferland & Rees 1988) provided it is sufficiently dense to be able to efficiently cool via bremsstrahlung.
— But can we say anything more about the site of reprocessing region ?
— In particular, can the reprocessor be the accretion disc ??
Here we explore the ionization structure of iron within a slab of matter illuminated from one side by an AGN-type X-ray spectrum, and from the other by a black body. Our findings are then compared to the conditions expected in the innermost regions of AGN accretion discs.

2. THE MODEL

The model consists of a spherical shell of matter with density n, radius R and total optical depth τ, illuminated from within by a central black body source (temperature T_s, luminosity L_s) and from the outside by a powerlaw X-ray spectrum (energy index α, luminosity L_h) between 4 eV and 100 keV. The shell is optically thick to scattering and contains elements at their cosmic abundances. The passage of radiation through the shell is is determined by solving a radiative transfer equation based on diffusion approximations for both the spatial transport and the dispersion in photon energy. Both Compton scattering and a variety of atomic processes are included. The method therefore permits the ionization and temperature structures of the shell to be self-consistently determined. Further details can be found in Ross (1979).

In this preliminary work we have chosen to explored the region of parameter space for which the iron is more recombined than Fe XXV. Iron Kα fluorescence from such material will therefore give an iron line at an energy <6.7 keV, consistent with the observations. We define the X-ray ionization parameter as $\xi_h = L_h/nr^2$ and τ^* as the optical depth from the outer edge of the shell

3. RESULTS

The relative abundance of Fe XXV is a good indicator of the ionization state of iron. When ~40% of the iron becomes more recombined than Fe XXV, further recombination is very likely to take place. We therefore adopt this value as a fiducial diagnostic of the ionization structure of iron within the slab.

Taking typical AGN input parameters .....

- The ionization curve is shown in Figure 1. It can be seen that for $\tau^* = 1$, a substantial fraction of iron is more recombined than Fe XXV for $\xi_h \lesssim 600$. The corresponding limit for $\tau^* = 0$ is $\xi_h \lesssim 400$.

- In Figure 2 we show the effect of varying the black body luminosity, L_s. It can be seen that the cooling rate, and hence the fraction of iron <Fe XXV is significantly increased for $L_s/L_h \gtrsim 10$. *i.e.* for $L_s/L_h \lesssim 10$, the ionization structure of the shell is primarily determined by the incident X-ray powerlaw.

- From Figure 3 it can be seen that the ionization structure of the cloud is sensitive to the energy index of the external radiation. As expected, as α increases, the Compton heating of outer layers of the shell decreases, giving rise to an increased fraction of iron in low ionization states.

- We find that these results are independent of the black body temperature T_s (until collisional ionization processes come into play — at $kT_s \gtrsim 0.3$ keV for iron >Fe XXIV), density ($10^{11} \lesssim n \lesssim 10^{15}$) and the assumed τ of the shell.

4. 'STANDARD' ACCRETION DISCS

The putative accretion disc fueling a central supermassive black hole is the most obvious site of the reprocessing material in AGN. As yet we have no detailed knowledge as to either the physical conditions within, or the location of the primary X-ray source. It is however likely that any disc-shaped accretion flow will intercept a large fraction of the emitted X-rays. Indeed the observed features in both Cyg X-1 and a number of AGN have already been shown to be consistent with the reprocessing expected from a simple semi-infinite, optically thick slab

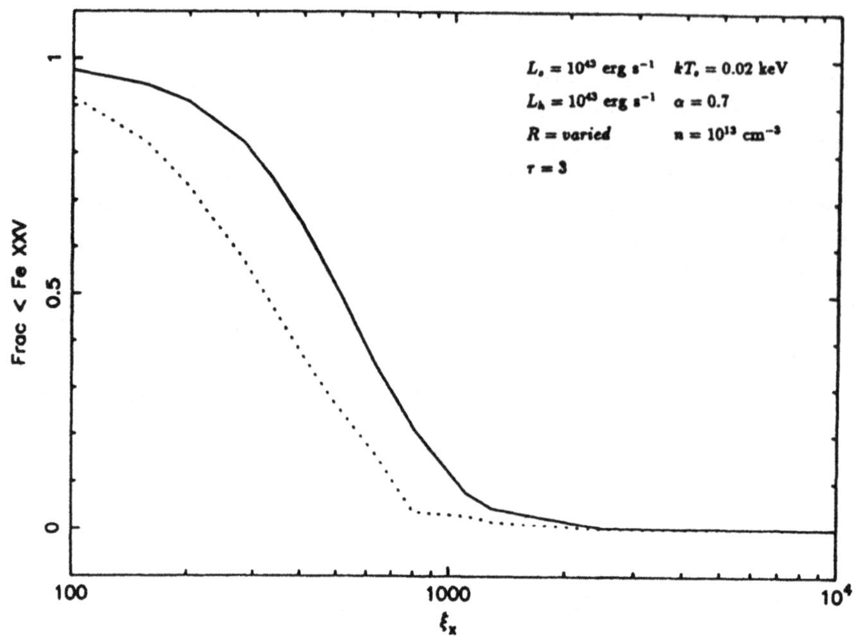

Figure 1. *The fraction of the total iron more recombined than Fe XXV as a function of the hard X-ray ionization parameter, ξ_h. In this and Figures 2–3, the solid curves indicate the conditions at $\tau^* = 1$, whilst the dotted curves show them at the outer edge of the shell ($\tau^* = 0$). The fiducial point where >40 % of iron is more recombined than Fe XXV is indicated.*

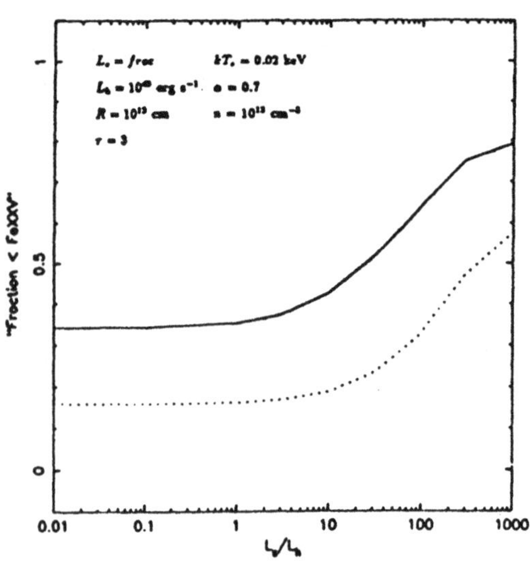

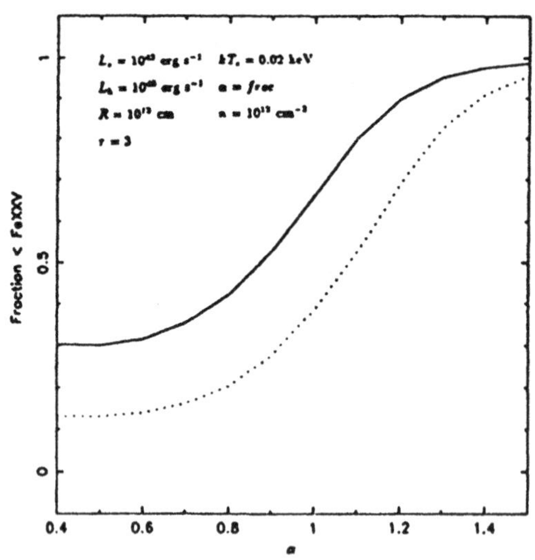

Figure 2.

Figure 3.

illuminated from a point source of X-rays above the centre (Fabian *et al.* 1989; George, Nandra & Fabian 1990; George & Fabian 1990). But are the physical conditions within standard accretion discs, in particular will the innermost regions of the disc be able to survive the intense external X-ray illumination at a temperature low enough that iron is not fully ionized ?

From Section 3 we require $\xi_h \lesssim 600$ in order that a sizable proportion of the iron in the illuminated matter is in a lower enough ionization state to emit the 6.4 keV line. In Figure 4 we show the vertically averaged density as a function of radius for the innermost ($P_r > P_g$) regions of a canonical AGN disc (using the expressions of Laor & Netzer 1989). Also shown is the region of the n–r plane in which $\xi_h \lesssim 600$. For the disc parameters considered it can be seen that for a standard 'α'-disc, for $r/r_g \gtrsim 20$ the density of the disc is sufficient such that > 40 % of the iron is more combined than Fe XXV. In Figure 5 we show the n–r plane for (a) $m_9 = 10^{-3}$ and (b) $m_9 = 10^{-1}$ black holes, with corresponding density curves for different accretion rates, $\dot{m}$. For each curve the critical radius beyond which $\xi_h \lesssim 600$ is indicated (and the points from the separate curves connected by the dashed line). Finally in Figure 5(a) we show the constraint imposed by a line—continuum lag of < 300 seconds. Thus it is clear that there may well be a region of parameter space in which standard accretion discs are consistent with providing the site of the reprocessing. Clearly for individual objects there are other constraints that can also be imposed by its luminosity, the temperature and τ_{es} of the disc *etc.*

It should be remembered that the vertical structure of accretion discs, particularly those externally illuminated by hard X-rays, is far from well understood. In addition the $P_r > P_g$ regions of α-discs are known to be unstable, and the effects of illumination on the instabilities is also unclear. It is possible that small dense clouds are able to condense within the unstable disc. These are then able to reprocess the primary X-ray emission (which itself may originate in the intercloud medium). The relative scale heights of the X-ray emission and blobs may then offer an explanation for different (time-dependent ?) fractions of direct:reflected components observed for various objects (Day *et al.* 1990). Further work addressing these problems is in progress.

REFERENCES

Bai, T., 1980. *Astrophys. J.*, **239**, 328.
Day, C.S.R., *et al.*, 1990. *Mon. Not. R. astr. Soc.*, in press.
Fabian, A.C., Rees, M.J., Stella, L. & White, N.E., 1989. *Mon. Not. R. astr. Soc.*, **238**, 729.
Ferland, G.J. & Rees, M.J., 1988, *Astrophys. J.*, **332**, 141.
George, I.M. & Fabian, A.C., 1990. *Mon. Not. R. astr. Soc.*, submitted.
George, I.M., Nandra, K. & Fabian, A.C., 1990. *Mon. Not. R. astr. Soc.*, **242**, 28P.
Guilbert, P.W. & Rees, M.J., 1988, *Mon. Not. R. astr. Soc.*, **233**, 475.
Hirano, T., *et al.*, 1987. *Publ. astr. Soc. Japan*, **39**, 619.
Kunieda, H., *et al.*, 1990. *Nature*, **345**, 786.
Laor, A. & Netzer, H., 1989. *Mon. Not. R. astr. Soc.*, **238**, 897.
Lightman, A.P. & White, T.R., 1988, *Astrophys. J.*, **335**, 57.
Nagase, F., 1989. *Publ. astr. Soc. Japan*, **41**, 1.
Nandra, K., 1990. This proceedings.
Nandra, K., *et al.*, 1989. *Mon. Not. R. astr. Soc.*, **236**, 39P.
Nandra, K., *et al.*, 1990. *Mon. Not. R. astr. Soc.*, in press.
Piro, L., Yamauchi, M. & Matsuoka, M., 1990. *Astrophys. J.*, **360**, L35.
Pounds, K.A., *et al.*, 1989. *Mon. Not. R. astr. Soc.*, **240**, 769.
Pounds, K.A., *et al.*, 1990. *Nature*, **344**, 132.
Ross, R.R., 1979. *Astrophys. J.*, **233**, 334.
Sato, N. *et al.*, 1986. *Publ. astr. Soc. Japan*, **38**, 731.
Suzuki, K. *et al.*, 1984. *Publ. astr. Soc. Japan*, **36**, 761.
White, N.E. *et al.*, 1986. *Mon. Not. R. astr. Soc.*, **218**, 475.

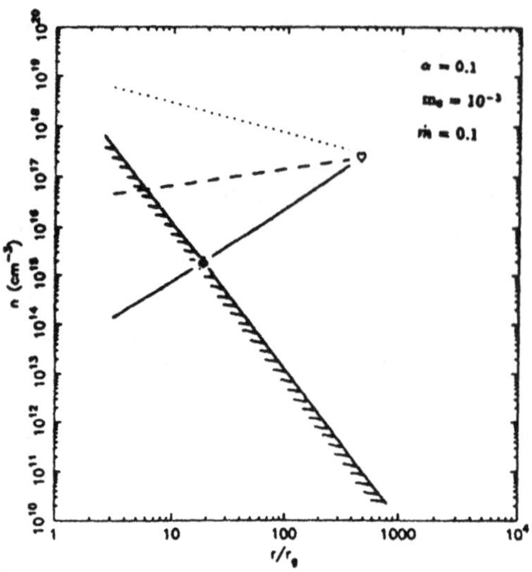

Figure 4. *The vertically averaged density vs radius (in units of r_g) in the radiation pressure supported region of an accretion disc where the viscous stress is proportional to (solid) $P_r + P_g$ (which we hereafter refer to as the α-disc), (dotted) P_g, and (dashed) $\sqrt{P_r P_g}$. The ♡ symbol in this and Figure 5 indicates where $P_r = P_g$. The hatched line delineates the region of the n–r plane such that $\xi_h > 600$. This line intersects the α-disc density curve at $r/r_g \sim 20$ (♣) indicating that beyond this radius the such a disc has a substantial fraction of ~neutral iron present within one Thomson depth of the surface despite the X-ray irradiation.*

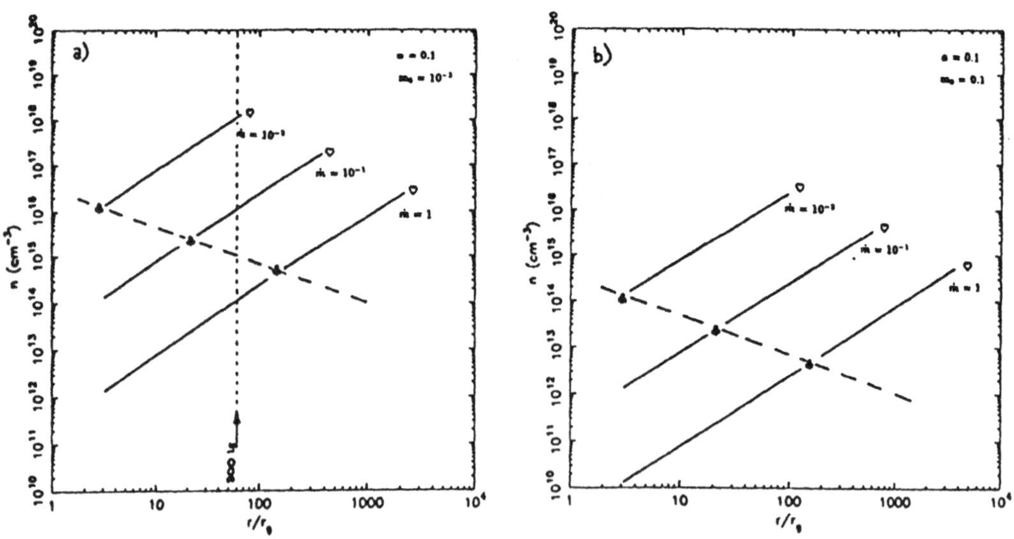

Figure 5. *Density as a function of radius for (a) $m_9 = 10^{-3}$, (b) $m_9 = 0.1$, and for several $\dot{m}$ (where m_9 is the black hole mass in units of 10^9 $M_\odot$ and $\dot{m}$ is the accretion rate in units of $\dot{m}_{Edd}$). Again the ♣ symbol on each curve indicates the radius beyond which $\xi_h > 600$. Figure 5(a) also shows the radius equivalent to 300 light seconds from the central object ($r_g = 1.5 \times 10^{14} m_9$).*

Iron K fluorescence and high energy bump from accretion disks in AGNs

G. Matt[1], *G.C. Perola*[1] *and L. Piro*[2]

[1]Istituto Astronomico dell'Universita' di Roma "La Sapienza", Via G.M. Lancisi 29, 00161 Roma, Italy
[2]Istituto di Astrofisica Spaziale, C.N.R., Via E. Fermi 21, 00044 Frascati, Italy

Recent *Ginga* observations have revealed the existence of both the iron K fluorescence line and a high energy bump in the X-ray spectra of some bright Seyfert 1 galaxies (Matsuoka et al., 1990; Piro et al., 1990; Pounds et al., 1990), which are currently attributed to the reprocessing of the primary X-ray photons by cold matter located in the neighbourhood of the putative black hole (Guilbert and Rees, 1988; Lightman and White, 1988). Assuming that the matter is distributed in an optically thick accretion disk, we have calculated, by means of Monte Carlo simulations, the intensities of the line and of the bump as a function of the disk inclination for different geometries of the disk surface. For the simplest case of a flat disk we have calculated the line profile taking into account also the kinematical and relativistic effects. The results obtained apply to matter which is neutral or mildly ionized ($T \lesssim 10^5$ K) and with a cosmic chemical abundance (Morrison and McCammon, 1983). A full account of this work is given in Matt et al. (1990); here some relevant results are briefly outlined.

The "effective" size of the reprocessing region

As an indicator of the "effective" size of the reprocessing region we have chosen the quantity $r_{1/2}$, that is the radius within which half of the reprocessed photons do come. This radius was computed for a geometry in which the source is represented by a point emitting isotropically a power law spectrum whit photon index -1.7, which is located on the simmetry axis at h=20 r_g above the surface of a flat disk ($r_g = GM/c^2$). The point source is a good approximation of an extended source, such as an optically thin corona, of size comparable to h.

In fig.1 $r_{1/2}$ for the line and the bump is given as a function of μ_i=cosθ_i, the cosine of the inclination angle, for a disk with inner radius r_i=10 r_g and outer radius r_o=10^4 r_g. The dependence on μ_i is due to the dependence in the angular distribution of the reprocessed radiation on the incidence angle of the impinging photons. The difference between the line and the bump derives from the difference in the phase matrices of the fluorescence and of the scattering (the latter being energy dependent). From the figure it appears that $r_{1/2}$ is comprised between about 40 and 90 r_g, that is a small fraction of

r_o. We would like to point out that this quantity can be associated with measurements of a time delay in the response of the reprocessed radiation to variation in the primary source, in order to estimate in a very first approximation the gravitational radius, and therefore the mass of the black hole. The mass (in units of solar masses) would then be of the order of a few 10^3 times the delay (in seconds). The exact relationship should take into account in detail the geometrical as well as the relativistic effects, as a function of θ_i, on the response to specific light curve profiles of the primary radiation, an item on which work is now in progress.

Line and bump properties for a flat disk

In order to take into account the kinematical and relativistic effects on the spectra we have used the weak-field approximation in the formulation of Chen et al. (1989), modified to include the anisotropy of the angular distribution of the reprocessed photons. The use of the weak-field approximation limits the validity of our results to inclination angles less than about 70°. The spectra between 5 and 9 keV of the reprocessed radiation for four inclination angles and three outer radii are shown in fig. 2. The profiles are very asymmetric, with the two characteristic horns due to the Doppler effect, and vary strongly both with the inclination angle and with the outer radius of the disk. In particular, the separation between the red and blu horns increases with the inclination angle and decreases with the outer radius. The K edge is broadened and its sharp indentation is blueshifted: for instance, this indentation for $\theta_i = 55°$ occurs at about 8.1 keV, corresponding to the rest energy of the edge in FeXX.

High resolution detectors are needed to derive the geometrical parameters from the line profiles. However, some information can be obtained also from low resolution data using integral quantities such as the equivalent width W_α of the line and the strength of the bump. Our computations show that W_α increases with μ_i, reaching the value of about 150 eV for a face-on disk and an outer radius of 10^4 r_g. W_α alone cannot be used, however, to estimate unambiguously θ_i, because it depends also on the abundance of iron Y_{Fe} with respect to the elements which contribute to the photoelectric absorption at 6.4 keV. An intermediate step is offered by the comparison of the line intensity with that of the bump, the latter being practically independent of Y_{Fe}. This can be done as in fig.3, where the quantity $R = W_\alpha N_{cd}/N_{cr}$ is given as a function of μ_i, N_{cr} (N_{cd}) representing the strength of the reprocessed (direct) continuum: as it can be seen, at energies above say 15 keV this quantity is fairly independent of μ_i (and of the other geometrical parameters over a wide range of them), and can be therefore used to estimate the iron abundance.

Non planar disk configurations

Large values of the equivalent width are sometimes interpreted as signature of a non planar configuration of the disk surface. Due to the dependence of this parameter on Y_{Fe}, the high energy bump is a more efficient indicator of strong deviations from the planarity.

We have considered a geometry where the source is again located at 20 r_g on the symmetry axis and the half-thickness Z of the disk is given by $Z = Ar_o(r/r_o)^\alpha$. It must be noted that in such a geometry the "secondary illumination", i.e. the illumination of the disk by the reprocessed photons themselves, plays an important role. The "secondary illumination" is much more important for the bump than for the line, because a high energy photon has a greater probability to be reflected than a 6.4 keV photon.

The equivalent width and the high energy bump at 20 keV (normalized to the direct continuum) for a conical configuration seen face-on is shown in fig. 4 as a function of A (r_i=10 r_g, r_o=10^3 r_g). The two quantities go through a broad maximum around A=1, corresponding to a cone half angle of 45°. We see that even strong deviations from the planarity do not produce a large increase in W_α, which remains less than 190 eV. Greater values can be obtained with a concave configuration, that is α greater than one. The maximum value of about 230 eV is reached with a parabolic configuration. We note that at least in one case, NGC 6814, a value of W_α significatively greater than 230 eV has been measured (Kunieda et al., 1990): in the framework of a "cold" disk model, the implication for this object is that Y_{Fe} is greater than the cosmic value, or, alternatively, that the continuum source is either beamed or partially hidden.

With regard to the bump, the dependence on A is similar in form but more pronounced, due to the greater importance for this quantity of the "secondary illumination". From the figure we see that the strength of the bump at maximum is about twice the value for the flat configuration and that it remains significatively greater also for smaller A, thus offering concrete prospects for the search of deviations from the planarity.

REFERENCES

Chen, K., Halpern, J.P., Philippenko, A.V.: 1989, *Astrophys. J.* **339**, 742

Guilbert, P.W., Rees, M.J.: 1988, *Monthly Notices Roy. Astron. Soc.* **233**, 475

Kunieda, H., Turner, T.J., Awaki, H., Koyama, K., Mushotzky, R., Tsusaka, Y.: 1990, *Nature* **345**, 786

Lightman, A.P., White, T.R.: 1988, *Astrophys. J.* **335**, 57

Matt, G., Perola, G.C., Piro, L.: 1990, *Astron. Astrophys.*, submitted

Matsuoka, M., Yamauchi, M., Piro, L., Murakami, T.: 1990, *Astrophys. J.* **361**, 440

Morrison, R., McCammon, D.: 1983, *Astrophys. J.* **270**, 119

Piro, L., Yamauchi, M., Matsuoka, M.: 1990, *Astrophys. J. Letters* **360**, L35

Pounds, K.A., Nandra, K., Stewart, G.C., George, I.M., Fabian, A.C.: 1990, *Nature* **344**, 132

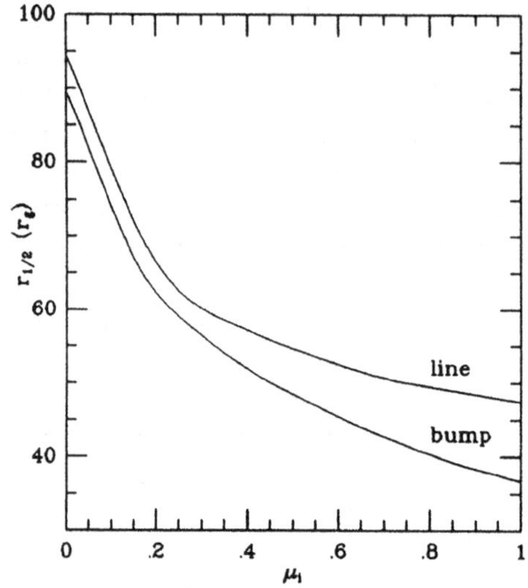

Fig. 1. The radius within which half of the reprocessed photons are emitted.

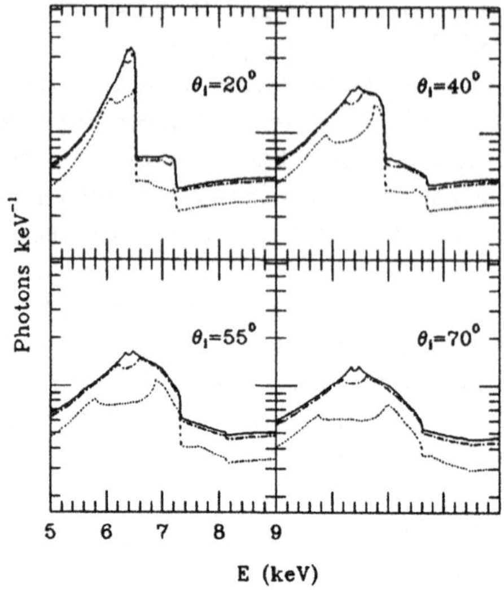

Fig. 2. The spectra of the reprocessed radiation for an inner radius r_i=10 r_g and an outer radius $r_o=10^2$ r_g (dotted line), $r_o=10^3$ r_g (dash-dotted line) and $r_o=10^4$ r_g (solid line).

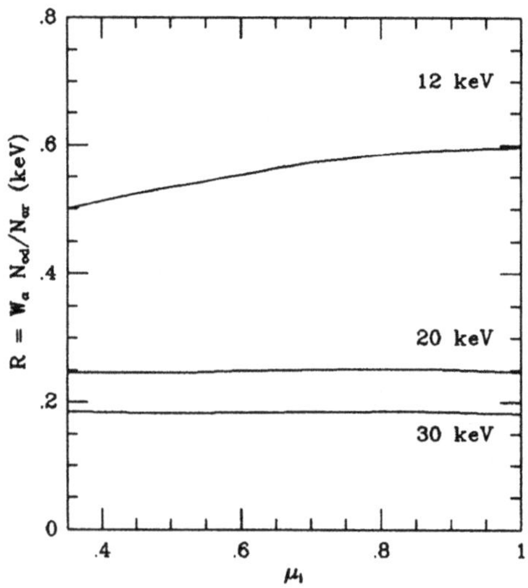

Fig. 3. The quantity R at three values of the energy.

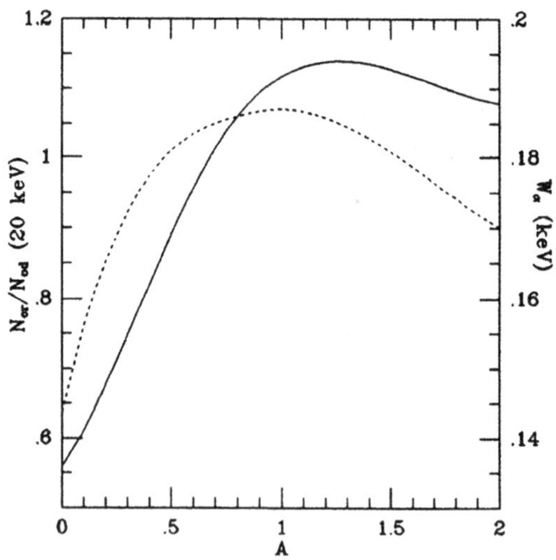

Fig. 4. W_α (dotted line) and the bump at 20 keV (solid line) for a disk with conical surface (seen face-on) vs. A=cotgψ where ψ is the cone half angle.

Line Profiles from a Disk around a Rotating Black Hole

Ari Laor
The Institute for Advanced Study
Princeton, NJ 08540, U.S.A.

1 Introduction

Accretion disks which are irradiated by a continuum X-ray source are a very likely place for the production of the Kα line in AGN and in galactic binary X-ray sources. They provide the high column density and covering factor required for efficient fluorescence, while allowing a very small column density on our line of sight, as implied by the detection of soft X-rays in most AGN. Further support for this was recently given by Pounds *et al.* (1990) who have detected, in a number of AGN, the spectral signature of a continuum component which is reflected from cold material. The line width, which is of the order of 1keV, is also naturally explained by accretion disks, as the line is significantly broadened by the high velocities in the inner regions of the disks. Electron scattering of a large enough optical depth can also significantly broaden the Kα line (e.g. Pozdnyakov *et al.* 1979), however the required conditions are not likely to occur if the line is produced by fluorescence in an accretion disk (George and Fabian 1990), and also not in the case where the line is produced in a X-ray heated corona (Kallman and White 1989).

Lines emitted from an accretion disc are strongly affected by relativistic effects. These effects were calculated by Fabian *et al.* (1989) by Stella (1990), and in the weak field limit by Chen and Halpern (1989), and George, Nandra, and Fabian (1990). These calculations were done for the case of a non-rotating black hole in which case the disk extends down to $6r_g$ ($r_g = GM/c^2$). However, the black hole is very likely to be rotating, in which case the disk extends down to about $1.23r_g$ (Thorne 1974), where the relativistic effects are significantly enhanced. This inner region could be important for the line emission since the very short variability time scales of the continuum, observed in some objects (e.g. Kunieda *et al.* 1990), indicate a compact source. This paper presents accurate calculations of line profiles for a line emitted from an accretion around a rotating black hole. These calculations were mainly intended for the interpretation of line profile measurements by future high resolution X-ray spectroscopy, in particular the BBXRT mission. However, as we describe below, the relativistic effects are strong enough, in some models, to have a detectable effect even with the currently available low resolution.

2 Method of Calculation

The relativistic effects on the emission from an accretion disk around a rotating black hole were first calculated by Cunningham (1975). The method applied here is somewhat different, and it is described in Laor, Netzer and Piran (1990).

The observed profiles were calculated for four different emissivity laws. The first two assume a power law dependence of the form $J(r_e) \propto r_e^\alpha$, with $\alpha = -2$ and $\alpha = -3$. These emissivity laws were used by Fabian *et al.* (1989), and were also calculated here to allow comparison with the previous calculations. Somewhat more plausible emissivity law can be deduced by assuming that a compact continuum X-ray source is situated at a height H above the disk center. A simple expression is obtained under the following simplifications; the line emissivity is proportional to the flux absorbed per unit area, the disk is flat, and relativistic effects on the flux from the continuum source are negligible. This results in an emissivity law of the form $J(r_e) \propto H/(r_e^2 + H^2)^{3/2}$. The line profiles were calculated for two cases $H = 3r_g$ and $H = 10r_g$. In the $\alpha = -2$ law the contribution from large r_e does not diminish, and it is therefore sensitive to the maximum radius assumed. All other emissivity laws have little emission beyond $400r_g$. A more realistic expression for the line emissivity as a function of r_e, θ_e and ν_e can be deduced from a detailed solution for the line production and transfer in the disk atmosphere, as was done by George and Fabian (1990), which is beyond the scope of this work.

3 The Line Profiles

Figure 1 presents the profiles obtained for the four emissivity laws described in section 2. In the case of a power law with $\alpha = -2$ the lines are strongly dominated by the emission from the outer part of the disk, while the power-law with $\alpha = -3$ represents the other extreme case were the lines are strongly dominated by the emission from the inner part. The other two emissivity laws, describing a continuum source above the disk with $H = 3r_g$ and $H = 10r_g$ represent intermediate cases. The main characteristics of these models are:

1. All lines are asymmetric. They are characterized by a steep blue wing and an extended red wing.

2. The line peak can be blueshifted by as much as 50% when a disk is seen at a high inclination. The maximum redshift is obtained for a disk seen face-on, however this redshift is not larger than 5%. Thus, for most inclinations and emissivity laws the peak of the line is blue shifted.

3. All emissivity laws which give a dominant contribution from the inner part ($r < 20r_g$) of the disk produce a single peaked line. A double peaked line is produced only by the $\alpha = -2$ emissivity law, where a significant fraction of the line flux comes from the outer part of the disk.

4. The typical line full FWHM at intermediate inclinations is $\sim \Delta\nu/\nu = 0.2$. Near edge-on ($\mu = 0.1$) it can reach about 0.5, while close to face-on it can be as low as 0.03.

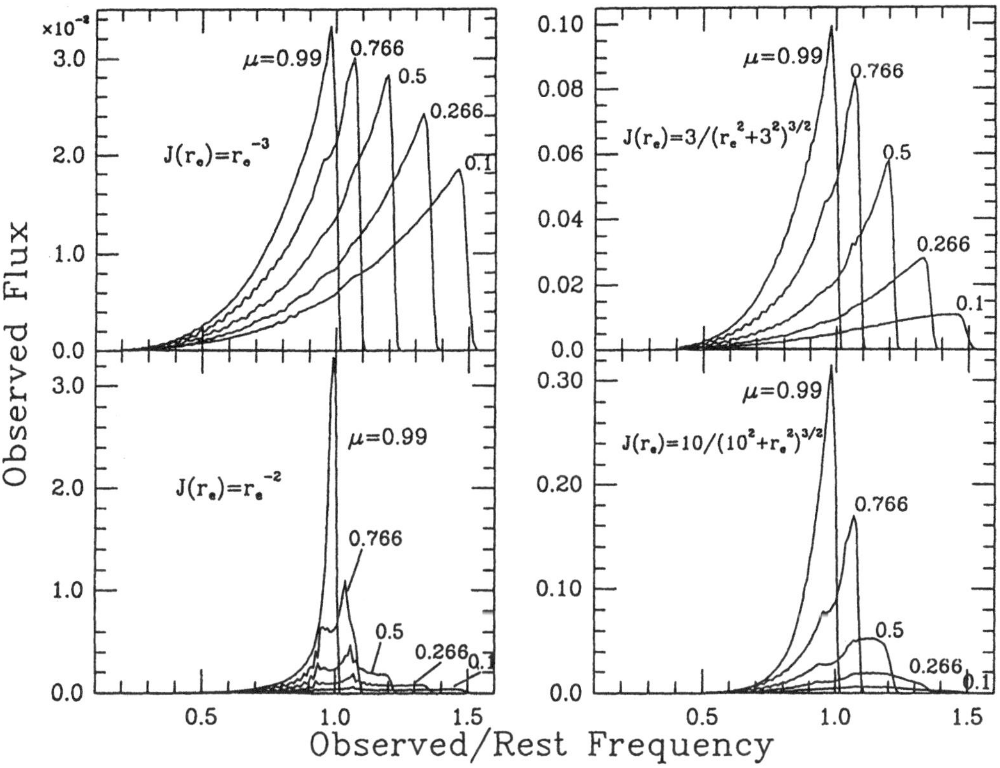

Figure 1: The line profiles for various emissivity laws. The emissivity law and the cosine of the observed inclinations are marked in each figure. The r_e^{-2} emissivity law gives least weight to the flux from the inner disk and thus displays the smallest relativistic effects, the r_e^{-3} law represents the other extreme case.

5. The line flux is strongly diminished at high inclinations in models were most of the flux comes from the outer disk ($\alpha = -2$ model). The line flux becomes nearly isotropic in the other extreme case ($\alpha = -3$ model) where the flux is dominated by emission from the inner disk.

4 Discussion

As already mentioned, similar calculations of line profiles by Fabian *et al.* (1989) and Stella (1990) assumed a non-rotating black hole. The profiles calculated here differ from previous results in that they reflect more enhanced relativistic effects, in particular the lines usually have a single, rather than a double peak, and the line can be strongly shifted in energy.

There have been two attempts to fit a specific disk model to two objects; Cygnus X-1 by Fabian *et al.* (1989), and NGC 3227 by George, Nandra, and Fabian (1990). These authors have shown that the observed line profile is consistent with fluorescence from an accretion disk, but the low spectral resolution did not allow to significantly constrain the model parameters.

The results presented above indicate that for most emissivity laws and inclinations the observed line is blue shifted. The predicted blue shift can be quite large. For example, a power law emissivity with $\alpha = -3$ implies that a significant number of objects should show a blue shift of 15% or more, such a high blue shift is in fact inconsistent with current observations, and one can deduce that such an emissivity law is unlikely to be common. Thus even though the currently available data is of a low resolution, one might be able to put interesting constraints on the form of the emissivity law, using only the distribution of line energy and EW of a large enough sample of objects.

The results presented above allow us to make the following predictions: 1. All lines should be asymmetric, having an extended red wing and a sharp blue wing. 2. The blue wing should extend in some objects up to an energy shift of 50%. Such a blue shift will not be common, as it is expected only for disks seen at a high inclinations, however its detection would serve as strong evidence that the black hole must be rotating. These predictions could be tested for individual objects in the near future with higher resolution X-ray spectroscopy missions (BBXRT, ASTRO-D etc.). Further details concerning this work are given in Laor (1990).

I would like to thank Tsvi Piran for his help in preparing this work. This work was supported in part by NSF grant PHY86-20266.

References

Chen, K., Halpern, J. P. 1989, *Ap. J.*, **344**, 115.

Cunningham, C. T. 1975, *Ap. J.*, **202**, 788..

Fabian, A. C., Rees, M., Stella, L., and White, N. E. 1989, *M.N.R.A.S.*, **238**, 729.

George, I. M., Nandra, K., and Fabian, A. C. 1990, *M.N.R.A.S.*, **242**, 28p.

George, I. M. and Fabian, 1990, *M.N.R.A.S.*, submitted.

Kallman, T., and White, N. E. 1989, *Ap. J.*, **341**, 955.

Kunieda, H., Turner, T. J., Awaki, H., Koyama, K., Mushotzky, R., and Tsusaka, Y. 1990, *Nature*, **345**, 786.

Laor, A. 1990, *Ap. J. Lett.*, submitted.

Laor, A., Netzer, H., and Piran, T. 1990, *M.N.R.A.S.*, **242**, 560.

Pounds, K. A., Nandra, K., Stewart, G. C., George, I, M., and Fabian, A. C. 1990, *Nature*, **344**, 132.

Pozdnyakov, L. A., Sobol, I. M., and Sunyaev, R. A. 1979, *Astr. Ap.*, **75**, 214.

Stella, L. 1990, *Nature*, **344**, 747.

Thorne, K. S. 1974, *Ap. J.*, **191**, 507.

X-RAY SPECTRA FROM CLOUDS IN ACTIVE GALACTIC NUCLEI

J. Felsteiner[1] and R. Opher[2]

[1] Department of Physics, Technion-Israel Institute of Technology, Haifa, Israel
[2] Instituto Astronômico e Geofísico, Universidade de São Paulo, Brazil

ABSTRACT

We present a cloud model which can explain certain basic characteristics of recently observed X-ray spectra from active galactic nuclei. The model includes cold opaque clouds and hot transparent clouds. Both nonisotropic power law emission from the central engine and nonsolar abundances of iron are considered.

1. INTRODUCTION

Besides coming directly to us, the X-ray power law continuum from AGN can be reflected back into our line-of-sight from opaque matter, together with characteristic emission and absorption features due to photoelectric, fluorescence and electron scattering processes. There can be substantial amounts of optically thick cold gas in the form of clouds or filaments close to the central engine (e.g. Guilbert and Rees 1988).

George *et al.* (1989, 1990) presented a model for the Fe line in NGC 3227. They considered a Keplerian disc illuminated by an external X-ray source. The results from Monte Carlo simulations were fitted to the Ginga data of NGC 3227. The model requires that NGC 3227 be viewed almost face on due to the large Doppler shifts from the Keplerian disc. The X-ray source was located a distance of 20 gravitational radii (R_G) above the center of the disc ($R_G = GM/c^2$).

Nandra *et al.* (1990) proposed a model for the Ginga data of MCG-6-30-15 of a power law X-ray source located above the central part of a disc. The disc is viewed almost face on and introduces a cold Fe edge by reflection and a 130 eV equivalent width fluorescent Fe line. The contributions of the Fe edge at 7.1 keV by the disc albedo and the 8.3 keV edge from a hot absorber combine to fit a measured edge energy of 7.8 keV. The excess of low energy photons indicates the need of a hot highly ionized absorber. Using a column density of 1.5×10^{23} H atoms cm^{-2} of hot matter with solar abundance and an ionization parameter of 3–10 gave an equivalent width for the Fe line of 50–170 eV, but with an unacceptable Fe K edge at 8.3–8.5 keV.

Pounds *et al.* (1990) studied the data from 12 Ginga observations of 8 Seyfert type AGN. The data were superimposed to produce a composite X-ray spectrum, Ginga-12. A simple power law fit to the Ginga-12 data yielded an energy index of $\alpha = 0.7$. Of particular interest is the data-minus-model residual from this fit which has a number of interesting features: 1) the Fe K line has an energy of 6.0–6.5 keV and an equivalent width of 75–150 eV; 2) a dip appears between 8–12 keV; and 3) an excess flux above 12 keV. A good fit to the Ginga-12 data was obtained with: 1) hot column densities of 0.06×10^{23} H atoms cm^{-2} and an additional column density at the Fe K edge of 0.9×10^{23} cm^{-2} (solar abundances); 2) a reflecting semi-infinite slab at an inclination angle of 9° (weakly constrained); and 3) an energy index of $\alpha = 0.91$. The two column densities imply an over abundance of Fe of 15 times the solar value. An equally good fit was also obtained without hot absorbers by increasing the reflected spectrum by a factor 1.5 and using an iron abundance twice the solar value.

The Fe K line is about 6.4 keV up to an ionization state of about XVIII, and for higher ionization states it rises to 6.7 keV for XXV (Koyama 1985). From Monte Carlo simulations, the equivalent width is about 100 eV for a column density of 10^{23} cm^{-2} and 200 eV for a column density of 2×10^{23} cm^{-2}. The intensity of the Fe line reaches a maximum at about 5×10^{23} cm^{-2}. Koyama (1985) noted that the equivalent widths of many pulsars such as Cen X-3, Vela X-1, and Her X-1 are about 5 times the expected values for an isotropic source. Similarly for AGN such as NGC 4151 the observed intensity of the Fe line exceeds the maximum expected value by a factor 5. The larger equivalent width and intensity of the Fe line can be explained if: 1) the X-ray emission is not isotropic, with less emission along our line-of-sight as compared with an isotropic source; or 2) the Fe line is overabundant by a factor of about 5.

2. MODEL

We study a cloud model which has: A) cold opaque reflecting clouds; and B) hot transparent clouds. Both the hot and cold clouds could be produced by a thermal instability, as discussed by Gouveia Dal Pino and Opher (1989, 1990). They showed that condensations can form in jets of AGN at a distance of about 1 kpc from the nucleus, and in the case of superluminal sources at a distance of about 1 pc from the nucleus. The indicated presence of condensations at a distance from the nucleus much less than 1 pc could also be due to a thermal instability.

Concerning the cold cloud contribution (A), similar to the observed asymmetry phenomenon of one sided jets in extragalactic nuclei, we assume that the central engine forms on its far side a hemispheric shell of opaque clouds which reflect the radiation from the central engine. The reflected spectrum from the opaque clouds (A) is not Doppler shifted as it would be if reflection occurs from a Keplerian disc. The hot transparent clouds (B) surround and are close to the central engine. The hot clouds (B) consist of two parts: B1) clouds reprocessing the power law continuum; and B2) clouds emitting thermal Fe lines and bremsstrahlung radiation. The clouds (B1) and (B2) may not necessarily be the same.

The reference model for the cold cloud contribution (A) is a hemispheric shell behind the central engine which reflects to us the power law radiation from the central engine. If the power law radiation is not isotropic, and less radiation (compared with an isotropic source) is emitted in our direction, then the albedo of the hemispheric shell is increased. We introduce a factor FA to include this possible effect, where $FA = 1$ for an isotropic source, and FA is greater than unity for a nonisotropic source. We also take into account the possibility that the hemispheric shell may not be exactly behind the central engine which could make FA less than unity. We investigate here the values $FA = 2$, 1.5, 1 and 0.7.

Concerning contribution B1, we investigate column densities $CD = 0.1$, 0.5, 1 and 2×10^{23} H atoms cm^{-2}. The temperature of B1 does not need to be very high; it just needs to be sufficiently high to ionize the elements of groups I and II discussed below.

The temperature also Doppler broadens the scattered radiation. To be specific, we used the same temperature T for B1 as we used for B2.

In general, the clouds B2 which emit the thermal lines and the bremsstrahlung radiation can be a hotter group of clouds than B1 with a smaller area. The B2 clouds can be closer to the central engine and the B1 clouds further out. We characterize the effective area of B2 by the ratio ($RBPL$) of the integrated emitted bremsstrahlung spectrum to the integrated power law spectrum. The values of RBPL studied were $RBPL = 0.005, 0.01, 0.02$ and 0.04. The high temperature clouds B2 (e.g., with $T = 5$ keV/k_B), besides being heated by photoionization, could be heated by Alfven waves, as discussed by Jatenco-Pereira and Opher (1989a,b,c). They showed that Alfven waves are of fundamental importance in explaining mass loss and heating in young and old stars and in the sun. The Alfven waves can also be of fundamental importance in the ejection and heating of matter in AGN.

The elements were divided into four groups. The groups used, and the log of the relative abundances of the elements, were: (I): H(0), He(-1.0); (II): C(-3.4), N(-4.0), O(-3.1), Ne(-4.0); (III): Mg(-4.5), Si(-4.5), S(-4.7); (IV): Fe(-4.4). For the hot clouds, groups I and II are ionized and we use for the present calculations $T = 5$ keV/k_B. We investigated also the possibility that the Fe abundance may be above normal. We introduced the factor FFE, with $FFE = 1$ corresponding to the above abundance of group IV. We investigated the values $FFE = 4, 2, 1.5$ and 1.

The Monte Carlo calculations made followed the calculations of Felsteiner and Opher (1976, 1989). The path of every photon is followed until it gets out of the region studied, or is otherwise absorbed. In case it is absorbed, we take into account K-shell X-ray fluorescence (both K_α and K_β). We use the Klein-Nishina formula modified to take into account the motion of the free electrons. This allows the photons not only to lose energy but also to gain energy. We evaluated the thermal Fe line emission from the results of Mewe *et al.* (1985). The thermal Fe line emission broadens the Fe peak as well as the broadening due to the taking into account of the Fe K_β emission.

3. RESULTS

A useful form of presenting results is in the form of a residual plot: calculated (or observed) spectrum minus the power law continuum. Such a residual plot is shown in Fig. 1b of Pounds *et al.* (1990). A power law with an energy index $\alpha = 0.7$ was used with their Ginga-12 data. Their residual plot shows a strong positive feature at 6.0–6.5 keV, a dip between 8–12 keV, and an excess above 12 keV. Above the 6.0–6.5 keV feature, the residual plot goes to zero near 7.1 keV. The Ginga-12 data was obtained from the large area counter (LAC) of the Ginga satellite (Turner *et al.* 1989). It consists of eight proportional counters which cover the range 1.5–35 keV. The energy resolution is 18% at 6 keV and scales as $E^{-0.5}$ over the full energy range.

We present our results in the form of residual plots, normalizing them to go to zero at an energy E_0. We use the relation to form the residual plot:

$$RP(E) = REF(E) - REF(E_0) \times (E/E_0)^{-(\alpha+1)},$$

where $REF(E)$ is the calculated spectrum, taking $E_0 = 7.1$ keV and $\alpha = 0.7$. Since LAC has an energy resolution of 18% at 6 keV, the residual plot (RP) of the observed data is not sensitive to narrow energy structures. We thus smoothed our calculated results with a Gaussian of energy width 5%, which is sufficiently large to smooth narrow energy structures but much smaller than the energy resolution of LAC.

Fig. 1 gives the RP of the cold clouds (A). Fig. 1a shows RP for $FA = 2$, 1.5, 1 and 0.7, with $FFE = 1$. Fig. 1b shows RP for $FFE = 4$, 2, 1.5 and 1, with $FA = 1$. We found for the equivalent widths in Fig. 1a $EW = 180$, 135, 90 and 63 eV, for $FA = 2$, 1.5, 1 and 0.7, respectively. In Fig. 1b $EW = 102$, 111, 107 and 90 eV, for $FFE = 4$, 2, 1.5 and 1, respectively.

Fig. 2 gives the RP from B1 (the reprocessing of the hot clouds). Fig. 2a shows RP (with $FFE = 1$) for $CD = 0.1$, 0.5, 1 and 2×10^{23} H atoms cm^{-2}, having the equivalent widths $EW = 9$, 52, 74 and 125 eV, respectively. Fig. 2b shows RP (with $CD = 0.1 \times 10^{23}$ H atoms cm^{-2}) for B1 for $FFE = 4$, 2, 1.5 and 1, having $EW = 34$, 17, 13 and 9 eV, respectively.

Fig. 3 gives the RP for B2 (emission of thermal Fe lines and bremsstrahlung radiation). Fig. 3a shows RP (with $CD = 0.1 \times 10^{23}$ cm^{-2} and $FFE = 1$) for $RBPL = 0.005$, 0.01, 0.02 and 0.04, having the equivalent widths $EW = 5$, 10, 20 and 40 eV, respectively. Fig. 3b shows RP (with $CD = 0.1 \times 10^{23}$ cm^{-2} and $RBPL = 0.02$) for $FFE = 4$, 2, 1.5 and 1, having $EW = 64$, 37, 29 and 20 eV, respectively.

The total equivalent width of the model and the residual plot are:

$$EW = EW(\mathrm{A}) + EW(\mathrm{B1}) + EW(\mathrm{B2}); \qquad RP = RP(\mathrm{A}) + RP(\mathrm{B1}) + RP(\mathrm{B2}).$$

We find from Figs. 1–3 that the present model can produce the basic characteristics of the residual plot of the Ginga-12 data (Fig. 1b, Pounds *et al.* 1990): 1) the positive feature at 6.0–6.5 keV; 2) the negative RP between 8–12 keV; and the positive RP above 12 keV. Equivalent widths in the observed range 75–150 eV are also obtained.

The present model can be easily generalized to include (besides other values of the parameters for FFE, FA, $RBPL$, CD, α, and T) the effects also of the shifts in energy of the iron edge and the Fe K lines due to high ionization of Fe.

ACKNOWLEDGEMENT

One of us (R.O.) would like to thank the Brazilian agency CNPq for partial support.

REFERENCES

Felsteiner, J. & Opher, R., 1976. *Astr. Astrophys.*, **46**, 189.
Felsteiner, J. & Opher, R., 1989. *Proc. 23rd ESLAB Symp. (ESA SP-296)*, **1**, 401.
George, I.M. *et al.*, 1989. *Proc. 23rd ESLAB Symp. (ESA SP-296)*, **2**, 945.
George, I.M. *et al.*, 1990. *Mon. Not. R. Astr. Soc.*, **242**, 28P.
Gouveia Dal Pino, E.M. & Opher, R., 1989. *Astrophys. J.*, **342**, 686.
Gouveia Dal Pino, E.M. & Opher, R., 1990. *Astr. Astrophys.* (in press).
Guilbert, P.W. & Rees, M.J., 1988. *Mon. Not. R. Astr. Soc.*, **233**, 475.
Jatenco-Pereira, V. & Opher, R., 1989a. *Mon. Not. R. Astr. Soc.*, **236**, 1.
Jatenco-Pereira, V. & Opher, R., 1989b. *Astr. Astrophys.*, **209**, 327.
Jatenco-Pereira, V. & Opher, R., 1989c. *Astrophys. J.*, **344**, 513.
Koyama, K., 1985. In *Galactic and Extragactic Compact Sources*, eds Y. Tanaka & W. Lewin (Tokyo, Inst. of Space & Astronautical Science).

Mewe, R. *et al.*, 1985. *Astr. Astrophys. Suppl.*, **62**, 197.
Nandra, K. *et al.*, 1990. *Mon. Not. R. Astr. Soc.*, **242**, 660.
Pounds, K.A. *et al.*, 1990. *Nature*, **344**, 132.
Turner, M.J.L. *et al.*, 1989. *Publ. Astr. Soc. Japan*, **41**, 345.

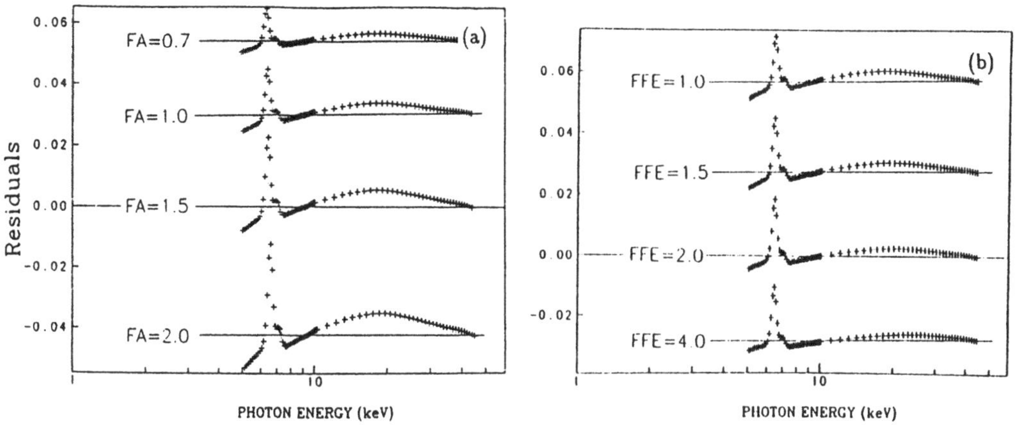

Fig. 1. Reflection from cold clouds. (a) FFE = 1; (b) FA = 1.

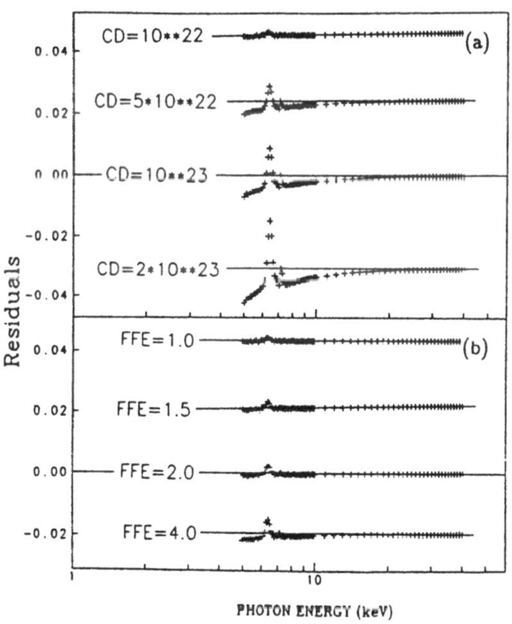

Fig. 2. Reprocessing by hot clouds. (a) FFE = 1; (b) CD = 10^{22} cm^{-2}.

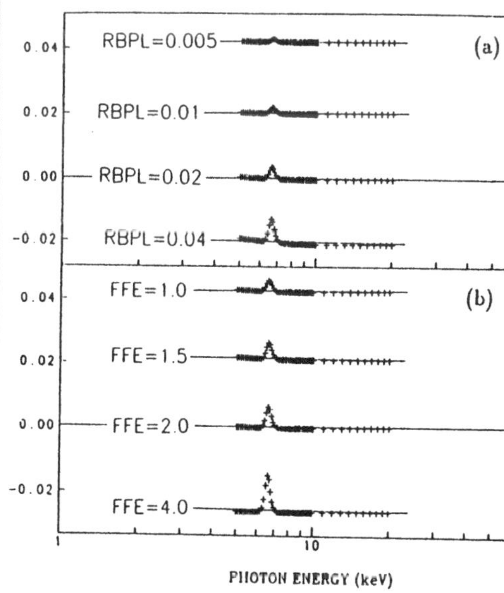

Fig. 3. Emission of thermal iron lines and bremsstrahlung, CD = 10^{22} cm^{-2}. (a) FFE = 1; (b) RBPL = 0.02.

THE ORIGIN OF THE X-RAY SPECTRA OF AGN

Gabriele Ghisellini[1], Andrzej A. Zdziarski[2], Ian M. George[3], Roland Svensson[4], A. C. Fabian[3] and Chris Done[5]

[1] Osservatorio Astronomico di Torino; [2] Space Telescope Science Institute; [3] Institute of Astronomy, Cambridge; [4] NORDITA, Copenhaghen; [5] NASA Goddard Space Flight Center.

ABSTRACT

The presence of a spectral component due to Compton reflection in the X-ray spectra of AGN, as recently observed by *Ginga*, give new clues to the properties of the central regions of these sources. These observations revive models in which much of the luminosity generated in X- and γ-ray energy bands in AGN is reprocessed via electron-positron pairs. In these models the X-ray spectral index is $\alpha_x \simeq 1$ for a wide range of parameters. If a substantial fraction of this radiation is incident upon cold material, the composite (directly observed + reflected) spectrum has an average spectral index of $\alpha_x \simeq 0.7$ in the 1–30 keV band, and steepens above ~ 1 MeV. Such spectra agree well with the X-ray and γ-ray observations of AGN and the constraints imposed by the cosmic γ-ray background.

WHY PAIRS ?

From the observed flux and the variability timescales, one can derive the compactness, ℓ, defined as $\ell \equiv L\sigma_T/(Rm_ec^3) \sim 4000(L/L_E)(3R_sR)$, where L/R is the luminosity to size ratio, L_E is the Eddington luminosity, R_s is the Schwarzschild radius, σ_T is the Thomson cross section, m_e is the electron mass and c is the speed of light.

Since approximately 10% of the bolometric luminosity of a typical AGN is emitted as X-rays, and since they vary with short timescales, the X-ray compactnesses are large, in the range 1–300 (Done & Fabian 1989).

Values of ℓ greater than unity indicate high photon densities, and therefore that collisions involving photons are important. Consequently, in a compact ($\ell > 1$) source there is a high probability that γ-rays collide with softer photons to produce electron positron pairs.

Assume that high energy electrons (or pairs) are continuously injected throughout the source at a rate corresponding to a compactness ℓ_h. These electrons produce, by Compton scattering off soft photons, a hard primary spectrum. The γ-rays so produced can be absorbed colliding with softer photons, producing relativistic electron–positron pairs.

For sufficiently high compactnesses ($\ell_h > 60$), all γ-rays produced in the center of the source get absorbed, and the reprocessing due to pairs 'saturates', in the sense that all the γ-rays are reemitted in the X-ray band. **In this regime the final X-ray spectral index tends to a value close to unity** (Svensson 1987).

The X-ray spectrum steepens from $\alpha \sim 0.6$ when pairs are only marginally important (for $\ell_h = 1$) to $\alpha \sim 1$ when the pair production process is saturated. For $\ell_h > 30$, up to 10 % of the injected electron luminosity is converted into rest mass of pairs, and ultimately into the annihilation line, broadened by scattering and by fast motions within the pair clouds.

At large ℓ_h γ-rays are efficiently absorbed, and the γ-ray spectral index α_γ is twice the spectral index of the target (X-ray) photons: $\alpha_\gamma \sim 2\alpha_x$ (Svensson 1984).

WHY PAIR MODELS DID NOT WORK.

Lightman & Zdziarski (1987) and Done & Fabian (1989) discussed the relevance of pair models for AGN, and concluded that a good agreement with observations was possible only in a small subset of the parameter space. In particular:

- To obtain $\alpha_x \sim 0.7$ the pair reprocessing has to be unsaturated, corresponding to intermediate values of ℓ_h, around a value of 30. With these relatively low values of ℓ_h, **too many γ-rays can escape the source, violating the limits imposed by the γ-ray background.**
- For higher values of ℓ_h, the spectrum steepens, and down-scattering becomes important, imprinting a break in the spectrum. Since this break has not yet been observed, we can derive a limit on the scattering optical depth of cold pairs inside the source, and hence on the maximum allowed compactness. This 'dead electron' problem is eased, but not completely solved, by reacceleration of the cold pairs (Done, Ghisellini & Fabian 1990).

WHY PAIR MODELS NOW WORK.

Pair models are now 'resurrected' not because of an improvement of the theory, but because of an improvement of the available data. *Ginga* has in fact allowed the decomposition of the of the 1–30 keV spectrum of bright AGN into a power law and a Compton reflection hump. **The spectral index of the power law component is $\alpha_x \sim 0.9$, and in general is steeper by a factor $\Delta\alpha \sim 0.2$ than the index derived with a single power law fit** (Pounds 1989).

Zdziarski *et al.* (1990) recently suggested that the source of the power law component is a compact region lying above (and below) an accretion disk. **The compactness $\ell_h > 60$ of this region means that pair production is saturated, and that the emerging X-ray flux is a power law of index $\alpha_x \sim 0.8 - 1$.**

Impinging onto the accretion disk, the hard X-rays are Compton down-scattered, and the soft X-rays are absorbed. The reflected spectrum then peaks at 20–30 keV, depending on the details of the incident spectrum (and hence on ℓ_h). An example is shown in Fig. 1.

As can be seen, the spectral index of the sum is $\alpha_x \sim 0.7$, the down-scattering break is hidden by the contribution of the reflected component, and the overall spectrum turns over at a few MeV, in full agreement with present observations. Note that:

- Saturated pair production is the reason for the universal spectral index of AGN.
- The region of the parameter space where the model works is large, the main requirement being $\ell_h > 60$.
- The model is easily testable. In fact we predict that the spectrum above ~ 30 keV should steepen, at least in the most compact sources, both because of the steepening of the incident spectrum (due to down-scattering by cold pairs) and because of the vanishing contribution of the reflected component.
- Spectral variability is allowed, but with small $\Delta\alpha$, as the luminosity of the compact source varies.
- The γ-ray spectrum of the compact source has $\alpha_\gamma \sim 2\alpha_x \sim 1.8$, similar to the shape of the γ-ray background. It seems therefore likely that the origin of the γ-ray background is emission from AGN.

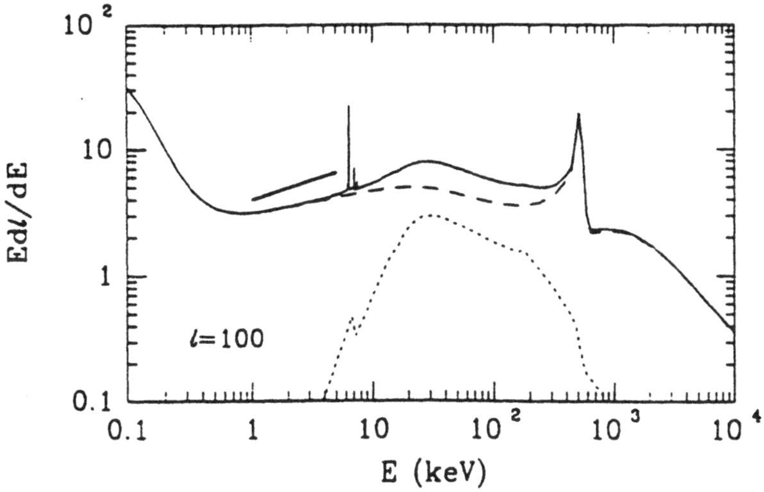

Fig. 1: *Spectrum from combined nonthermal pair cascades and Compton reflection, for a compactness* $\ell = 100$. *The dashed curve is the incident pair cascade spectrum. The dotted curve is the reflected spectrum, and the solid curve is the total observed spectrum. For comparison an* $\alpha = 0.7$ *power law is also shown as a thick solid line. For clarity, the iron* K_α, K_β *and nickel* K_α *fluorescence lines are shown in the total spectrum only. The normalization of the spectrum corresponds to* $\ell/2$ *directly radiated away and* $\ell/2$ *reprocessed by the slab. From Zdziarski et al. 1990.*

References

Done, C., & Fabian, A.C., 1989, *Mon. Not. R. astr. Soc.*, **240**, 81.

Done, C., Ghisellini, G. & Fabian, A.C., 1990. *Mon. Not. R. astr. Soc.*, **245**, 1.

Lightman, A.P. & Zdziarski, A.A., 1987, *Astrophys. J.*, **319**, 643.

Pounds, K.A., 1989, in *AGN and the X-ray background*, 23rd ESLAB Symposium, ESA SP-296, p. 753.

Svensson, R., 1984, in: *X-ray emission from Active Galactic Nuclei*, p.152, eds. Trumper, J., & Brinkmann, W., Max Planck, Garching.

Svensson, R., 1987. *Mon. Not. R. astr. Soc.*, **227**, 403.

Zdziarski, A.A., Ghisellini, G., George, I.M., Fabian, A.C., Svensson, R., & Done, C., 1990, *Astrophys. J.*, **363**, L1.

TWO PHASE ACCRETION DISKS AS A MODEL FOR THE X-RAY EMISSION FROM SEYFERT GALAXIES

F. Haardt and L. Maraschi

Dipartimento di Fisica, Universitá di Milano

ABSTRACT

A thermal two-phase accretion disk model is considered. A substantial fraction f of the gravitational power is assumed to be dissipated via buoyancy and reconnection of magnetic fields in a hot tenuous "corona" surrounding the main body of the disk. The main cooling mechanism of the hot layer is Comptonization of soft photons thermally produced in the underlying cold phase. About 50% of the hard Comptonized photons are backscattered into the thick phase where they are in part reprocessed into the soft blackbody power and in part reflected. The fraction of backscattered, absorbed and reflected photons is estimated by Montecarlo simulations. Coupled thermal balance equations for the two phases yield the temperature of the hot phase and the slope of the Comptonized component as a function of f and τ, the optical depth of the hot phase. We find that for $f \geq 2/3$, and $\tau < 1$, the temperature of the hot phase adjusts so as to maintain $\alpha \simeq 1$. For small τ the temperature of the hot phase is sufficient for pair production to be important. Montecarlo simulations with parameters satisfying the balance equations, show that the model can account in a self-consistent way for the components present in the X-ray spectra of Seyfert galaxies, for their average properties and for the observed short term spectral variability.

I. INTRODUCTION

The origin of the X-ray emission from Seyfert galaxies and Quasars is an important open problem in extragalactic astronomy. Here we discuss a model in which an accretion disk has a two phase structure: an optically thick inner layer (phase 1) surrounded by a hotter tenuous phase (phase 2). The latter is directly heated by dissipation processes and cooled via Comptonization of the soft photons provided by the optically thick component. The proposed physical picture may correspond to accretion disk models where dissipation occurs through magnetic field amplification, buoyancy and reconnection as discussed by various authors (e.g. Burm 1986, Shibata *et al.* 1990).

We shall simply assume that a fraction f of the gravitational power P_G is directly dissipated in the hot corona, while $(1-f)P_G$ is the power dissipated within the optically thick phase. Using f and τ, the optical depth of the corona, as input parameters we derive coupled energy balance equations for the two phases taking into account the heating of the thick layer by the Comptonized photons scattered downwards (Section II). Part of the downdriven X-ray photons actually experience direct Compton reflection, and consequentely a broad bump peaked at ~ 30 keV is present in the overall outgoing spectrum.

The thermal balance equations determine what fraction of the gravitational power is emitted respectively in 1) a blackbody component 2) a power law component due to unsaturated thermal Comptonization 3) a reflection bump. The global subdivision of the emitted power into different spectral components implicitly determines the slope α of the power law and the temperature T_C of the hot phase. In Section III we derive α and T_C adopting analytic approximations for the Comptonized spectrum and solving the implicit equation by an iterative procedure. These results allow us to discuss the dependence of the spectral slope on the model parameters, f, τ, and T_{BB} . The spectral variability associated with a variation of f is examined in detail (Section IV). In Section V realistic emission spectra are computed by a Montecarlo technique for the model parameters derived in Section III. The simulations confirm the analytically determined value of α for the angle averaged photons but show that the upward scattered ones have a systematically flatter energy distribution in the X-ray range. The results are discussed in Section VI.

II. ENERGY BALANCE EQUATIONS

We approximate the two phase disk as two uniform adiacent slabs. The two main parameters are f and τ , while the thickness of the cool phase does not enter explicitly, provided it is sufficiently large to reprocess the hard photons. The hot phase is assumed to be tenuous ($\tau < 1$) so that the main cooling mechanism is Comptonization of the soft photons emitted by the thick phase.

Let us call L_s the total soft photon luminosity emerging from the cool phase of the disk. The total luminosity radiated (in all directions) by the hot phase can be expressed as AL_s where A is the amplification factor due to Comptonization. The luminosity added by the Compton process is

$$L_C = (A-1)L_s \tag{1.a}$$

The total Compton luminosity L_C can be broken down into the upward and the downward components L_{uC} and L_{dC}, so that

$$L_C = L_{uC} + L_{dC} \tag{1.b}$$

We have estimated directly through Montecarlo simulations (see Section III) the fraction of down driven Compton power (see also Pozdniakov *et al.* 1983) obtaining, for the interesting range of parameters:

$$L_{dC} = \eta L_C, \quad \text{with } \eta \simeq 0.5-0.6 \tag{1.c}$$

The hard photons directed downwards are partly absorbed and partly reflected by the optically thick layer. This process has been discussed by Lightman and White (1988), and Guilbert and Rees (1988), who have estimated the fraction of reflected photons showing that their energy spectrum is broadly peaked around 30 keV. The effective albedo a integrated over energy can be obtained by a semi-analytical procedure. We have verified by direct Montecarlo simulations that $a \simeq 0.15$ is a value appropriate for the range of parameters to be considered. We thus set

$$L_{rfl} = aL_{dC} \tag{1.d}$$

$$L_{abs} = (1-a)L_{dC} \tag{1.e}$$

Figure 1: Power flow chart of the model. See text for definitions.

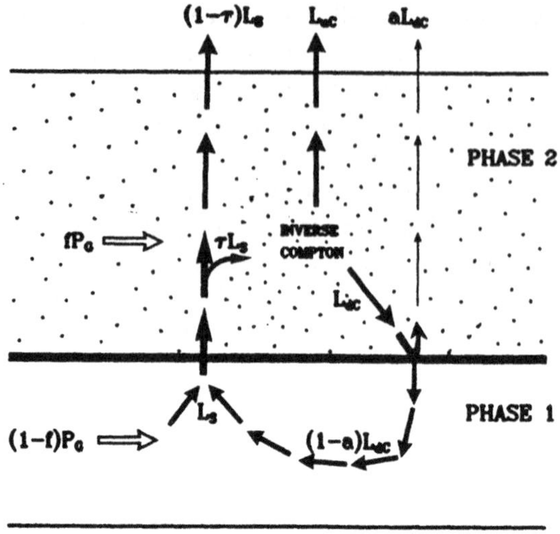

The absorbed power $(1-a)L_{dC}$ will contribute to L_s, while the reflected one aL_{dC} will contribute to the hard component of the emitted spectrum. Second order reflections from the hot layer can be shown to be unimportant for $\tau < 1$. An energy flow chart is depicted in Fig 1.

The above considerations can be summarized in the following two equations:

Energy balance for the thick phase:

$$(1-f)P_G + (1-a)L_{dC} = L_s \tag{2}$$

Energy balance for the hot phase:

$$fP_G + L_s = AL_s \tag{3}$$

Using eqs.(1) one can solve for L_s and A, obtaining:

$$L_s = [1 - f(1-(1-a)\eta)]P_G \tag{4.a}$$

$$A = 1 + \frac{f}{1 - f[1-(1-a)\eta]} \tag{4.b}$$

The fractional term in (4.b) represents the mean gain in energy of the soft photons in crossing the hot layer. For small values of f the amplification is $\simeq 1+f$, while for $f = 1$ (*i.e.* in the case of negligible dissipation occurring in the thick phase) we have the limit

$$A(1) = 1 + \frac{1}{(1-a)\eta} \tag{4.c}$$

The outgoing luminosity L_{out} can be written as:

$$L_{out} = L_s + L_{uC} + L_{rfl} \tag{5}$$

whith

$$L_{uC} = \frac{(1-\eta)f}{1-f+f\eta-fa\eta} L_s \tag{6.a}$$

$$L_{rfl} = \frac{a\eta f}{1-f+f\eta-fa\eta} L_s \tag{6.b}$$

It can be easily verified that $L_{out} = P_G$. The three contributions correspond to 1) a blackbody component, 2) a power law due to Comptonization 3) a reflection bump. It is interesting that their relative intensities depend mainly on f, as a and η can be computed and in any case do not vary substantially.

III. SPECTRAL SHAPE OF THE COMPTONIZED COMPONENT

Since the energy balance equations (4a,b) determine the value of the amplification factor A required for the Comptonization process, they implicitly determine a relation between the optical depth and the temperature and therefore the spectral shape of the Comptonized component. In the following we employ results from the Comptonization theory to derive explicitly the dependence of T_C and α on τ. In the case of $\tau < 1$ one can approximate the Comptonized spectrum with a power law (cfr. Rybicki Lightman 1979)

$$I(\epsilon) = I(\epsilon_0)\left(\frac{\epsilon}{\epsilon_0}\right)^{-\alpha} \tag{7}$$

where ϵ_0 is the average energy of the soft photons and α is given by:

$$\alpha = \frac{-\ln\tau}{\ln A_1} \tag{8}$$

where A_1 is the average energy gain in a single scattering. We can thus write the total Compton luminosity L_C as

$$L_C = (A-1)L_s = \int_{\epsilon_1}^{3kT_C} I(\epsilon)d\epsilon \tag{9}$$

where $\epsilon_1 = A_1\epsilon_0$. Thus, assuming $L_s \simeq I(\epsilon_0)\epsilon_0$

$$A - 1 = (1-\alpha)^{-1}\left[\left(\frac{3kT_C}{\epsilon_0}\right)^{1-\alpha} - \left(\frac{\epsilon_1}{\epsilon_0}\right)^{1-\alpha}\right] \tag{10}$$

We solve eq. (10) by an iterative procedure using (8) and an expression for A_1, valid at relativistic and subrelativistic temperatures, obtained from polynomial fitting of Montecarlo simulations and checked with the results of Podznyakov *et al.* (1983). The derived values of α and T_C are plotted vs. τ for various values of f in Fig 2. The most important result apparent from the figure is that for a range of three orders of magnitude in τ the spectral index α varies between 1.4 and 1.1 (for $f = 1$). Thus it is extremely insensitive to this parameter. The physical reason is that the temperature adjusts so as to keep α nearly constant, which is the only way to produce a constant amplification factor.

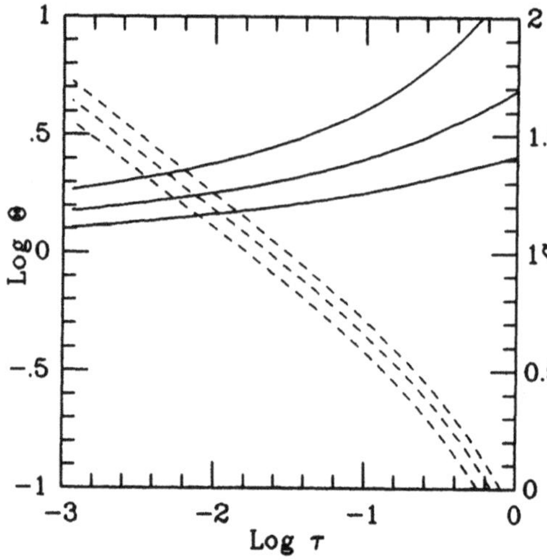

Fig. 2: Coronal equilibrium temperature (dotted line, left scale) and average X-ray spectral index (solid line, right scale) as functions of τ. Higher $\theta = kT_C/mc^2$ and lower α correspond to increasing values of f. Different curves refer to f=0.5, 0.75, 1.

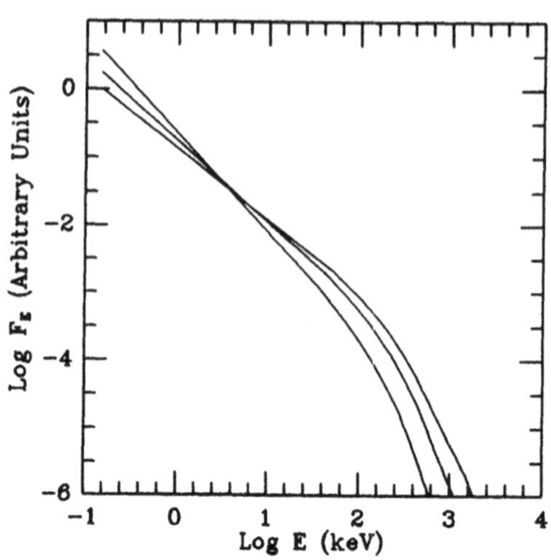

Fig. 3: X-ray spectra computed for varying soft photon input and constant total output.

The second result evident from the figure is that α is always larger than 1 and is larger for smaller values of f. We will show in the following through Montecarlo simulations that the spectral index of the upward scattered photons is systematically flatter than the average one computed above. Good agreement with the observed "universal" value of α =0.7-0.8 (e.g. Pounds *et al.* 1990) can be however reached only for values of f close to 1.

In the solution T_{BB} was fixed at 50 eV. The dependence of the results on this value is very small: the change in α is less than 0.1 for T_{BB} varying between 5 and 500 eV.

For low τ the temperature reaches values close to mc^2, at which pair production may become important. Preliminary work on these regimes indicates that pairs, at least in an equilibrium approximation, have a stabilizing effect, providing an increased optical depth and cooling and thus limiting the maximum attainable temperature.

IV. SPECTRAL VARIABILITY

Suppose the total output from the system is constant on short timescales, since it is implausible that the accretion rate may change substantially on these timescales. It is however possible that the equilibrium described in the previous section is verified in an "average" sense but is violated on timescales on which the magnetic structures, responsible for the dissipation in the hot layer, may change.

We model this situation computing Comptonization spectra with constant power output but varying the soft photon input. The temperature is determined from the condition that the luminosity is constant: thus a higher soft photon input must correspond (for fixed τ) to a steeper spectrum and to a lower temperature. To this end we adopt the approximations discussed in Zdziarski (1985) which take into account deviations from a

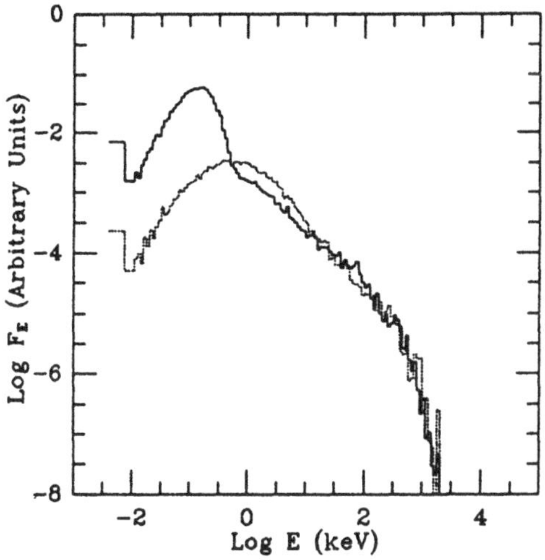

Fig. 4a: Comptonization spectra produced in the hot phase in the upward (solid line) and downward (dotted line) directions. The hot phase parameters are: $\tau = 0.1$, $\theta = 0.5$. The soft photon source is a blackbody with $kT = 50$ eV.

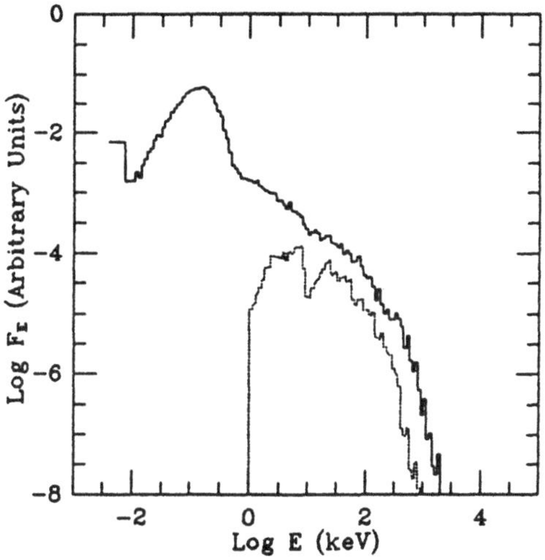

Fig. 4b: Sum (solid line) of the upward Comptonized spectrum and of the reflected part of the downward Comptonized spectrum. The purely reflected component is also shown (dotted line). Parameters are the same of Fig. 4a.

power law due to the Wien peak. The results are illustrated in Fig 3. An increase of the power in the soft band by a factor 1.5 causes a significant reduction of temperature and a steeper spectrum, which has higher intensity in the soft and medium X-ray ranges. We propose that this effect accounts for the observed spectral variability of Seyfert 1 galaxies in the X-ray range. In several cases the spectrum is observed to steepen with increasing intensity in the 2-20 KeV band (e.g Treves et al 1990). It is interesting to mention that in the case of 3C120, where the steepest spectrum has $\alpha \simeq 1$, we observed a pivoting of the spectrum around few keV (Maraschi et al. 1991), as indicated by the present computation.

V. REALISTIC EMISSION SPECTRA

The quasi-analytic approximations used above have been necessary in order to derive average spectral properties which in turn allowed us to determine the physical parameters of the Comptonization region in a consistent way. Once these parameters are known we can take a closer look at the emission spectra by performing Montecarlo simulations. The technique follows closely the method outlined by Pozdnyakov, Sobol and Sunyaev (1983).

We choose values of τ and T_C , satisfying the balance equations, and assume the photon source on one side of the slab, with a blackbody spectrum at T_{BB} =50 eV. We determine separately the energy distributions of photons emerging from the two sides of the slab: the side opposite to that of incidence (transmitted photons) and the same side (backscattered photons).

The spectra of transmitted and backscatterd photons differ significantly in shape in the energy range corresponding to the first scattering, which is clearly due to the

Table 1

τ_{es}	Θ	$EW(i=0^o)$	α_{av}	α_{up}	α_{up+rfl}	$(A-1)_{MC}$
0.03	1.	350 eV	1.2	0.85	0.65	1.9
0.05	0.76	300 eV	1.25	0.9	0.7	2.0
0.1	0.5	250 eV	1.25	0.85	0.7	2.2
0.25	0.28	160 eV	1.3	1	0.85	2.3
0.5	0.16	145 eV	1.35	1.1	0.9	2.1

α_{av}: "average" spectral index derived from the balance equations for f=1.
α_{up}: spectral index derived from the best fit of the Montecarlo simulations of the upward Comptonized component in the [2-30] keV range.
α_{up+rfl}: spectral index derived from the best fit of the Montecarlo simulations of the sum of the upward Comptonized component plus the reflected one in the [2-30] keV range.
$(A-1)_{MC}$: Compton amplification factor derived from integration of Montecarlo spectra. The analytical value for f=1 is A=2.

higher probability and energy transfer of head on collisions. Thus the transmitted component is flatter and contains slightly less power than the backscattered one. We can verify that the analytically estimated value of α closely corresponds to that of the angle averaged distribution. In order to characterize the spectral shape we performed best fits with power laws in the 2-20 keV range of the Montecarlo distributions. In Table 1 we compare the estimated spectral indices α_{up} of the transmitted component with the "average" for various values of the Comptonization parameters. In a second step we produce a Montecarlo simulation of the interaction of the backscattered photons with the thick layer, computing the integrated albedo and the energy distribution of the "reflected" photons. The sum of the transmitted and reflected components which build up the observed spectrum is shown in Fig 4b. Spectral indices for the "observed" distribution estimated by best fits in the 2-30 keV range are reported in Table 1.

Finally we estimate the equivalent width of the Fe line for an isotropic distribution of ionizing photons, and assuming that photoionization of the K_α level occurs at an effective optical depth ~ 1. The effective optical depth is defined as in Rybicki and Lightman (1979), considering the thick layer opacity as due to iron bound-free absorption and electron scattering. Due to the enhanced flux of backscattered photons with respect to the transmitted ones the E.W. can be somewhat in excess of the usual 2π case. A clear prediction of the present model is that the line will be broadened by electron scattering in the hot layer. Quantitative estimates of this effect are underway.

VI. DISCUSSION AND CONCLUSION

The above treatment is limited in several ways. At high temperatures the contribution of pair production in the spectral formation and opacity needs to be included. Preliminary work along these lines indicates that the results reported above are not substantially modified. More physics could be put in the study of the actual dissipation processes and of the vertical structure of both the hot and cool layers. This would however introduce substantial complications, probably requiring sophysticated numerical methods.

The new aspect of the model is that with f as a single fundamental parameter it reproduces very closely the three components observed in the X-ray spectra of Seyfert galaxies. In particular, the strength of the reflection bump and of the Fe line are accounted for due to the 2π solid angle and to the backward asymmetry of the Comptonized flux. A very interesting consequence, which should be worked out in detail, concerns the correlation of time variability between the different components. One can anticipate that the reflected and Fe line components should vary together with the continuum in the same bands, with small delays, while the relation with the blackbody component depends on what physical mechanism actually causes the variability. In Section IV above we have shown an example in which soft and medium X-ray intensities are correlated, but anticorrelated with the hard X-ray intensity.

The model is tightly constrained: in order to reproduce the average slope observed in Seyfert galaxy spectra in the medium X-ray range it is required that practically all the power be dissipated directly within the hot layer. It is possible to relax this requirement introducing a covering factor of the hot phase with respect to the thick one. In this way part of the blackbody luminosity would emerge freely and the soft photon flux available for Comptonization would be smaller, allowing larger values of the amplification and flatter spectra.

In conclusion we have shown that a simple, essentially one parameter model of a twophase disk can account for the X-ray spectral properties of Seyfert galaxies. This remarkable success encourages further work along this line.

References

Burm,H. 1986, *Astr. Ap.*, **165**, 220.

Guilbert, P. W. and Rees, M. J. 1988, *M.N.R.A.S.*, **233**, 475.

Lightman, A. P. and White, T. R. 1988, *Ap. J.*, **335**, 57.

Maraschi, L., Chiappetti, L., Falomo, R., Garilli, B., Malken, M., Tagliaferri, G., Tanzi, E. G., and Treves, A. 1991, *Ap. J.*, **368**, 138.

Maraschi, L. and Molendi, S. 1990, *Ap. J.*, **352**, 452.

Pounds, K. A., Nandra, K., Stewart, G. C., George, I. M., and Fabian, A. C. 1990, *Nature*, **344**, 132.

Pozdnyakov, L. A., Sobol, I. M., and Syunyaev, R. A. 1983, in *Sov.Scien.Rev./Sec.E: Astr. and S.Phys.Rev./Vol.2*, ed R. A. Syunyaev (Moscow), p 189.

Rybicki, G. B., and Lightman, A. P., 1979, *Radiative Process in Astrphysics* (New York: Wiley).

Shakura, N. I. and Sunyaev, R. A. 1973, *Astr. Ap.*, **24**, 337.

Shibata, K., Tajima, T., and Matsumoto, R. 1990, *Ap. J.*, **350**, 295.

Turner, T.J. and Pounds, K.A., 1989, *M.N.R.A.S.*, **240**, 833.

Vishniac, E. T., Jin, L., and Diamond, P. 1990, *Ap. J.*, **365**, 648.

Zdziarski, A. A. 1985, *Ap. J.*, **289**, 514.

COMPTON BROADENING OF THE IRON LINE IN NGC 3227

Bożena Czerny and Magda Zbyszewska
Nicolaus Copernicus Astronomical Center
Bartycka 18, 00-716 Warsaw, Poland
and
D.J.Raine
University of Leicester, Department of Physics and Astronomy
Leicester LE1 7RH, UK

1. INTRODUCTION

The presence of the reflected component in the X-ray spectra of AGN is the principal evidence for the existence of cold matter deep in the nuclei of AGN (see Fabian, these proceedings). The iron K-fluorescence line has been detected in a number of sources (*e.g.* Nandra; Piro; Kunieda – these proceedings).

An emission feature at ≈ 6.5 keV with equivalent width of 130 ± 60 eV in the Ginga spectrum of the Seyfert 1 galaxy NGC 3227 has been reported by Pounds et al. (1989). The shape of the line was modelled by George, Nandra and Fabian (1990) assuming that the line originates in the inner part of an accretion disk and that the broadening is caused by the Doppler effect. However, in such a model there are difficulties in reproducing a line which is both broad and prominent because of the character of the dependence of the luminosity and Doppler shifts on the inclination angle.

Here we propose that the cool gas producing the iron line is embeded in a hot corona. The existence of such a corona seems to be required since it can effectively smear out the Lyman edge expected in the intrinsic disk spectrum (Czerny and Zbyszewska 1990). The Comptonization by a hot corona results also in a broadening of the iron line even if the disk is observed from the top and Doppler shifts are almost negligible.

2. LINE PROFILE

We assume that the hot corona at the top of the cold matter is completely ionized so electron scattering is the only source of opacity. The iron line is formed at the basis of the corona as a result of irradiation of the gas by hard X- and γ-rays. We do not discuss the dilution of the ingoing hard photon flux. Instead we concentrate on the shape of the iron K line emitted by such a stratified medium. The line is broadened as a result of Compton scattering in the corona. All Doppler shifts are ignored in this study.

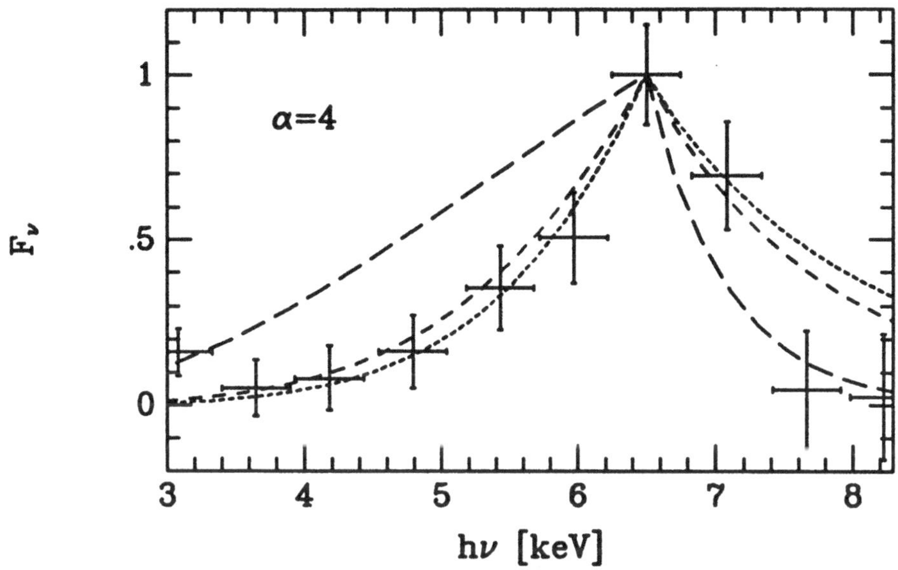

Fig.1 The comparison of data points of NGC 3227 with theoretical profiles for $\tau = 3$ – dotted line, $\tau = 5$ – dashed line, $\tau = 10$ – long dashed line and $kT = 4.46$keV, $kT = 1.87$keV, $kT = 0.53$keV respectively.

We estimate the shape of the iron line using simple formulae for Comptonization of the monochromatic photons at frequency ν_o by a spherically symmetric hot cloud of optical depth τ and temperature T. If $h\nu_o < kT$ the emerging radiation is well described by the simplified formulae (Sunyaev and Titarchuk 1980; see also Rybicki and Lightman 1979):

$$F_\nu = \frac{\alpha(\alpha+3)}{(2\alpha+3)}\left(\frac{\nu}{\nu_o}\right)^{\alpha+3} \quad \text{for} \quad \nu < \nu_o$$

$$F_\nu = \frac{\alpha(\alpha+3)}{(2\alpha+3)}\left(\frac{\nu}{\nu_o}\right)^{-\alpha} \quad \text{for} \quad \nu > \nu_o$$

The energy index α is related to the optical depth and temperature of the plasma through the Compton parameter y:

$$\alpha = -\frac{3}{2} + \sqrt{\frac{9}{4} + \frac{4}{y}}$$

The parameter y describes the relative change of the photon energy due to scatterings. The full-width-half-maximum of such a line can be estimated as:

$$(\Delta\nu)_{FWHM} = 0.72\ \nu_o \frac{2\alpha+3}{\alpha(\alpha+3)}$$

This means that a line with width $\sim$ 1.5keV ($\sigma \sim$ 0.75keV) can be formed as a result of Comptonization by a medium with $y \sim 0.1$.

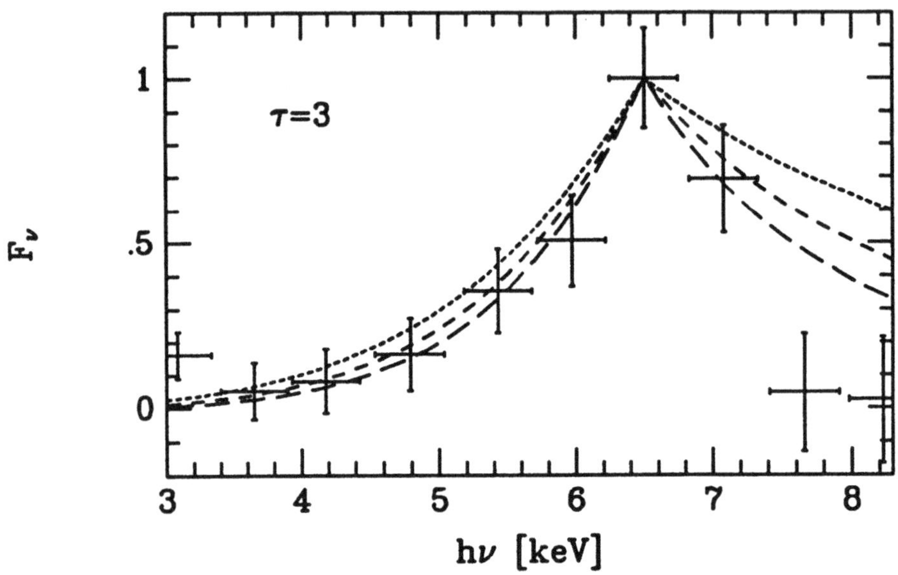

Fig.2 The comparison of data points of NGC 3227 with theoretical profiles for $\alpha = 2$ – dotted line, $\alpha = 3$ – dashed line, $\alpha = 4$ – long dashed line and $kT = 12.50$keV, $kT = 6.94$keV, $kT = 4.46$keV respectively.

In our calculations we used more general formulae for the energy flux given by Sunyaev and Titarchuk (1980; Appendix) to cover the cases $h\nu_o > kT$ as well. The numerical results are shown in Figures 1 and 2, together with the observational points for NGC 3227.

3. CONCLUSIONS

1) The low frequency wing of the iron line in NGC 3227 is in good agreement with the model for $\tau \leq 5$ and $\alpha \sim 3 \div 4$ ($y \sim 0.2 \div 0.14$). The temperature of gas with such parameters should not be lower than a few keV. These plasma parameters are consisted with the non-detection of the Lyman edge. Since the presence of the hot

corona seems to be required by the data any future models should incorporate the stratification of the temperature at the surface layers of the irradiated cool gas.

2) The high frequency wing is not well reproduced by the theoretical curves. Therefore, the iron edge at $\sim$ 7.5keV seems to be required in spite of the asymmetry of the Comptonized line profile (Pounds *et al.* (1989) suggested that such edge was an artefact of their spectral fitting based on symmetrical Gaussian profile).

REFERENCES

Czerny, B. & Zbyszewska, M., 1990. *Month. Not. R. astr. Soc.*, (submitted).

George, I.N., Nandra, K. & Fabian, A.C. 1990. *Month. Not. R. astr. Soc.*, **242**, 28P.

Pounds, K. A., Nandra, K., Stewart, G. C. & Leighly, K., 1989. *Month. Not. R. astr. Soc.*, **240**, 769.

Rybicki, G.B., Lightman, A.P., 1979. In:*Radiative Processes in Astrophysics*, eds. John Wiley & Sons, Inc..

Sunyaev, R.A. & Titarchuk, L.G., 1980. *Astr. Astrophys.*, **86**, 121.

Iron K-lines from Accretion Disks around Schwarzschild Black Holes

L. Stella [1,*] and S. Campana [2]

[1] Osservatorio Astronomico di Brera, Via Brera 28, 20121 Milano, Italy
[2] Università degli Studi di Milano, Via Celoria 16, 20133 Milano, Italy
* affiliated to I.C.R.A.

Abstract

Broad Fe K-shell emission lines at $\sim 6-7$ keV are observed in a number of bright accreting galactic X-ray sources. In the case of the black hole candidates Cyg X-1 and 4U1543-47 the line centroid energy is significantly lower than 6.4 keV, indicating a substantial redshift. Blending of different lines from different ionisation stages of iron and thermal broadening produce widths smaller than observed. The observed widths and redshifts are also unlikely to arise from Comptonisation. The iron K-line profile is most likely dominated by bulk plasma motions in the vicinity of the accreting collapsed object. A fully relativistic model is summarised for the emission-line profile produced by a Keplerian disk orbiting a Schwarzschild black hole. It is shown that the model predictions are in good agreement with the low resolution ($E/\Delta E \leq 10$) line profiles that have been obtained so far. The response of the line intensity to variations at the central source is investigated. A detailed comparison with the recent Ginga results on the variability of the Fe-line intensity of the Seyfert Galaxy NGC 6814 is carried out and constraints are derived on the disk geometry and mass of the central object. Finally a method to accurately measure the mass of disk accreting black holes in active galactic nuclei is outlined.

1. Introduction

Among galactic X-ray sources, spectral features around 6-7 keV are present in virtually all classes of bright X-ray binaries. The spectra of many bright low-mass X-ray binaries (LMXRB) show also iron K-line emission with centroid energies of 6.4-6.7 keV and EWs between 30 and 150 eV (White, Peacock and Taylor 1985; White et al. 1986; Suzuki et al. 1984: Hirano et al. 1987). In the black hole candidate binary Cyg X-1 an emission line of EW $\sim$ 100 eV is observed with a centroid energy of $\sim$ 6.2 keV (Barr, White and Page 1985; Makishima 1986; Balucinska-Church and Barr 1990). In the ultra-soft X-ray transient 4U1543-47, which is considered a potential black hole candidate on the basis of its X-ray spectral properties, an emission line feature has been revealed at an energy of $\sim$ 5.9 keV (van der Woerd, White and Kahn 1989). The X-ray spectra of a number of unobscured Seyfert galaxies also display emission line features around 6.4 keV (Nandra et al. 1989, 1990; Pounds et al. 1990; Pounds et al. 1989, 1990; Piro et al. 1990).

Our capability of exploiting the information contained in these lines is still hampered by the modest energy resolution (up to $E/\Delta E \sim 10$ for gas scintillation proportional counters: GSPCs) of the X-ray detectors covering the $6-7$ keV range which have been flown on X-ray astronomy satellites. Nevertheless in several cases (e.g. Sco X-1, Sco X-2, 4U1705-44, 4U1608-52 among LMXRB; Cyg X-1, 4U1543-47 among black hole candidates) it has been possible to establish that the iron K-lines are significantly broader than the detector resolution, and reach a full width at half maximum (FWHM) of 1 keV or more.

2. Properties of Fe K-lines and Current Scenarios

The centroid energy of Fe Kα-lines is a function of the degree of ionisation and varies from about 6.4 – 6.5 KeV for Fe I-XX, to 6.6 – 6.7 keV for Fe XXIII-XXV and 6.9 keV for Fe XXVI. If the X-ray flux from the vicinity of the collapsed object irradiates the surrounding plasma, the temperature of the plasma depends on the ionisation parameter $\Xi = F/P_{gas}c$, where F is the incident flux (in the 1-1000 Ry range) and $P_{gas}c$ is the gas pressure (Krolik, McKee and Tarter 1981). For ionisation parameters greater than a critical value (roughly $\Xi \geq 10$) the plasma temperature is determined by the effect of Compton scattering and the plasma temperature is high and comparable to the Compton temperature $T_{IC} = <\epsilon>/4$ or $T \sim 10^7 - 10^8$K , where $<\epsilon>$ is the mean photon energy. For low values of the ionisation parameter the effects of atomic processes (photoionisation heating, recombination and collisional excitation cooling) are dominant and the plasma temperature is $\sim 10^4$K. Depending on which of the two situations occurs, K-line emission will come predominantly from recombination onto Hydrogenic or Helium-like iron, around $\sim 6.7 - 6.9$ keV, or from fluorescence of weakly ionised atoms at energies of ~ 6.4 keV. The centroid line energies measured in accreting X-ray sources are usually in the range 6.4 – 6.7 keV, although the black hole candidates Cyg X-1 and 4U1543-47 have centroid energies significantly lower than 6.4 keV. This indicates that a substantial redshift, in addition to line broadening, is present in several cases.

The apparent line width is affected by 4 main broadening mechanisms (cf. Kallman and White 1989 and references therein):

i) Blending of different narrow line components, arising from different ionisation stages: this can produce a brodening of $\Delta E \sim 0.5$ keV at the most.

ii) Thermal broadening is approximately given by $\Delta E \sim Ev_{thermal}/c < 0.05$ keV , and is therefore negligible even for the highest plausible temperatures.

iii) Compton broadening is a result of the energy shift $\Delta\epsilon = \epsilon(4kT - \epsilon)/(m_ec^2)$ due to each Compton scattering. The detailed profile, centroid energy and width of a Comptonised line profile depend on the Thomson depth and the temperature of the electron gas. A Comptonisation width of ~ 1 keV can be achieved for a Thomson depth of ~ 3.

iv) Doppler broadening due to bulk motions of the plasma in the vicinity of the collapsed object. In the case of Keplerian motion $\Delta E \sim (r/50r_s)^{-1/2}$ keV, where $r_s = 2GM/c^2$ is the Schwarzschild radius and M is the mass of the central object.

Much of the modelling of the accretion flow towards collapsed objects assumes the presence of an accretion disk. In X-ray binaries containing a mass donor star overflowing its Roche lobe there is incontrovertible evidence that the accretion flow is mediated by a disk. Accretion disks around the collapsed object are also likely to be present in some wind-fed X-ray binaries and in the nuclei of active galaxies. Only those systems which are likely to possess an accretion disk extending to very small radii, comparable to the dimensions of the collapsed object, are considered in this paper (X-ray pulsar systems will therefore be excluded).

Two basic accretion disk pictures have emerged over the last decade to explain the characteristics of bright disk-accreting X-ray sources. In the *hot corona accretion disk model*, a central hot corona ($kT \gg 10^8$ K) forms, perhaps as a result of instabilities of the radiation pressure dominated region of standard α-viscosity disks. A cold "standard" disk surrounds the hot inner corona, and perhaps also extends to within the hot corona (Shapiro, Lightman, Eardley 1976; Galeev, Rosner and Vaiana 1979).

Low temperature thermal (or nearly thermal) emission ($T < 10^7$K) is emitted by the cold disk, which can explain the soft X-ray components observed in Galactic black hole candidates and the XUV bumps in Seyfert galaxies and Quasars. The hard X-ray power-law components observed from these sources is supposed to originate in non-thermal emission processes in the inner hot corona. Part of the hard X-ray radiation emitted by the hot corona is intercepted by the surrounding cold disk, giving rise to $\sim 6.4 - 6.5$ keV fluorescence lines and to a characteristic reflected spectral component at higher energies (Guilbert and Rees 1988; Lightman and White 1988; George and Fabian 1990; Fabian and George 1990; Matt et al. 1990a, 1990b).

In the other accretion disk scenario, the X-ray heating from the central source will affect the vertical structure of the disk by generating a *Compton heated accretion disk corona* (ADC) in (almost) hydrostatic equilibrium above the disk surface and extending over a wide range of radii (White and Holt 1982; Begelman, McKee and Shields 1983; Begelman and McKee 1983; Fabian et al. 1982). This model has been developed to intepreted the orbital modulations and partial X-ray eclipses observed in a number of bright LMXRB, and has been discussed mostly in relation to the relatively low temperature spectra (equivalent bremsstrahlung temperatures of < 10 keV) observed from these sources. The local emission-line profiles for this model have recently been worked out in some detail by Kallman and White (1989). The 6.7 and 6.9 keV lines from Hydrogenic and Helium-like iron dominates the emission for radii $< 30\ r_s$. At larger radii fluorescence emission around ~ 6.4 keV from lower ionisation stages becomes gradually dominant.

It is worth emphasising that the two scenarios outlined above are not fully developed yet, nor are they fully independent of each other. The point relevant to the present discussion is that a ~ 1 keV width of the iron K-line is likely to be produced by dynamical effects in both cases.

3. Emission Line Profiles from a Relativistic Disk

Our idealised approach assumes that a narrow iron K-shell line at a fixed energy is produced close the surface of the accretion disk and that the line photons can freely propagate outwards (Fabian et al. 1989). The disk emission regions which are relevant to this problem extend to radii of $\leq 100\ r_s$, such that a number of relativistic effects, including gravitational and transverse shifts, are important. The line profile produced by an accretion disk orbiting a Schwarzschild black hole is calculated using a fully relativistic treatment.

We assume a thin disk extending from r_i to r_o, which is observed at an inclination i. Using the disk coordinates r_{em} and ϕ_{em} ($\phi_{em} = 0$ at the line of nodes, where the disk crosses the plane of the sky) to identify an emitting point, the ratio $(1+z)$ of the emitted energies of photons from that point to the observed energies is given by

$$(1+z) = E_{em}/E_{obs} = \left(1 - \frac{3r_s}{2r_{em}}\right)^{-1/2} \left\{1 + \frac{\cos\beta}{\left[2r_{em}(1+\tan^2\xi)/r_s - 2\right]^{1/2}}\right\}, \qquad (1)$$

where $\xi + \pi/2$ is the angle between the direction of emission of the photon and the line connecting the emitting point to the black hole and β the angle between the disk plane and the plane of the photon trajectory (defined by the black hole, the emitting point and the observer). β can be expressed analytically as a function of ϕ_{em} and the system inclination i. To calculate ξ a numerical integration of the photon geodesic equation in a Schwarzschild metric is, in general, required (see the appendix in Fabian et al. 1989 for details).

The first term on the right hand side of Eq. 1 depends only on the radius at which a line photon is emitted and contains the gravitational and transverse parts of the general relativistic shift. This term always produces a redshift. The second term contains the Doppler part of the general relativistic shift. As usual, this term can be either "blue" or "red". The relative contribution of the gravitational and transverse redshifts increases for decreasing radii.

The specific line intensity $I_{E_{em}}$ in the disk comoving frame, is approximated by a delta function $I_{E_{em}} = \varepsilon\ \delta(E_{em} - E_o)$, where E_o is rest energy of the line. The geometry of the inner accretion disk surrounding a black hole and the corresponding line emissivity as a function of radius are poorly understood. It is customary to assume that the surface line emissivity ε varies as r_{em}^q. Alternatively, ε can be derived self-consistently by numerical methods once a disk geometry is prescribed. In all cases, for radii much larger than the dimensions of the central

X-ray emitting region (where the radiation from the central X-ray source intercepts the disk only obliquely) we expect that $\varepsilon \propto r_{em}^{-3}$. Here we assume that the line intensity in the comoving frame is isotropic.

By integrating Eq. 1 over the line emitting portion of the disk the emission line profile seen by an observer at infinity is calculated (see the appendix in Fabian et al. 1989 for details). The important parameters that determine the line profile are the inner radius, r_i, the outer radius, r_o, the inclination, i, and the radial dependence of the line emissivity. Radii are expressed in units of the Schwarzschild radius $r_s = 2.95 \times 10^5 (M/M_\odot)$ cm , such that all figures can easily be rescaled for any mass. The line is roughly similar to the two-horned shape familiar from observations of the disks of cataclysmic variables and spiral galaxies.

In Fig. 1 we show the line profile stepping through a range of each parameter. At each step the other parameters are fixed at $r_i = 10\ r_s$, $r_o = 100\ r_s$, $i = 30°$ and $q = -2$. In the first panel the inner radius is varied: as r_i decreases the red wing of the line becomes more elongated, whereas the blue side of the line remains almost unchanged. This is due to the increasing importance of gravitational and transverse redshifts for small radii. As the outer radius increases the line profile becomes narrower and the shift of the blue and red horn of the line decreases (see the second panel of Fig. 1). This is a consequence of the lower velocities in the outer disk regions.

Aberration, time dilation and blueshift, modified by the fraction of the disk which is relevant and the disk inclination, make the blue horn brighter than the red horn. The effect is more pronounced for high inclinations: the blue part of the line is dominant over a wide range of parameters and the mean photon energy of the line is higher than the rest energy, despite gravitational and transverse redshifts. This is clearly seen in the third panel of Fig. 1. As the inclination decreases the gravitational and transverse redshifts become dominant and the mean photon energy of the line is lower than the rest line energy.

For different values of q the importance of the line wings relative to the line core changes, due to a different contribution of the inner disk region to the integrated line profile (see the fourth panel of Fig.1). It is evident that for $q = -2$ the line profile is most sensitive to the outer radius and the inclination.

The overall shape of the line can be used as a sensitive diagnostic to determine the inclination of the system, as well as the location of the line emission region. The plots are given in terms of r/r_s and are, therefore, independent of the mass of the central object.

4. Comparison with Observed Line Profiles

The relativistic disk line profiles discussed in the previous section could be relevant to the iron K-lines of a wide range of accreting compact objects including low mass X-ray binaries, galactic black hole candidates and active galactic nuclei. Contrary to the other broadening mechanisms discussed in section 2, line widths of $\sim$ 1 keV or more and substantial redshifts are easily obtained from a relativistic accretion disk. However, instrumentation flown so far has only a resolution of at best $E/\Delta E \sim 10$ in the 6 to 7 keV range and the distinctive profiles shown in Fig. 1 will be considerably smeared. In Fig. 2 we show the profile as it would be seen using a GSPC for $r_i = 20\ r_s$, $r_o = 200\ r_s$, $i = 30°$ and $q = -2$. The double peak has been smoothed into a single broad peak, that is red-shifted from the original rest energy. This is very similar to the feature reported from Cyg X-1 by Barr, White and Page (1985) and by Makishima (1986). The EXOSAT GSPC spectrum of Cyg X-1 is easily fitted with the relativistic line model (Fabian et al. 1989). Despite the large uncertainty, the derived inclination is consistent with the constraints from optical measurements and from the lack of an X-ray eclipse. The line rest energy is constrained to be between 6.4 and 6.8 keV and is consistent both with fluorescence from low ionisation stages of iron at 6.4 keV and $\sim$ 6.8 keV emission from Hydrogenic or Helium-like iron in a hot plasma. By analogy with the case of Cyg X-1 and, possibly, 4U1543-47 we can predict that a broad iron feature showing a inclination-dependent red- or blue-shift of up

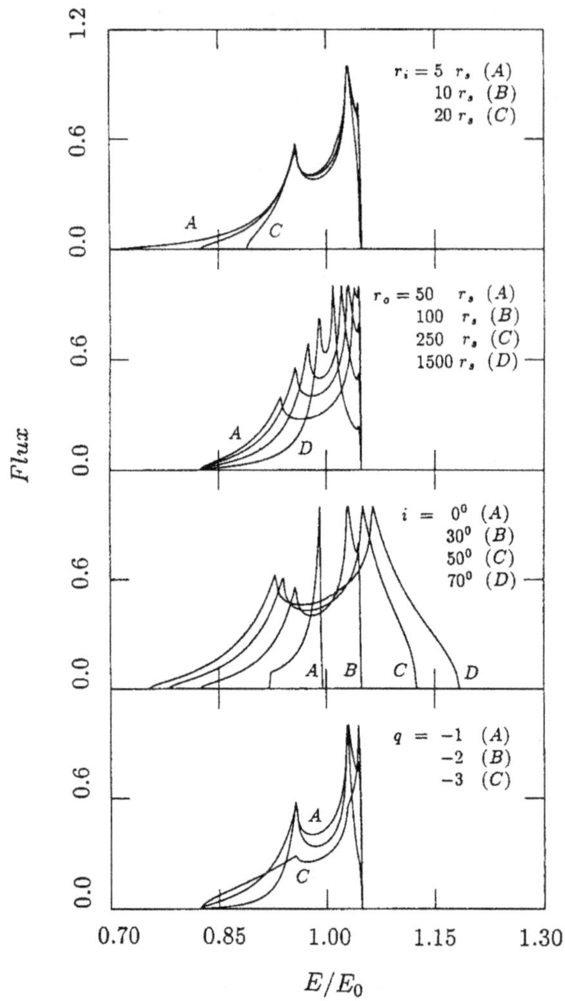

Figure 1. Line profiles from a relativistic accretion disk. When not specified, the other parameters are fixed at $r_i = 10\ r_s$, $r_o = 100\ r_s$, $i = 30°$ and $q = -2$.

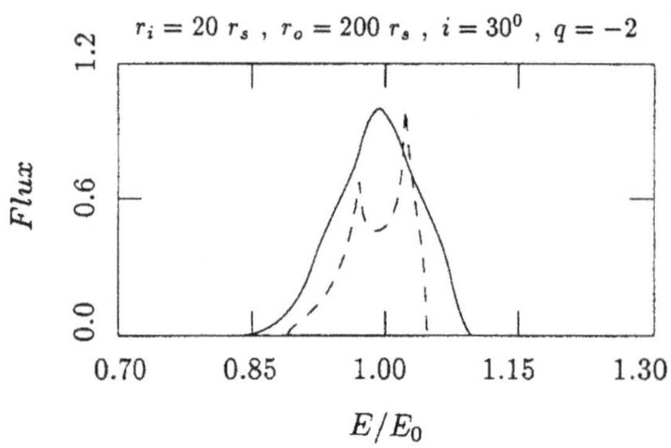

Figure 2. Line profile from a disk with $r_i = 20\ r_s$, $r_o = 200\ r_s$, $i = 30°$ and $q = -2$, convolved with a 10 per cent wide rectangular function for comparison with the GSPC observation of Cyg X-1 reported by Barr, White and Page (1985). The dashed curve gives the corresponding full resolution profile.

to 1 keV with a FWHM of $\sim$ 1 to $\sim$ 2 keV and an EW of 100 to 200 eV should be found in all accreting black holes X-ray sources involving a disk and corona. The X-ray transients ASM 2000+25 and GS 2023+338 (Tanaka 1989, 1990) are important targets in this respect.

Similar line broadening and asymmetries are also to be expected from LMXRB where the low surface magnetic dipole moment of the neutron star allows the disk to approach close to the neutron star, and in disk-accreting AGN (George and Fabian 1990; Fabian and George 1990; Matt et al. 1990a, 1990b).

5. Line Intensity Variations

Variability in disk accreting X-ray sources is likely produced in the vicinity of the collapsed object, where the bulk of the X-ray flux is emitted. Flux variations from the central X-ray source induce changes in the disk line emissivity. Light travel time effects cause the innermost line emitting regions of the disk to respond first followed by the outer regions. To give a quantitative description of the disk "reverberation" we consider the case in which the dimensions of the central source of radiation are smaller than the inner radius of the line emitting portion of the disk. For an observer at infinity any variation in the flux from the central source will be "echoed" by different locations in the disk at different times. In the limit of a point-like central source, neglecting all general relativistic delays in light propagation, a flux variation from the central source reaching a distant observer at a time t_o will be echoed after a time $t - t_o$ by the disk region defined by

$$t - t_o = r(1 - \sin i \sin \phi)/c \quad . \tag{2}$$

The locus of constant t on the disk (the "echo front") is an ellipse of eccentricity $\sin i$, with the major axis along the projection of the line of sight on the disk plane and the most distant focus at $r = 0$. As in the stationary case the line profile at the observer's time t is determined by r_i, r_o, i and ε; in this case, however, the disk region at different r and ϕ respond to flux variations from the central source with a time delay given by Eq. 2 and therefore the line emissivity ε contains and additional dependence on r and ϕ. The relevant profile integrations can be carried by specifying a model for the variability of the central X-ray source. Fig. 3 shows the response of the line intensity from an accretion disk to a rectangular impulse lasting for a short time $\Delta t = r_s/c$, (times are expressed in units of the Schwarzschild radius light crossing time $r_s/c = 0.98 \times 10^{-5}(M/M_{\odot})$ s). The inner disk radius is taken to be $r_i = 25\ r_s$ and the disk line emissivity to vary as r^{-3}, consistent with a central source of X-ray radiation with radius $\ll r_i$. For relatively low inclinations ($i < 30^{\circ}$) the line intensity response shows a single peak, centered around a time delay of $\sim 25 - 40\ r_s/c$. These values are consistent with simple-minded light travel time arguments for a disk with $r_i = 25\ r_s$. For higher inclinations, however, an additonal peak forms at shorter time delays, whose relative amplitude increases for increasing inclination. For $i > 50^{\circ}$ the first peak dominates, and the maximum in the response is reached for delays of $\sim$ 10, 6, 2 and 1 r_s/c, respectively for $i = 50^{\circ}, 60^{\circ}, 70^{\circ}$ and 80°. This simply reflects the fact that for increasing inclinations the disk regions closest to the line of sight respond faster and faster (the disk "echo front" becomes more and more elongated towards the observer). Given the short duration $\Delta t = r_s/c$, of the variation (in fact the shortest that might be expected from an emitting region encompassing the black hole), the curves in Fig. 3 approximate the transfer function Ψ of the line intensity for the system, i.e. the line intensity produced in response to a δ-function variation in the central source at $t = 0$. The disk line intensity for sufficiently high inclinations, therefore, can respond much faster than r_i/c.

By definition of transfer function $I_l(t) = \Psi * I_c(t)$, where $I_c(t)$ and $I_l(t)$ represent the intensity of the central source and the line intensity, respectively, and "$*$" indicates a convolution. Therefore, in the presence of long variations at the central source, the line intensity remains insensitive to any narrow features of Ψ.

Delays between the variations at the central source and line variations can be studied through the cross-correlation function $C(\tau)$ of $I_c(t)$ and $I_l(t)$. In particular the peak of $C(\tau)$ is often used to measure a fiducial delay. It is easy to show that $C(\tau) = \Psi * A_c(\tau)$, where $A_c(\tau)$ represents the auto-correlation function of $I_c(t)$ (cf. Penston 1990). This relation shows that narrow features of Ψ can be seen in the cross-correlation function only in the presence of a comparably narrow autocorrelation function of the central source variations. We conclude that in the case of the disk of Fig. 3 the fast response of the line intensity for $i > 50^{\circ}$ (corresponding to the first peak) can be detected if comparably fast variations characterise the variability of the central source.

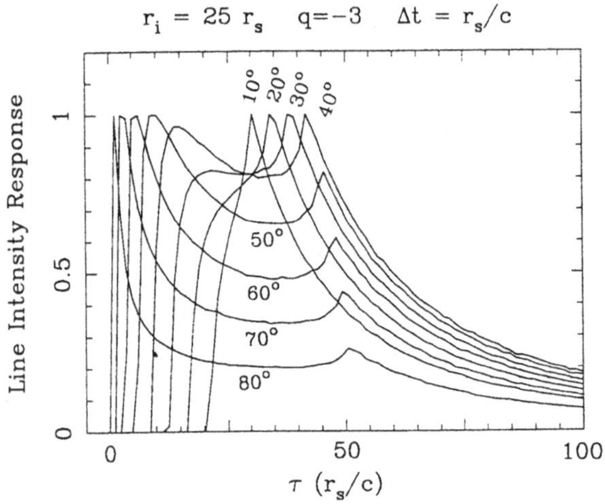

Figure 3. Response of the line intensity from a relativistic disk to a short rectangular flux variation at the central source, as a function of time delay (τ). Different curves correspond to different inclinations.

6. The Case of NGC 6814

Recent Ginga observations have revealed a dramatic variability in the Seyfert galaxy NGC 6814 (Kunieda et al. 1990a, 1990b). In particular the 2 – 20 keV flux, corresponding to an average luminosity of $L_x \sim 10^{43}$ erg s^{-1}, was observed to vary by a factor of 2 in $\sim$ 50 s. An intense iron line was detected with a centroid energy of 6.4 ± 0.1 keV, an average EW of $\sim$ 300 eV and a FWHM of < 1 keV. The Fe-line intensity variations were strongly correlated with variations of the X-ray continuum. No asymmetry or lag was found in the cross-correlation function, setting an upper limit of $\sim$ 250 s to any delay of the Fe-line intensity.

In this section we show how the available observations can be used to derive constraints on an idealised accretion disk scenario for the Fe-line formation in NGC 6814. We assume an infinite flat disk illuminated by a spherical central source of hard radiation. In order to maximise the fraction of radiation intercepted by the disk and, therefore, the EW of the line we assume that the radius of the central source r_c coincides with the inner radius r_i. (Even in this case, however, it is difficult to achieve an EW as high as $\sim 300\ eV$, cf.George and Fabian 1990; Matt et al. 1990a, 1990b). The disk line emissivity for this geometry is calculated using the prescription of Chen and Halpern (1989). The stationary line profile observed at infinity is determined in this case only by the inner radius r_i and the inclination i. The delay of the maximum response of the line intensity depends also on the variability timescale at the central source (see previous section). In the absence of the autocorrelation function of the continuum X-ray flux, we consider the shortest possible variability timescale consistent with the assumed geometry, *i.e.* r_c/c.

Fig. 4a shows the centroid energy of the line profiles obtained in this way as a function of i and for a range of different r_i. The centroid energy is calculated from the ratio of the 1st and 0th moments of the energy distribution in the line and is normalised to the line rest energy. It is apparent that the centroid energy is blueshifted (redshifted) for inclinations higher (lower) than $\sim 50^\circ$. As expected, disks extending to smaller radii r_i are characterised by more pronounced shifts. An observed energy of 6.4 ± 0.1 keV can either result from a low ionisation fluorescence line (rest energy $E_o \sim 6.4$ keV) with only little blueshift or redshift, or from a substantially redshifted line with $E_o \sim 6.7 - 6.9$ keV from the highest ionisation stages of iron. The points in Fig. 4a consistent with the former (latter) interpretation are indicated by filled (empty) symbols.

Fig. 4b shows the width of the line profiles, calculated from the ratio of the square root of the 2nd moment and 0th moment of the energy distribution. A normalisation of $2.35/E_o$ is used, which allows to reproduce the FWHM in the case of a Gaussian profile. The width of the line depends strongly on r_i and i. Note that the increasing contribution of the gravitational and transverse redshifts for decreasing radii reduces the dependence of the line width on the inclination. Filled symbols are consistent with the observed FWHM < 1 keV.

Fig. 4c shows the delay in the line intensity variations, as evaluated from the maximum of the response to a rectangular impulse from the central source. The impulse duration r_c/c is in this case comparable to the light crossing time of the inner emitting disk regions and the first peak in the disk transfer function (see Fig. 3) is substantially broadened. Nevertheless a factor of ~ 3 decrease of the delay for increasing inclinations is apparent for $r_i \geq 10$. For smaller values of r_i the behaviour is complicated by light deflection. In addition to the constraints provided by the line centroid energy and width (Figs. 4a and 4b), the following considerations apply.

(A) The assumed flat geometry and Keplerian velocity distribution of the disk requires that the luminosity is in the sub-Eddington regime. This implies $M > 10^5 M_\odot$ and, therefore, the scale of Fig. 4c must convert to seconds at least. By imposing that the line delay is < 250 s, we rule out all profiles characterised by $\tau > 250\ r_s/c$.

(B) The shortest delays ($\tau \sim 7\ r_s/c$) compatible with the constraints in Figs. 4a and 4b, are for $r_i = 3\ r_s$ and $i < 25^\circ$. By imposing that this delay is < 250 s the mass is constrained to be $M < 3 \times 10^6 M_\odot$.

(C) A somewhat stronger constraint on the mass is obtained by requiring that the minimum variability timescale of the central source $r_c/c = r_i/c$ is shorter than 50 s. This implies $M < 5 \times 10^6 (r_s/r_i) M_\odot$. Therefore, all profiles with $r_i > 50 r_s$ are ruled out, as they would require a mass inconsistent with the sub-Eddington regime.

Fig. 4c summarises all available constraints. Empty symbols correspond to the range of parameters allowed for a high ionisation Fe line with $E_o \sim 6.7 - 6.9$ keV. In this case small values of r_i and i are required in order to produce a substantial redshift. The range of parameters allowed in the case of a low ionisation fluorescence line (6.4 keV) is marked by filled symbols. In this case $10\ r_s \leq r_i \leq 50\ r_s$ and a large range of inclinations is allowed. In either interpretations, the mass of the central black hole must be $10^5 M_\odot < M < 5 \times 10^6 (r_s/r_i) M_\odot$. The latter inequality provides a stronger constraint in the case of a 6.4 keV line.

7. Measuring Black Hole Masses

Flux variations at the central source induce also characteristic profile changes of the line from the accretion disk. Owing to light travel time effects, the innermost disk regions, corresponding to the highest Keplerian velocities, respond first. This causes a faster response of the line wings than the core.

Figure 5 shows three examples of line profiles and their response to a step-like doubling of the flux from the central object, which lasts for a short time Δt. Once the entire echo front is included in the line emitting portion of the disk two features form on the wings of the line profile. As time increases, the echo front propagates outwards through the disk and the two features

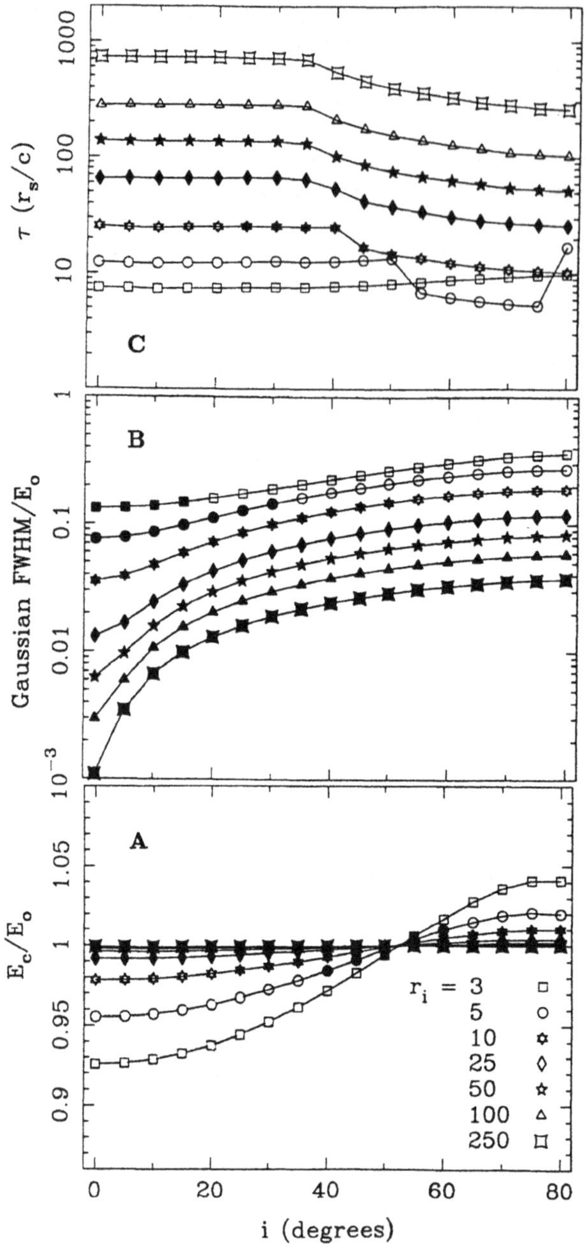

Figure 4. (a) Normalised line centroid energy vs inclination for the idealised accretion disk geometry discussed in the text. Different symbols correspond to different inner radii r_i. The points consistent with the interpretation in which the 6.4 ± 0.1 keV line centroid energy observed in NGC 6814 results from a line with rest energy of $E_o \sim 6.4$ keV ($E_o \sim 6.7 - 6.9$ keV) are indicated by filled (empty) symbols. (b) Normalised Gaussian FWHM vs inclination. Filled symbols are consistent with a FWHM of < 1 keV. (c) Delay in the response of the line intensity vs inclination. Filled (empty) symbols represent solutions consistent with a line rest energy of $E_o \sim 6.4$ keV ($E_o \sim 6.7 - 6.9$ keV). Constraints from panels a and b have been taken into account, together with the additional constraints discussed in the text.

drift toward the blue and red horns of the line profile as the Keplerian velocities decrease. Once the front echoing the final part of the source flare has propagated beyond the outer disk radius, the line will have returned to its original profile. The position of the peak of the drifting features at a time $t - t_o$ is determined by the outermost echo front within the line-emitting portion of the disk (i.e. the disk region "echoing" the onset of the source flare), and in particular by the two zones at radius r_f where blueshift and redshift are highest. The drifting blue and red features provide a sensitive diagnostic of the inner disk regions. The information contained in them is exploited by expressing their shift as a function of the disk parameters and time (Stella 1990).

By measuring the observed position of the blue and red drifting features at any given time the accretion disk inclination i can be determined together with r_f. These measurements of radii involve scale lengths in units of the Schwarzschild radius inferring nothing about the mass of the central object. However, by monitoring the position of the blue or red drifting features at two different times t and t', the difference of the corresponding radii $r'_f - r_f$ is evaluated also in absolute units. The combination of the relative and absolute measurements of $r'_f - r_f$ provides the mass of the central object (Stella 1990).

The number of X-ray photons at the Earth per Schwarzschild radius light crossing time is a factor of $\geq 10^4$ higher for bright AGN than for X-ray binaries. Therefore, prospects for observing the drifting features from individual source flares on the wings of X-ray lines are far more promising in AGN.

The Balmer emission lines in the optical spectra of several AGN show a broad double-horned profile which is likely to be produced by matter orbiting the central massive black hole in a disk-like geometry. Observation of the characteristic optical line profile changes induced by flux variations at the central source is within reach of current instrumentation.

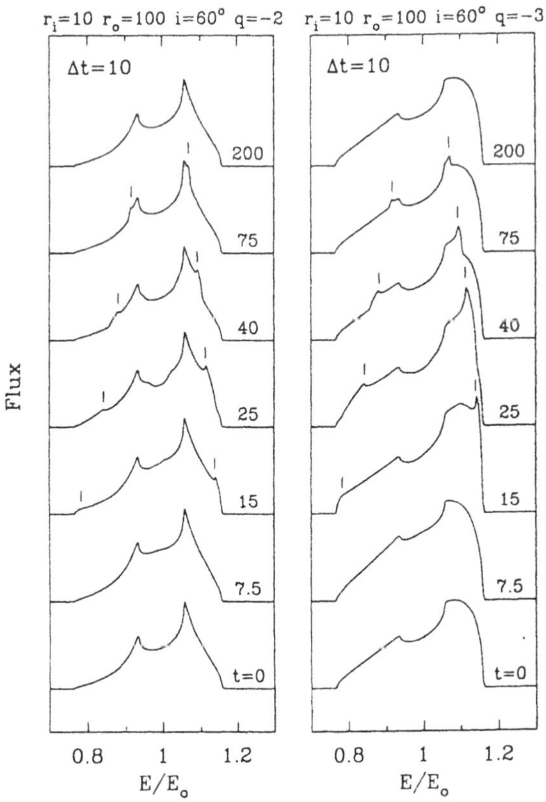

Figure 5. Response of the line profile from a relativistic Keplerian disk to variations of the flux from the central source, as a function of energy. Two examples are shown, corresponding to different choices of the parameter q. Radii are in the range of values inferred from the widths of Fe Kα lines in accreting X-ray binaries. A rectangular doubling of the central source flux, starting at $t_o = 0$ and lasting for Δt, is assumed. The profiles are shown for seven different times t after the flare onset. The vertical lines represent the predicted position of the drifting features (cf. Stella 1990).

8. Conclusions

The iron K-line profiles of accreting collapsed objects in X-ray binaries and AGN should be resolved by the next generation of X-ray detectors, such as the solid state spectrometers and CCD detectors to be flown on BBXRT, Astro-D, Spectrum X-Gamma, AXAF and XMM (resolution of $E/\Delta E \geq 50$ at 6-7 keV). If future observations confirm the relativistic accretion disk picture outlined above, the dimensions (in units of r_s), inclination and emissivity distribution can be directly measured from the iron K-line profiles. In addition the characteristic variability expected in the line profiles can provide far more reliable measurements of black hole masses in AGN than those available to date. Iron K-lines may be the observational key to the study of the innermost regions of accretion flows around black holes.

References

Balucinska-Church, M., and Barr, P., 1990, this volume.

Barr, P., White, N.E., and Page, C.G., 1985, *Mon. Not. R. ast. Soc.*, **216**, 65P.

Begelman, M.C., and McKee, C.F., 1983, *Astrophys. J.*, **271**, 89.

Begelman, M.C., McKee, C.F., and Shields, G.A., 1983, *Astrophys. J.*, **271**, 70.

Chen, K., and Halpern, J.P., 1989, *Astrophys. J.*, **344**, 115.

Fabian, A.C., and George, I.M., 1990, this volume.

Fabian, A.C., Guilbert, P.W., and Ross, R.R., 1982, *Mon. Not. R. ast. Soc.*, **199**, 1045.

Fabian, A.C., Rees, M.J., Stella, L., and White, N.E., 1989, *Mon. Not. R. ast. Soc.*, **238**, 729.

Galeev, A.A., Rosner, R., and Vaiana, G.S., 1979, *Astrophys. J.*, **229**, 318.

George, I.M., and Fabian, A.C., 1990, *Mon. Not. R. ast. Soc.*, in press.

Guilbert, P.W. and Rees, M.J., 1988, *Mon. Not. R. ast. Soc.*, **233**, 475.

Hirano, T., et al., 1987, *Publ. Astron. Soc. Japan*, **39**, 619.

Kallman, T., and White, N.E., 1989, *Astrophys. J.*, **341**, 955.

Krolik, J.H., McKee, C.F., and Tarter, C.B., 1981, *Astrophys. J.*, **249**, 422.

Kunieda, H. et al., 1990a, *Nature*, **345**, 786.

Kunieda, H. et al., 1990b, this volume.

Lightman, A.P. and White, T.R., 1988, *Astrophys. J.*, **335**, 57.

Makishima, K., 1986, in *The Physics of Accretion onto Compact Objects*, Eds. K.O. Mason, M.G. Watson and N.E. White, (Springer-Verlag, Berlin), p. 249.

Matt, G., Perola, G.C., and Piro, L. 1990a, *Astron. Astrophys.*, in press.

Matt, G., Perola, G.C., and Piro, L. 1990b, this volume.

Nandra, K., et al. 1989, *Mon. Not. R. ast. Soc.*, **236**, 39P.

Nandra, K., Pounds, K.A., and Stewart, G.C. 1990, this volume.

Penston, M.V., 1990, in *Variability in AGN*, Ed. H.R. Miller and P.J. Wiita, (Cambridge Univ. Press), in press.

Piro, L., Yamauchi, M., and Matsuoka, M., 1990, *Astrophys. J.Letters*, **360**, L35.

Pounds, K.A., et al., 1989, *Mon. Not. R. ast. Soc.*, **240**, 769.

Pounds, K.A., et al., 1990, *Nature*, **344**, 132.

Shapiro, S.L., Lightman, A.P., and Eardley, D.M., 1976, *Astrophys. J.*, **204**, 187.

Stella, L., 1990, *Nature*, **344**, 747.

Suzuki, K. *et al.* , 1984, *Publ. Astron. Soc. Japan*, **36**, 761.

Tanaka, Y., 1989, in *23rd ESLAB Symposium: Two Topics in X-ray Astronomy*, Ed. J. Hunt, B. Battrick (ESA-SP296), p. 3.

Tanaka, Y., 1990, this volume.

van der Woerd, H., White, N.E., and Kahn, S.M., 1989, *Astrophys. J.*, **344**, 320.

White, N.E., and Holt, S.S., 1982, *Astrophys. J.*, **257**, 318.

White, N.E., Peacock, A., and Taylor, B.G., 1985, *Astrophys. J.*, **296**, 475.

White, N.E., et al., 1986, *Mon. Not. R. ast. Soc.*, **218**, 129.

Rapid Variability of Iron line from NGC 6814

H. Kunieda, H. Awaki, K. Koyama, Y. Tawara, and Y. Tsusaka
Department of Physics, Nagoya University
Nagoya, Japan

C. Done, R. Mushotzky, and T. J. Turner
NASA/Goddard Space Flight Center
Greenbelt, MD 20771. USA

K. Leighly, and S. Tsuruta
Montana St. University
Bozeman, MT 59717, USA

GINGA observational results of the fastest variable Seyfert I galaxy NGC 6814 is presented. The shortest variable time scale of 50 sec was found. It suggests the central mass of $10^6 M_\odot$. Strong Iron line at 6.4 keV is observed with the equivalent width of up to 500 eV. The Iron line flux changed together with the total X-ray flux. The delay time seems to be less than 256 sec. If the line were attributed to cold Iron, this result might imply the cold accretion matter comes closer than 10^{13} cm to the central source. Quasi-periodical time behavior of ~ 12000 sec is suggested with dips and sharp drops. This time scale corresponds to the Keplerian orbit at R ~ 10^{13} cm around the central object of $10^6 M_\odot$.

1. X-ray Light Curves of NGC 6814

NGC 6814 has been observed with Ginga at three occasions; April 28-30, 1989, April 15-17, 1990, and October 6-7, 1990. The first observation was performed as a collaborative observation of Mushotzky and Nagoya group. The second one was conducted along with the proposal by Tsuruta and Kunieda. The third one is one of the target-of-opportunity observations of Ginga team. In this proceeding, data taken in the first two observations are presented. When Ginga observed this source, it was in a moderate level of 5 - 40 c/s, which corresponds to 0.5 - 4 mCrab. X-ray light curves (2 - 20 keV, 1 point = 64 sec) in Fig.1 a, b show rapid variability by a factor of 5 or more.

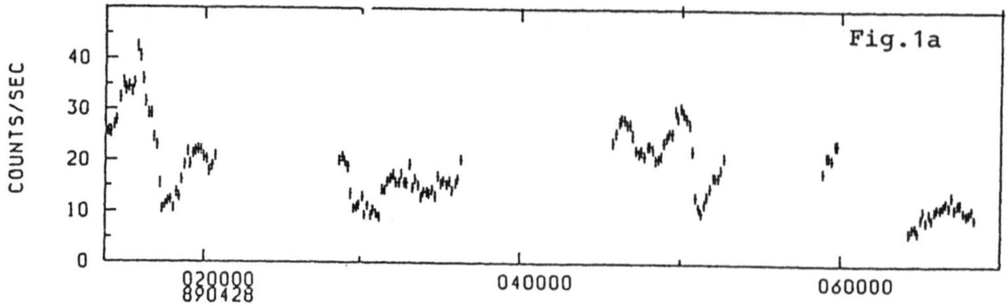

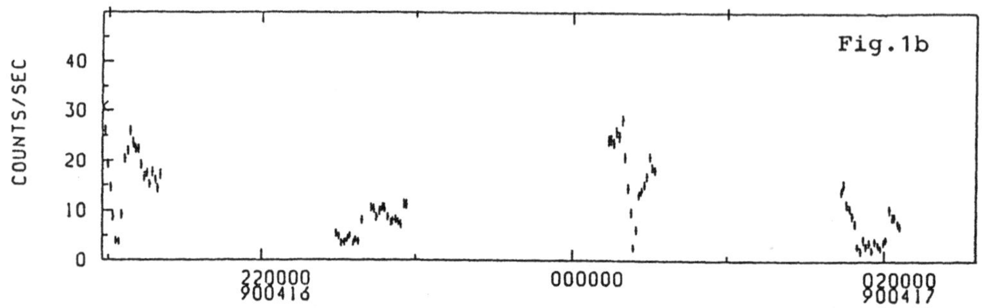

Fig. 2 a, b are the blow up version with 16 sec time resolution of the X-ray light curves at the fastest variable parts of Fig. 1a and b. In 1989, flux decreased down to quarter of the maxima, while it dropped to almost zero at the minima in 1990. The shortest doubling (halving) time scale is found to be ~ 50 sec in both cases.

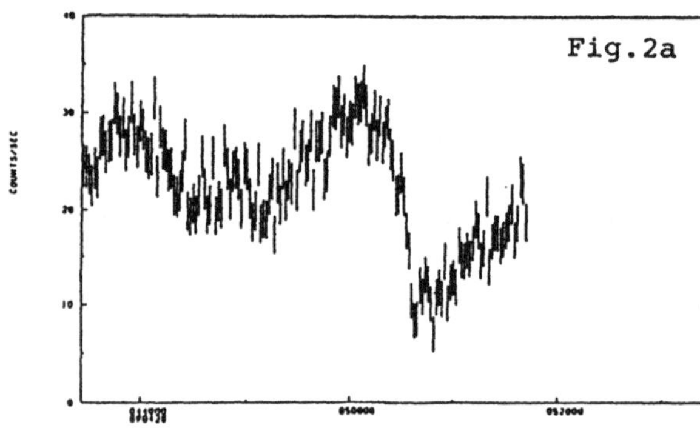

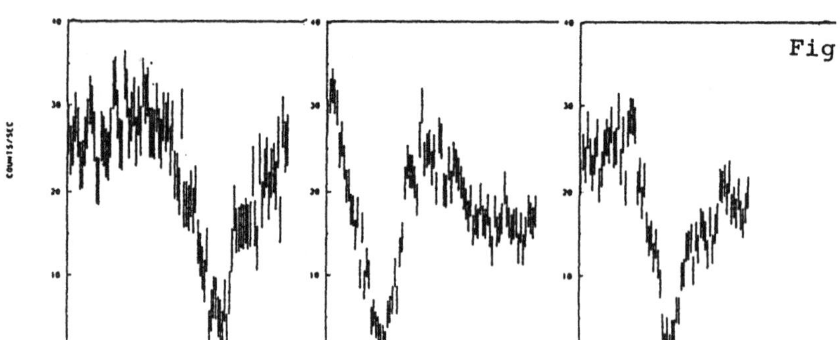

2. X-ray Spectra

X-ray data are sorted into 4 groups according to the total X-ray flux. They are well described by a single power law model with an Iron line. Fig. 3 is the spectrum of the brightest group in 1990. Best fit parameters are summarized in Table 1. The photon index Γ of NGC 6814 is about 1.4, which is much smaller than the so-called canonical value of 1.7. Such a small Γ value seems to be constant among the four spectra at different flux levels.

The most striking result is the strong Iron lines. The line flux increases as the total flux increases. The linear correlation can be clearly seen in Fig. 4. If the Iron line flux changed slower than the total X-ray flux, such a linear relation should be smeared out. Therefore, we could imagine that the Iron line flux changes as fast as the total X-ray flux.

Fig.3

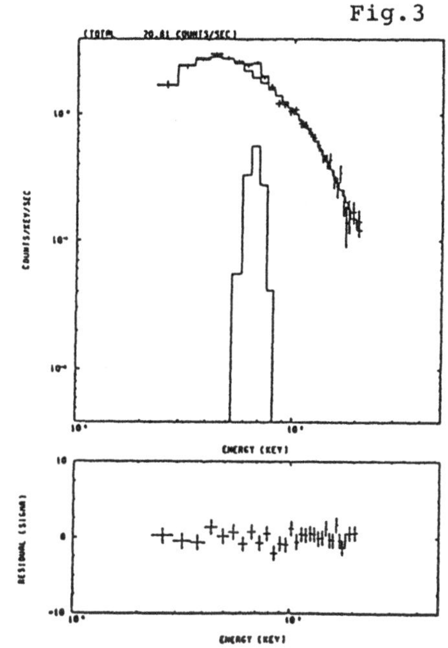

Table 1. Power Law Model Parameters for NGC 6814

	Total Flux* (c/s)	Γ	I_{Fe} (c/s)	E_{Fe} (keV)	N_H (log/cm^2)	EW (eV)	χ_v^2
1989							
1	8.72±0.13	1.45±0.11	0.14±0.14	6.40±0.55	22.93±0.08	125±127	1.318
2	14.55±0.13	1.39±0.06	0.51±0.14	6.34±0.16	22.77±0.06	305±82	1.010
3	19.52±0.15	1.40±0.05	0.76±0.15	6.38±0.12	22.70±0.06	344±69	1.693
4	25.80±0.15	1.45±0.04	1.03±0.16	6.39±0.09	22.66±0.04	359±56	1.492
1990							
1	6.82±0.13	1.61±0.12	0.12±0.13	6.48±0.64	22.74±0.10	149±165	0.943
2	14.74±0.22	1.37±0.10	0.80±0.20	6.48±0.15	22.45±0.15	528±133	1.256
3	20.81±0.31	1.45±0.09	0.87±0.26	6.40±0.17	22.59±0.09	389±117	0.839
4	26.98±0.31	1.42±0.05	1.33±0.20	6.42±0.09	22.53±0.06	467±70	2.321

* 2.3 - 20.9 keV (top + mid layrs)
Errors for 90 % confidence level are shown, except the total flux.

3. Iron Line Flux

Each energy spectrum accumulated every 256 sec is fit with a (power law + one line) model. The light curves of the Iron line flux shown in the lower panels in Fig. 5a, b are thus obtained. The large area at 6 -7 keV of the GINGA detector and the large equivalent width of 300 - 600 eV enable us to examine the Iron line line flux with high time resolution. The sharp peak of the total flux in Fig.5 a (~ 6 hour) is well followed by a sharp peak of the Iron line flux, as is the case of the sharp drop in Fig.5 b (~ 3.5 hour). Cross-correlation functions and timing analyses in detail are performed by Done et al.(1990).

Fig.4

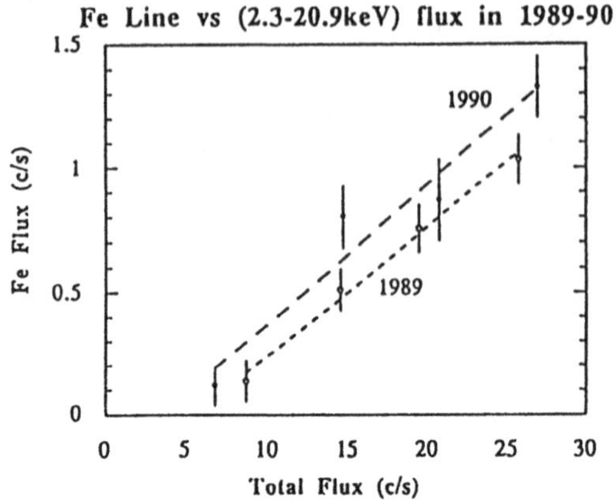

Fig.5a

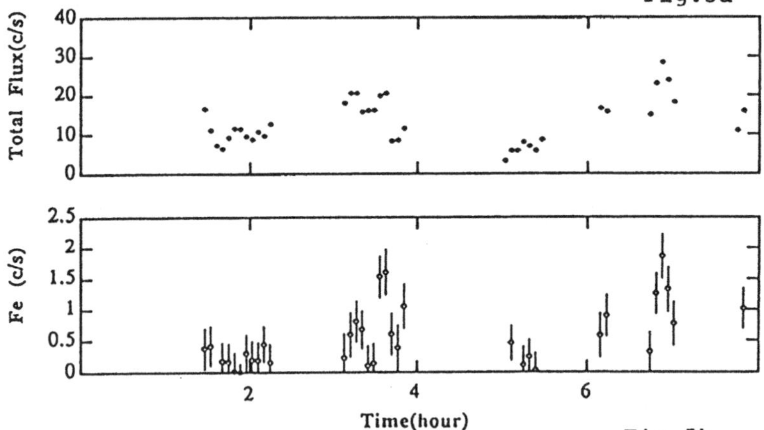

Fig.5b

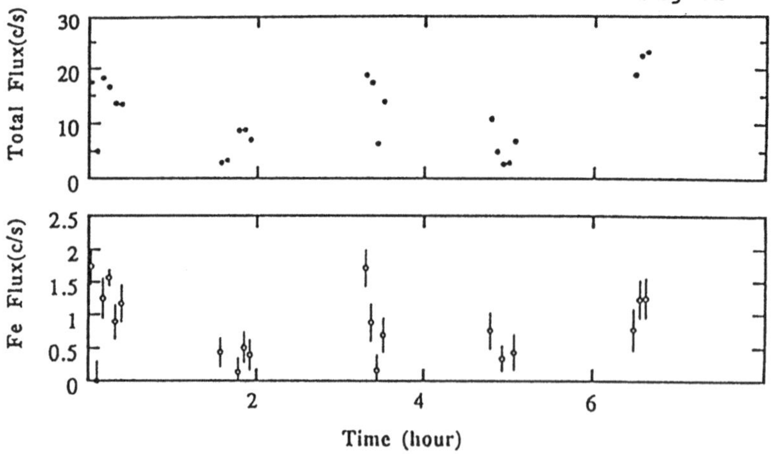

4. Quasi-periodicity

As is shown in Fig. 2, dips are found in April 1990, in which X-ray flux drops down to zero. During the observations, three similar dips are observed, being separated with 1415 min and 200 min. It reminds us the 12000 sec (200 min) periodicity reported by EXOSAT group (Mittaz et al. , 1989). If the light curves in 1989 are examined carefully, one can find the sharp drops with arrows in Fig.1a are also separated with the period of ~ 200 min. The power spectral analysis of the X-ray light curve in 1989 done by Done (1990) also showed that the power is peaked at about 12000 sec. Similar results from the data in 1990 will be shown in the future by Leighly .

5. Discussion

(1) Time Variability --- Geometry

The fastest variable time scale of the X-ray continuum was observed to be $\delta t \sim$ 50 sec. It implies the size of the emitting region to be smaller than $c \times \delta t \sim 1.5 \times 10^{12}$ cm. If we assume that the X-rays are mainly emitted from a region as large as 5 R_s (R_s: Schwarzschild radius), the central object is estimated to be $10^6 M_\odot$. It is the smallest value among the central mass estimated for AGN. On the other hand, the upper limit of delay time of the Iron line flux to the continuum X-rays is 256 sec, which means the reprocesser responsible for the Iron line is located within 10^{13} cm from the central object. This is the first evidence of (cold) accretion matter so close to the central engine of AGN.

Another time scale found from NGC 6814 is the quasi-periodicity of 12000 sec. The power spectrum by Done and folded light curves in 1989 may confirm the quasi-periodicity. More over, the sharp drops found in April 1989 and the dips in April 1990 were separated with ~ 12000 sec and its multiple. If we assume that such periodicity is caused by a Keplerian motion around the central object, the radius of orbit is estimated to be 10^{13} cm with the central mass of 10^6 M. Assuming the dip is due to a cloud in this orbit, the size of the central source is derived from the fastest variable time scale of $\delta t = 50$ sec as $< (\delta t/\tau_{orbit}) \times 2\pi R \sim (50 \text{ sec}/12000 \text{ sec}) \times 2\pi \times 10^{13}$ cm $\sim 3 \times 10^{11}$ cm. The cloud size of $\sim 10^{12}$ cm is derived from the dip width of 200 - 300 sec. Since the column density of more than 10^{24} cm^{-2} is required to stop all X-rays, volume density is estimated to be $>10^{12}$ cm^{-3}, if a spherical shape of cloud is assumed.

(2) Iron Line Profile --- Geometry and Ionization State

◊ Line Energy at ~ 6.4 keV

As is listed in Table 1, the line energy is always consistent with 6.4 keV, which can be attributed to a fluorescence line from cold Iron. It implies that cold material is accreting closer than $R \sim 10^{13}$ cm (c x 256 sec) from the central source. Since the cold matter responsible for the Iron line is expected to be illuminated by strong radiation ($L \sim 10^{42-43}$ erg/s) from the central source, cloud density has to be as high as 10^{14-15} cm^{-3} to keep the ionization parameter ($\xi = L/nR^2$) smaller than 100.

There is another possibility that the Iron line at 6.7 keV from highly ionized ion (> XVII) is shifted down to 6.4 keV due to the gravitational red shift. It happens when the Iron line is emitted from the region at a distance of 11 - 20 R_S.

◊ Line Broadening < 1 keV

The upper limit of the Iron line width is 1 keV. If a wide line results from the Doppler broadening due to the Keplerian motion, the radius of orbit has to be larger than 10 R_s for the edge on geometry.

◊ Low energy tail of the line profile

There is an indication in the spectrum of whole data that the Iron line profile exhibits an asymmetry with a tail on the low energy side. It could be explained by additional component of redshifted lines. It may be observed from an accretion disk, which extends into a small radius of 10 R_S.

◊ Large equivalent width = 150 ~ 500 eV

The line equivalent width represents relative ratio of line and continuum fluxes. It positively correlates to the column density of surrounding matter, if the gas is spherically distributed. The absorption in low energy range suggests N_H of $10^{22.7}$ cm^{-2} on the line of sight. Since the observed column is too small to create large equivalent width up to 500 eV, large amount of reprocessing matter is required around the central source.

The other explanation is the non-uniform radiation model. In other words, strong ionizing flux is emitted toward other directions than the line of sight.

Last possibility is the reprocess in highly ionized plasmas (> Fe XXIV). It is because the fluorescence yield of Iron becomes 0.75 at an ionization state of XXVI, which is twice as large as that of cold Iron (0.34).

6. Conclusion

Rapid variabilities of the total X-ray flux and the Iron line flux clearly indicate the size of 10^{12-13}cm with the central object of 10^6 M. This is one of the important results originated from the large area of GINGA detectors. On the other hand, the Iron line diagnostics in detail on the line energy, width and profile is just started by GINGA. The future missions such as BBXRT and ASTRO-D will enable us to examine the diagnostics with their high energy resolution.

IRON LINE VARIABILITY AND GEOMETRY IN SEYFERT GALAXIES

Roland Walter
Geneva Observatory
Chemin des Maillettes 51, CH-1290 Sauverny, Switzerland

Abstract

Significant Iron features were recently found in several bright active galactic nuclei (AGN) with Tenma, Exosat and Ginga. When detected, the 6.4 keV fluorescent line equivalent width (EW) is equal to a few 0.1 keV. Such large equivalent widths are the signature of optically thick matter reprocessing X-ray photons. This matter largely covers the hard X-ray source independently of geometry (Makishima, 1987). We show here that variable Iron line equivalent widths are found in four Seyfert galaxies. For these objects, the Iron line equivalent widths are decreasing with increasing X-ray continuum flux. In NGC 5548 and NGC 3516, these variations are correlated with those of the ratio of hard and soft X-ray fluxes. Such behavior could be understood with geometrical variations where the soft X-ray luminosity is proportional to the square of the distance separating the X-ray source and the reprocessing matter. Such geometrical variations cannot be fitted with an extended reprocessing thin disk. Possible geometries for the soft X-ray source include the central part of a geometrically thick disk.

Geometry

The Iron fluorescence line provides a diagnostic on the distribution of optically thick cold matter in the nucleus of Seyfert galaxies and therefore on the geometry of these objects. Ginga observations of Iron features in Seyfert galaxies give evidences for large amount (Ω of a few and $N_{II} > 5\ 10^{23}$ cm^{-2}) of cold ($T <$ 0.1keV) matter reprocessing the hard X-ray continuum (Pounds, 1989).

We have to consider three different geometries for the reprocessing region : (a) a large scale reprocessing region, where its size in unit of time is many time larger than the time scale of X-ray variation (as in reprocessing torus, broad emission line region or galaxy), (b) an intermediate reprocessing region, which size is many times larger than its distance to the X-ray source (as in a thin disk), but where this distance is comparable or shorter than the time scale of variation of the X-ray continuum and (c) a local reprocessing region, which size in unit of time is comparable or shorter than the time scale of X-ray variation (as in a thick disk). In a large scale reprocessing region, the Iron line flux is constant and the equivalent width is inversely proportional to the X-ray flux (Fx). In the intermediate geometry, the slope of the decrease of the equivalent width $-\log EW / \log Fx$ is comprised between 0 and 1, even if the relative geometry of the X-ray source and of the reprocessing region is variable. In a local reprocessing region, both Iron line flux and equivalent width may be variable (depending on the relation between geometry and X-ray flux) with a slope of decrease maybe larger than one if geometrical variations occur. Iron line variability measurement on several time scales provide therefore valuable clues on the geometry of Seyfert galaxies.

Possible red-shifted Iron lines and rapid variabilities of soft X-ray sources provide some clues for optically thick matter within a few hundred gravitational radius (Rg) from the hard X-ray source. The positive correlation found between the Iron fluorescence line and the X-ray lightcurves in NGC 6814 (Kunieda et al, 1990) suggests an upper limit of 100 Rg for the distance separating the X-ray source and the reprocessing region. Such observations rule out models with large scale reprocessing region.

Iron line variability

Figures 1 and 2 show the variations of the Iron line equivalent width with the continuum flux in four Seyfert galaxies : NGC 5548, NGC 3516, NGC 4151 and MCG-6-30-15. In all of them, there is some evidence that the Iron line equivalent width is inversely correlated to the 2-10 keV flux. The slope of the decrease is given in the following table 1 :

Object	NGC 5548	NGC 3516	NGC 4151	MCG-6-30-15
–log EW / log Fx	2.0 ± 0.8	2.0 ± 0.8	1.0 ± 0.5	1.0 ± 0.8

There is also some evidence for Iron line flux variability in NGC 5548 and NGC 3516. On the other hand, this flux remains constant in NGC 4151 and MCG-6-30-15. This may indicate that the reprocessing region is very large (as it is also indicated by the extended soft X-ray source in NGC 4151) or that we have a local reprocessing region with a particular link between geometrical and X-ray flux variations. But it may also indicate that we are not yet able to detect Iron line flux variations (measured in NGC 6814) in the majority of the sources, as the error bars on the Iron line flux are always very large. In other sources where the Iron line was detected, the continuum flux did not show sufficient variations (NGC 5506, Pounds et al, 1989 ; NGC 7469, Piro et al, 1990) or the signal to noise ratio (3C273, Turner et al, 1989) was to low to infirm or confirm a correlation between equivalent width and continuum flux.

In the following, we will assume a local reprocessing region even if our data are not excluding a large scale geometry. A local geometry is supported by some other observations pointed out above. Furthermore, if the reprocessing region is too large, the covering factor of the reprocessing matter cannot show large variation and the slope of the decrease of the equivalent width with the continuum (table 1) cannot be larger than 1, contrasting with our observations for NGC 5548 and NGC 3516. In the local model, the variability of the equivalent width suggests a variation of the covering factor of the reprocessing region above the X-ray source. In NGC 5548 this geometrical variation is also suggested by the simultaneous smooth variation of the optical depth of the X-ray source (figure 3).

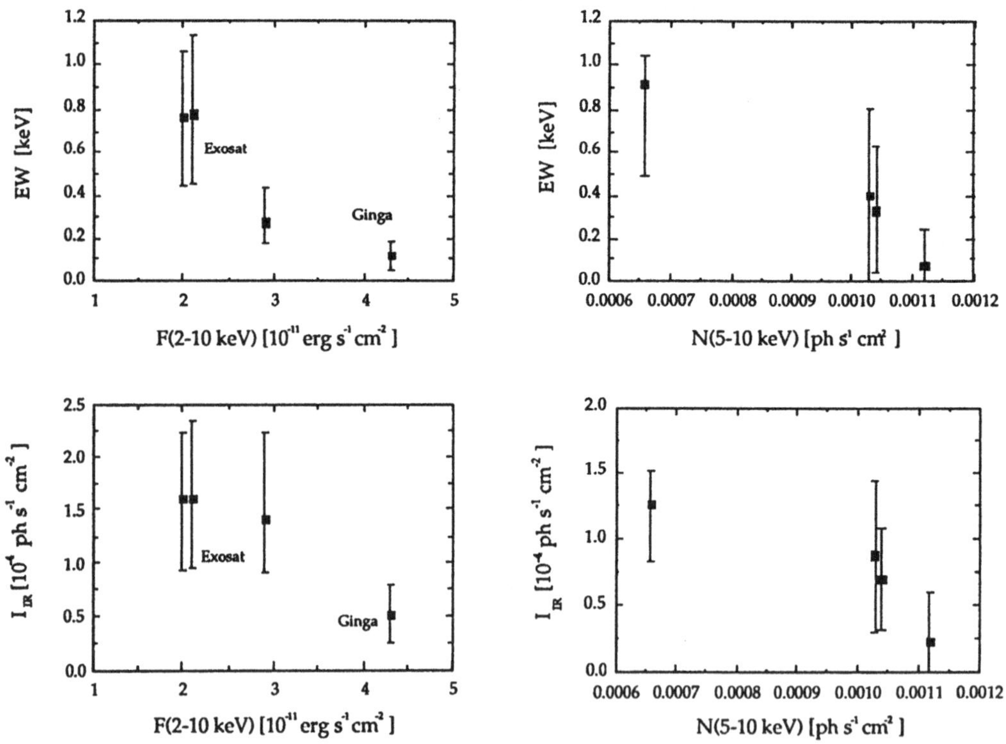

Figure 1 : Iron line equivalent width and flux variability in NGC 5548 (left) and NGC 3516 (right). Ginga observation of NGC 5548 is described in Pounds et al (1989). Other datas are from Exosat archive observations.

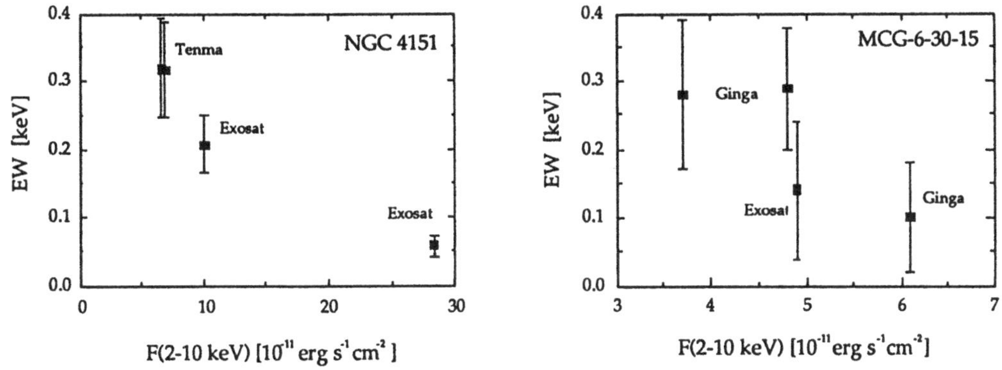

Figure 2 : Iron line equivalent width variability in NGC 4151 (left) and MCG-6-30-15 (right). Datas are from Perola et al (1986) and Matsuoka et al (1986) and respectively from Matsuoka et al (1990) and Nandra et al (1989).

If the distance separating the X-ray source and the reprocessing region is not larger than a few hundred Rg, as it is suggested by the correlation of the X-ray continuum and Iron line lightcurves in NGC 6814 and by the time lag measured between hard and soft X-ray lightcurves in NGC 5548 (see below), *the mean equivalent width measured on a few ten hours observation is an indicator of the covering factor of the reprocessing region at the epoch of observation*, even in luminous sources. As the mean Iron line equivalent width is decreasing with increasing continuum flux, the covering factor of the cold matter must decrease in the same time. This implies that *the distribution of cold matter is not spherical but highly anisotropic*. These variations can only be understood if a major portion of the reprocessing matter is located beyond the X-ray source in a solid angle smaller than 2π.

NGC 5548

The soft excess Seyfert galaxy NGC 5548 is a very interesting object in the X-ray energy band because it shows variations up to a factor of ten in the soft X-ray energy band and because the soft X-ray source is not covered by a large or variable amount of absorbing matter. On the other hand, the hard X-ray source is geometrically separated and covered by a larger amount of absorbing matter (Walter and Courvoisier, 1990). There is also some evidence in NGC 5548 for a lag between soft and hard X-ray lightcurves, which suggests that the distance separating the X-ray source and optically thick matter is one light hour. This time lag is ten times more than the upper limit measured in NGC 6814, which is in agreement with the fact that NGC 5548 is 8 times more luminous than NGC 6814.

Exosat observations of this source (Walter and Courvoisier, 1990) show that the X-ray powerlaw index is variable between 0.3 and 0.9 with a poor correlation with the 2-10 keV continuum flux. The powerlaw index is well correlated to the ratio of hard to soft X-ray fluxes as it is expected in inverse comptonization. This ratio is a measure of the optical depth of the hot ($kT_e \approx m_e c^2$) cloud of electrons comptonizing the soft X-ray photons. The optical depth is smoothly variable, suggesting a geometrical variation and is also related to the soft X-ray luminosity. Furthermore, the variations of the Iron equivalent width arise simultaneously with the X-ray spectral variations.

In NGC 5548, we are observing correlations between the hot matter optical depth τ and the soft X-ray luminosity L_{bb} and between the Iron line equivalent width and the 2-10 keV luminosity L_x. These correlations may be written $EW \sim L_X^{-2.0\pm0.8}$ and $\frac{L_X}{L_{bb}} \sim \tau \sim L_{bb}^{-0.56\pm0.1}$ which gives $EW \sim L_{bb}^{-0.9\pm0.5}$.
The covering factor of the reprocessing region is therefore inversely proportional to the soft X-ray luminosity. A geometry where *the soft X-ray luminosity is proportional to the square of the distance separating the X-ray source and the reprocessing matter* will fit this behavior. Possible geometries for the soft X-ray source include the central part of a geometrically thick disk.

If the hot comptonizing electrons are heated in a shock arising in a collimated wind, as in the wind and shock model (Courvoisier and Camenzind, 1989), the distance separating the disk and the shock in

NGC 5548 may be specified with our observations and we find that the Thomson optical depth of the wind at the location of the shock is independent of the activity of the nucleus and is approximately equal to one. Furthermore, the cooling rate associated with the comptonizing electrons gives a few tenth for the optical depth of the hot region, as observed in NGC 5548.

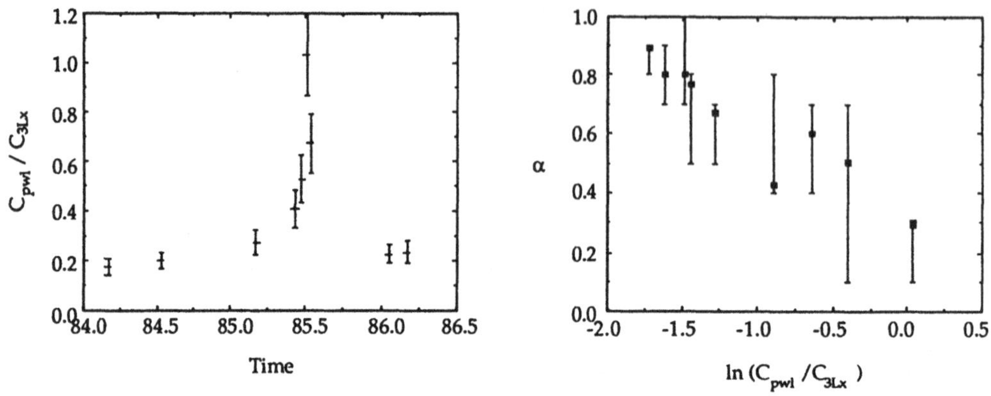

Figure 3 : Smooth variation of the optical depth of the comptonizing cloud in NGC 5548 (left) and correlations between the powerlaw index and the logarithm of this optical depth (right).

Conclusion

We show that the *mean* 6.4 keV Iron line equivalent width is variable in four Seyfert galaxies, namely NGC 5548, NGC 4151, NGC 3516 and MCG-6-30-15. Measurements in other active galactic nuclei do not infirm these observations. Even if our data are not excluding a large reprocessing region, a local reprocessing region is supported by other observations and is suggested in NGC 5548 and NGC 3516 by the large slope of the decrease of the Iron line equivalent width with increasing continuum flux. In this local reprocessing region model, the variation of the mean equivalent width measured on these two objects suggests a geometrical variation.

This geometrical variation is also supported by the simultaneous variation of the optical depth of the hot comptonizing cloud in NGC 5548. Furthermore this variation indicates that the distribution of the cold reprocessing material is highly anisotropic around the hard X-ray source. The correlation between the optical depth of the hot comptonizing cloud and the soft X-ray luminosity show that the covering factor of the reprocessing matter is inversely proportional to the soft X-ray luminosity. Such a behavior may be explained if the soft X-ray luminosity is proportional to the square of the distance separating the reprocessing matter and the X-ray source. Such geometrical variation cannot be explained with a simple thin disk model with a radius larger than several hundred of Rg. On the other hand, the overall variability is consistent with a variable position of the X-ray source in the funnel of a thick accretion disk.

References

Courvoisier, T.J.L. and Camenzind, M. : 1989, *Astron. Astrophys.*, **224**, 10

Kunieda, H. et al : 1990, *Nature*, **345**, 786

Makishima, K. : 1987, *in Physics of Accretion onto Compact Objects*, Mason, Watson and White (eds), Springer

Matsuoka, M. et al : 1986, *Pub. Astr. Soc. Japan*, **38**, 285

Matsuoka, M. et al : 1990, *Astrophys. J.*, **361**, 440

Nandra, K. et al : 1989, *Mon. Not. Roy. Astron. Soc.*, **236**, 39p

Perola, G.C. et al : 1986, *Astrophys. J.*, **306**, 508

Piro, L. et al : 1990, *Astrophys. J. (Letters)*, in press

Pounds, K.A. : 1989, *in Proc. 23rd ESLAB Symp. (ESA SP-296)*

Pounds, K.A. et al. : 1989, *Mon. Not. Roy. Astron. Soc.*, **240**, 769

Walter, R. and Courvoisier, T.J.L. : 1990, *Astron. Astrophys.*, **233**, 40

X-ray spectral changes in rapidly variable Seyfert galaxies

F. Fiore[1,2], G.C. Perola[1], L. Piro[3], M. Yamauchi[2] and M. Matsuoka[2]

[1]Istituto Astronomico dell'Universita' di Roma "La Sapienza", Via G.M. Lancisi 29, 00161 Roma, Italy; [2]RIKEN, The Institute of Physical and Chemical Research, Saitama 351-01, Japan; [3]Istituto di Astrofisica Spaziale, C.N.R., Via E. Fermi 21, Frascati, Italy

Introduction

It has been recently recognized that spectral variability above 1 keV is a quite common phenomenon among Active Galactic Nuclei (AGN). In most cases it can be roughly described by a softening of the spectrum as the source brightens even if there is often a large scatter around this general trend. The discovery by *Ginga* of a new hard spectral component in several AGN (e.g. Pounds *et al.* 1990, Matsuoka *et al.* (1990) and Piro *et al.* 1990), whose intensity becomes comparable to that of the main power law above $\sim$ 15 keV, suggests a natural interpretation of that behaviour: it can arise from non-simultaneous variations of the two components. The hard component is probably produced, as the iron K fluorescence line at 6.4 keV, by reprocessing of the main one onto optically thick matter surrounding the central non-thermal source. Therefore, it is possible to investigate the spatial distribution of the thick gas by studying the temporal variations of the iron line intensity and of the spectral shape at high energies. The variations of the latter, particularly in faint sources, can be conveniently investigated through a hardness ratio. If the hard component stays constant while the intensity of the main power law varies, we shall observe a smooth correlation between the hardness ratio and the intensity. More realistically, the temporal behaviour of the "reprocessed" component will be related to the size of the "reprocessing region" and it will be a smoothed and delayed version of that of the main one. In this case we shall observe also a scatter around the correlation between hardness ratio and intensity.

We have analyzed *Ginga* observations of the Seyfert 1 galaxies MCG-6-30-15 and NGC 4051, which are known to be highly and rapidly variable. In this paper we present some results of the analysis of the spectral variations above 4 keV. A complete and detailed discussion can be found in Fiore *et al.* (in preparation). Results of a different analysis of these observations have been presented by Matsuoka *et al.* and Nandra *et al.* (1990, MCG-6-30-15) and by Done *et al.* (1989, NGC 4051).

Observations and results

MCG-6-30-15 was observed from September 9 to 10, 1987 with a mean count rate $C_T \sim 15$ cts s^{-1} in the 1.7-18 keV energy band. The source varied during the observation by a factor $\sim$ 2 in a few hours. NGC 4051 was observed from May 13 to 16, 1988 with a mean count rate C_T 7 cts s^{-1} and it varied by up to a factor 5 in a few hours. In Fig. 1 a, b the Hard Color (HC = C(8.1-18.0 keV)/C(3.5-5.8 keV)) is plotted against C_T. Each point represents the data accumulated in one satellite orbit (90 m). The probability of a constant value is 0.0033 % ($\chi_r^2 = 3.1$) for MCG-6-30-15 and 0.25 % ($\chi_r^2 = 2.0$) for NGC 4051. These probabilities have been evaluated in the hypothesis of pure statistical fluctuations.

The strong time variability of the Ginga internal background, together with the diffuse X-ray background spatial fluctuations, introduce, however, an additional "systematic" error in the background evaluation, which could generate an artificial scatter in the HC diagrams. A detailed analysis of the *Ginga* background and a quantitative estimate of the magnitude of the systematic error shows that the case of intrinsic null HC variation cannot be excluded for the fainter of the two sources, namely NGC 4051.

The HC of MCG-6-30-15 shows both an inverse trend with C_T and a scatter around it. This behaviour can be reproduced in the framework of the two components model mentioned above. This model, described by

$$F(E) = [C_\gamma E^{-\alpha}(1 + rA(E,\alpha)) + I_{Fe}]e^{-\sigma(E)N_H}, \tag{1}$$

was tested against the pulse height spectra of each orbit. Here α and N_H are the spectral index and absorbing column, r is the ratio between the "reprocessed" and the direct component, $A(E,\alpha)$ is the X-ray albedo of the optically thick gas, as described by Lightman and White (1988), and I_{Fe} is the intensity of the iron line. This model can reproduce the data with α fixed at 1.9, the value that gives the best results in terms of χ^2. The variations in r are significant, being 0.016 % the probability of a constant value, and reflect those of HC. The two dashed lines in Fig. 1a represent the expectation values of a model with $\alpha = 1.9$, $N_H = 8 \cdot 10^{21}$ cm^{-2} and $r = 0$ or 6. This result would imply that the reprocessed component changed on time scales longer than 90 m and, because of the scatter, shorter that about 2 days. If, however, the systematic error in the backgroud subtraction is taken into account and, conservatively, most of the scatter in HC is ascribed to it, the quantity r turns out still to be variable, but the implication now would be that the reprocessed component remained practically unchanged over the two days of observation. The other parameter representing this component, namely I_{Fe}, was consistent with a costant value, but it is too poorly determined for a statistically significant test of the above mentioned possibilities.

The HC of NGC 4051 (Fig. 1 b) is consistent with a constant value, if the systematic error in the background subtraction is taken into account. The modest spectral variations above 4 keV suggests that the reprocessed component varies nearly simultaneously with the direct one. To check this suggestion we looked for variations in the other parameter representing the reprocessed radiation, namely the intensity of the iron line. The line count rate was estimated by subtracting from the counts in a narrow interval centred at the line energy a continuum estimated by interpolating the counts in two adjacent bands. The histogram in Fig. 2 represents the average counts in the line, accumulated in bins of 256 s, in three flux intervals. The errors represent the dispersion of the points in each interval. The mean count rate is 0.14 ± 0.02 cts s^{-1}, and the constant value hypothesis can be excluded at the 99 % confidence level.

We estimated the cross-correlation function between the line and the 3.5-5.8 keV count rates by using the algorithm of Edelson and Krolik (1988). The result is plotted in Fig. 3, where each point represents the mean correlation in 5120 s. The only point with a cross-correlation coefficient significantly different from zero is that centred at zero lag, thus suggesting that the line intensity responds to the rapid variations of the continuum on a time scale of a few thousands seconds at most. A detailed analysis of the short time variations of the iron line is in progress and will be published elsewhere. Here we would only like to emphasize that the analysis of both the parameters chosen to study the behaviour of the reprocessed component indicate that it varies and follows the direct component down to time scales of at most a few thousands seconds.

Conclusions

An analysis of 2 *Ginga* observations of the Seyfert galaxies MCG-6-30-15 and NGC 4051 yield the following new results:

a) The spectral variations exhibited by these sources above 4 keV can be explained by a model with two components whose relative normalization is variable.

b) The main component can be described by a power law with α constant and steeper (~ 2) than the mean spectral index of $\alpha = 1.7 + / - 0.15$ found by Mushotzky (1982) and Turner and Pounds (1989). The value found here is, on the other hand, in agreement with that predicted in non-thermal models with pair production (e.g. Zdziarski *et al.* 1990).

c) The second component can be identified with the "reflection bump" theoretically predicted by Lightman and White (1988) and Guilbert and Rees (1988). It is natural to explain the observed spectral variability at high energies in terms of a delay in the response of the reprocessed continuum to the variations of the main power law. The temporal behaviour of the reprocessed component would be, therefore, related to the size of the "reprocessing region". Our results indicate that the reprocessed component varies on a time scale t_{RC} longer than 90 m in MCG-6-30-15 and at most of the order of 90 m in NGC 4051. Because most of the "reprocessed photons" come from a region whose size, in unit of gravitational radii, should be the same from one source to the other (Matt *et al.* 1990), one would conclude that the difference in t_{RC} reflects a difference in the central black hole mass. This conclusion is also supported by the different mean X-ray luminosity, that is 10^{43} erg s^{-1} for MCG-6-30-15 and $2 \cdot 10^{41}$ erg s^{-1} for NGC 4051.

References

Done, C. et al.: 1989, *M.N.R.A.S.* **243**, 713

Edelson, R., Krolik, J.: 1988, *Astrophys. J.*, **333**, 646

Guilbert, P.W., Rees, M.J.: 1988, *M.N.R.A.S.*, **233**, 475

Lightman, A.P., White, T.R.: 1988, *Astrophys. J.*, **335**, 57

Matt, G., Perola, G.C., Piro, L.: 1990, *Astron. Astrophys.*, in press

Matsuoka, M., Yamauchi, M., Piro, L., Murakami, T.: 1990, *Astrophys. J.*, **361**, 440

Mushotzky, R. F.: 1982, *Astrophys. J.*, **254**, 22

Nandra, K., Pounds, K. A., Steward, G. C.: 1990, *M.N.R.A.S*, **242**, 660

Piro, L., Yamauchi, M., Matsuoka, M.: 1990, *Astrophys. J. Letters*, **360**, L35

Pounds, K.A., et al.: 1990, *Nature*, **344**, 132

Turner, T.J., Pounds, K.A.: 1989, *M.N.R.A.S.*, **240**, 833

Zdziarski, A.A. et al.: 1990, *Astrophys. J. (Letters)*, **363**, L1

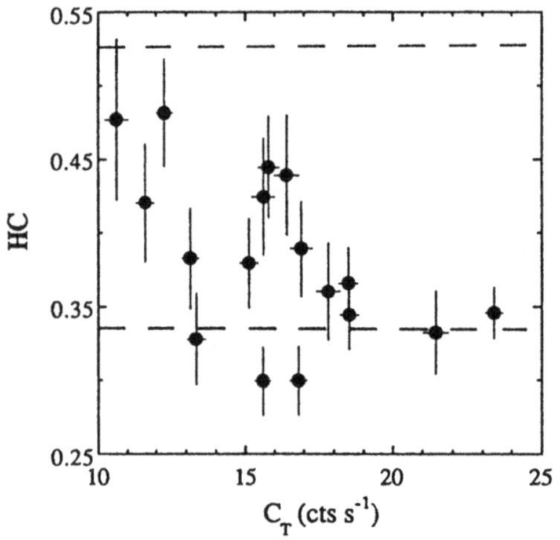

Figure 1 a. MCG-6-30-15: Hard Color - Intensity diagram. The two dashed lines represent the expectation values of the two components model in (1) with $\alpha = 1.9$ and $N_H = 8 \cdot 10^{21} cm^{-2}$ and $r = 0$ or 6.

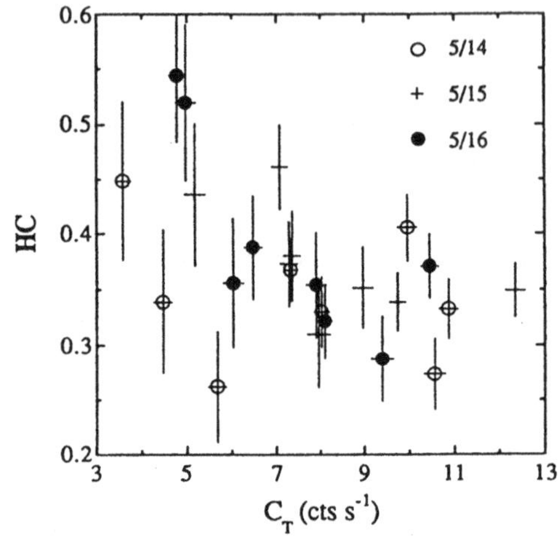

Figure 1 b. NGC 4051: Hard Color - Intensity diagram. Different symbols represent the three days of observations.

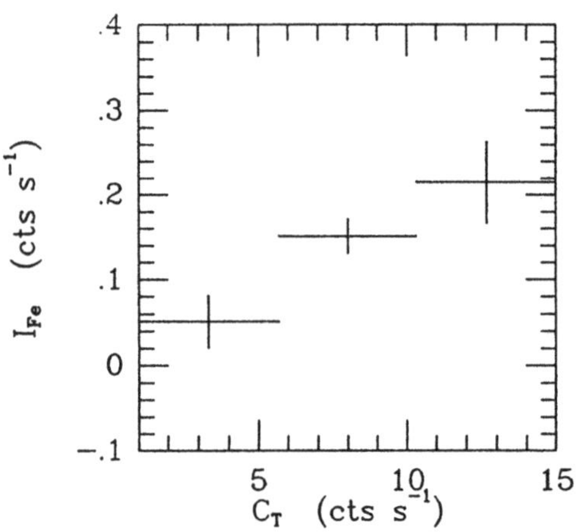

Figure 2. NGC 4051: Fe line counts - Intensity diagram. The histogram represents the average count rate of the iron line in three flux intervals.

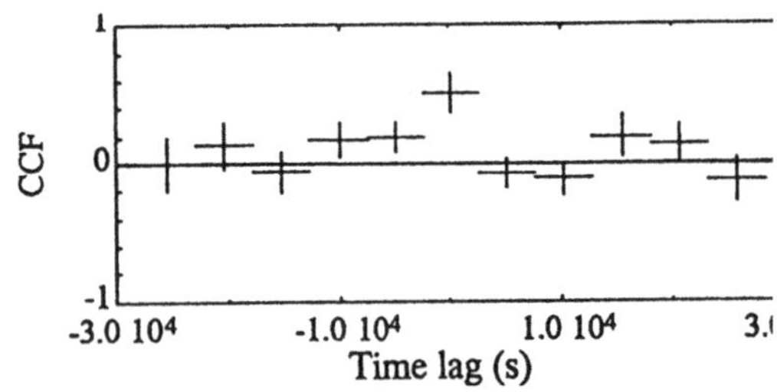

Figure 3. NGC 4051: the Cross Correlation Function between the light curve of the iron line and of the 3.5-5.8 keV continuum. Each point represents the mean correlation in 5120 s.

Soft Excess Emission in *Einstein* SSS+MPC Observations of Seyfert Galaxies - a Possible Interpretation as a Blend of Soft X-ray Lines

T.J. Turner[1], K.A. Weaver[2], R.F. Mushotzky, S.S. Holt & G.M. Madejski[1]

NASA/Goddard Space Flight Center
[1]Universities Space Research Association
[2]University of Maryland

We have re-investigated the X-ray spectra of 26 Seyfert galaxies for which simultaneous Solid State Spectrometer (SSS) and Monitor Proportional Counter (MPC) data were available. This analysis has improved over the previous analysis of the SSS and MPC data by a) using a predictive model for the amount and geometry of ice on the SSS detector (based on calibration observations), b) use of a more sophisticated background subtraction algorithm for the MPC data, and c) simultaneous fitting of the 2 datasets using XSPEC, giving an effective bandpass of ~ 0.6 - 20 keV.

Most of the data were adequately described by an absorbed power-law model, the distribution of spectral slopes and absorbing columns in good agreement with previous observations by EXOSAT (Turner & Pounds 1989).

EXOSAT and Einstein IPC observations showed that an excess of counts over the hard X-ray spectrum, is a common feature of AGN (Wilkes & Elvis 1987; Turner & Pounds 1989). The EXOSAT low energy telescope and the Einstein IPC were very sensitive at soft X-ray energies but had poor energy resolution. The combination of the moderate energy resolution SSS (160 eV) and high energy (slope constraining) MPC data have provided the best detector combination to date for the determination of the form of soft excess emission components.

Simple absorbed power-law fits revealed that for some sources the allowed parameter space lay in a 'forbidden' region where the upper limit on absorption was below the Galactic value. That, combined with the strong evidence for soft spectral components in the EXOSAT and Einstein IPC data, led us to try more complex models to fit our data. The addition of a soft continuum component significantly improved the fit in ~ 30% of cases (at > 95% confidence), but the shape of the residuals (over a simple power-law fit) in many of the sources strongly suggested that a more suitable parametrization of the soft excess may be an emission line feature (Figure 1). We used a broad Gaussian model as a parametrization for this soft emission line component, which improved ~ 60% of all fits (at > 95% confidence).

To ensure that our results were not affected by systematics, we excluded observations within 100 days of launch (after which the amount of ice on the SSS detector could be determined accurately). We also examined the distribution of detections of soft excesses and found them to be random across the Seyfert galaxy observations. The results are robust within the 40% uncertainty quoted as the maximum error in the current ice model (for details, see Turner *et al.* 1991). This complex form is not observed in any other type of object observed with the SSS detector.

The best-fit line energies were typically ~ 0.8 keV and the equivalent widths ranged from 50 eV to a few hundred eV. The line energies imply an origin as iron L emission with possible contributions from other elements such as oxygen and neon. One possible site for the soft X-ray line production is an equilibrium plasma. Alternatively, the shape of the SSS residuals could be explained as an absorption feature superimposed on a continuum soft excess component.

Line luminosities were ~ 10^{42} erg/sec, comparable to luminosities in the iron K lines detected in *Ginga* Seyfert observations (*e.g.* Pounds 1989). If these soft lines are produced in the same reprocessing region as the iron K lines, then to get comparable luminosities in iron K and the soft X-ray regime we require that the reprocessing region is highly ionized with $\xi > 100$. This ionization parameter is consistent with the line blend centroid energy of ~ 0.8 keV for a slab of solar abundance material illuminated form above (see Kallman, these proceedings). Thus, these data support the hypothesis that at least some AGN have ionized reprocessing material in the active nucleus.

How do these components fit into the general picture of soft excess emission in Seyfert galaxies? Now we have evidence for 4 different types of soft excess emission:

1) A steep, ultra-soft, rapidly variable excess (*cf.* Wilkes & Elvis 1987; Turner & Pounds 1989),

2) Leakage of continuum emission through a patchy absorber, *e.g.* NGC4151 (*e.g.* Pounds *et al.* 1986; Holt *et al.* 1979),

3) A blend of soft emission lines ~ 0.8 keV,

4) Extended soft emission, *e. g.* NGC4151 (Elvis, Briel and Henry 1983; Fassnacht *et al.* 1988, 1990.

It is of course possible that the soft X-ray line emission originates in the patchy absorber or within the extended emission region. Further constraints on the origin of these lines may be obtained by a study of their time variability.

References

Elvis, M., Briel, U.G., & Henry, J.P. 1983, *Ap.J.*, **286**, 105.

Fassnacht, C., Elvis, M., Wilson, A.S., & Briel, U. 1988, *B.A.A.S.*, 20, **4**, 970.

Fassnacht, C., Elvis, M., Wilson, A.S., & Briel, U. 1990, *Ap.J.*, **361**, 459.

Holt, S.S. 1979, in *HEAO Science Symposium,* (MSFC : NASA CP-2113), p 292.

Pounds, K.A., Warwick, R.S., Culhane, J.L., & de Korte, P.A.J. 1986, *M.N.R.A.S.*, **218**, 685.

Pounds, K.A. 1990, in *Proceedings of the 23rd ESLAB Symposium,* Ed. J. Hunt and B. Battrick (Paris: ESA publications), p 753.

Turner, T.J. & Pounds, K.A. 1989, *M.N.R.A.S.*, **240**, 833.

Wilkes, B. & Elvis, M. 1987, *Ap. J.*, **323**, 243.

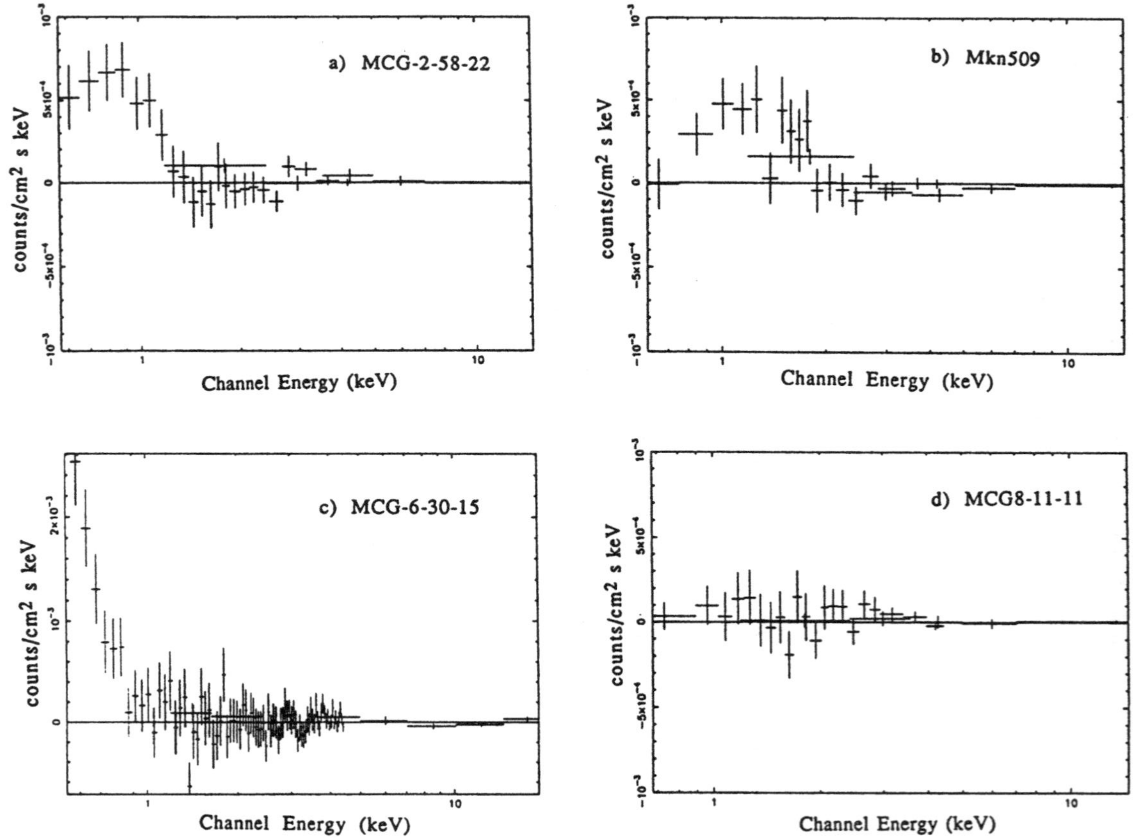

Figure 1 shows the residuals from fitting absorbed power-law models to 4 of the SSS+MPC observations. In each case the data above 2 keV were fit, and the best-fit model extrapolated to lower energies. The residuals from MCG-2-58-22 and Mkn509 suggest the presence of soft X-ray line emission. The excess in MCG-6-30-15, however, is adequately described by a continuum model such as thermal bremsstrahlung. MCG8-11-11 shows no evidence for soft excess emission and is shown for comparison.

IV. Future Missions

THE X-RAY ASTRONOMY MISSION SAX

G.C. Perola

Istituto Astronomico, Universita' di Roma "La Sapienza"

OUTLINE OF THE MISSION

SAX, acronym of "Satellite for Astronomy in the X-rays", is a project of the Italian Space Agency (ASI) in collaboration with the Netherlands Agency for Aerospace Programs (NIVR). The program entered the phase C/D (a continuous phase to completion) in April 1989, and the launch date is planned for the end of 1993.

The SAX mission will be devoted to spectral and variability studies of celestial sources by means of a set of instruments covering a wide energy band, from about 0.1 keV to about 200 keV. The mission was conceived with the following capabilities:

- Imaging and broad-band spectroscopy in the range 0.1-10 keV, with angular resolution of 1 arcmin and $E/\Delta E \sim 12$ at 6 keV.
- Continuum and line spectroscopy with $E/\Delta E \sim$ 5-20 over the range 3-200 keV.
- Time variability studies of bright source spectra on short (msec) and long (days to months) time scales.
- Systematic studies of the long term variability of sources brighter than 1 mCrab through repeated surveys of selected regions of the sky.

The instruments chosen to obtain these capabilities are briefly described in Table 1 (for more details, see Butler and Scarsi, 1990). The Narrow Field Instruments (NFI) are coaligned to provide a simultaneous coverage of the celestial targets over the energy band 0.1-200 keV, and comprise:

- the Concentrator/Spectrometer (C/S) - 4 units;
- the High Pressure Gas Scintillation Proportional Counter (HPGSPC);
- the Phoswich Detector System (PDS).

The Wide Field Camera (WFC) operates in the 2-35 keV range and is made of two units pointing in opposite directions and at 90° to the NFI.

The satellite is a free flyer of the one ton class, stabilized on three axis and capable of pointing at any source in the sky within every single year of its operational lifetime - the main constraint on the portion of sky accessible at any one time being posed by the requirement that the solar array should normally remain within 30° of the Sun line. The pointing stability will be better than 2 arcmin, and the attitude reconstruction

Table 1. The SAX scientific instruments

Concentrator/Spectrometer - 4 units	
Optics (per unit):	
Focal length	185 cm
Number/length of double cone mirrors	30/30 cm
On axis resolution (HPR)	1 arcmin
Field of view (FWHM)	30 arcmin
GSPC focal plane detectors:	
Energy resolution ΔE/E	$0.08/\sqrt{E/(6\,keV)}$
Position resolution	$0.5/\sqrt{E/(6\,keV)}$ mm = $0.9/\sqrt{E/(6\,keV)}$ arcmin
ME-GSPC (3 units)	
Energy range	1-10 keV
Effective area *per unit* at 3 keV	48 cm^2
7 keV	50 cm^2
LE-GSPC (1 unit)	
Energy range	0.1-10 keV
Effective area at 0.25 keV	56 cm^2
0.60 keV	28 cm^2
1.0 keV	61 cm^2
7.0 keV	50 cm^2
High Pressure GSPC	
Exposed geometrical area	450 cm^2
Field of view (FWHM)	1° (collimated)
Energy range	3-120 keV
Energy resolution ΔE/E	$0.03/\sqrt{E/(60\,keV)}$
Phoswich Detector System	
Exposed geometrical area	800 cm^2
Field of view (FWHM)	1.5° (collimated)
Energy range	15-200 keV
Energy resolution ΔE/E	$0.18/\sqrt{E/(60\,keV)}$
Wide Field Camera - 2 units	
Effective area of proportional counter per unit	250 cm^2 (through mask)
Field of view per unit	20°x20°
Angular resolution	5 arcmin
Energy range: centre of field	2-30 keV
full field	2-10 keV
timing (no image)	2-35 keV
Energy resolution	20% at 6 keV

post facto better than 1 arcmin. The orbit will be circular and low altitude (600 km), with an inclination of about 2° with respect to the Equator and a period of 94 minutes. This type of orbit offers the advantage, particularly valuable for the high energy non-imaging instruments, of limiting the particle induced background. Consequently the Ground Station for uplink telecommand and downlink telemetry will be located near the Equator (Malindi, Kenya) and the minimum contact time will be ten minutes per orbit, while the Operational Control Centre, connected to the Station by a communication satellite link, will be located in Italy. The minimum (design) lifetime of the mission will be two years, but its estimated duration is over three years.

The scientific objectives and the instrumental payload have been conceived and planned under the responsibility of a consortium of Institutes in Italy, together with Institutes in Holland and with the partecipation of the Space Science Department of ESA. The composition of the consortium is listed below:

- Istituto per le Tecnologie e lo Studio delle Radiazioni Extraterrestri (ITESRE/CNR), Bologna
- Istituto di Fisica Cosmica e Tecnologie Relative (IFCTR/CNR), Unita' GIFCO dell'Universita', and Osservatorio Astronomico di Brera, Milano
- Istituto di Fisica Cosmica e di Applicazioni dell'Informatica (IFCAI/CNR), and Unita' GIFCO dell'Universita', Palermo
- Istituto di Astrofisica Spaziale (IAS/CNR), Frascati, and Unita' GIFCO dell' Istituto Astronomico, Universita' "La Sapienza", Roma
- Space Research Laboratory, SRU, Utrecht
- Space Science Department, SSD, of ESA, Noordwijk

In addition, a collaboration with the Max Planck Institute for Extraterrestrial Physics in Munich was set up for the prototype testing and the calibration of the C/S mirrors.

STRATEGY AND OBJECTIVES OF THE SCIENTIFIC PROGRAM

SAX will be operated like an observatory in space, capable of about 2000 pointings in its minimum lifetime of 2 years. While the NFI will be prime for most of the time, the WFC will be periodically used to scan the galactic plane to monitor the temporal behaviour of sources above 1 mCrab and to detect transient phenomena; thanks to their large field of view, the WFC will also be used for long term monitoring of bright selected sources while the NFI will carry out their sequence of pointed observations. The observational program will be held flexible enough to accomodate targets of opportunity, in particular those arising from the WFC surveys. The observations will be organized on the basis of a "core program", mainly devoted to systematic studies, and of a "guest observer program".

Fig. 1 shows the broad band sensitivity of the NFI in a typical exposure time of 10^4 sec. Given the instrumental capabilities and the ample spectral coverage, SAX can be profitably used in practically every one of the numerous astrophysical areas where the previous missions have demonstrated the importance of X-ray measurements. What follows is merely a list of these areas, with just a mention of some of the prominent aspects on which SAX can give significant contributions:

- Compact galactic sources, containing a degenerate or collapsed stellar object: continuum shape and variability; temporal behaviour of spectral features (such as Fe fluorescence, cyclotron line emission and absorption, photoelectric absorption) as a function of rotation and orbital phases; study of transient phenomena.
- Active galactic nuclei: spectral dynamics of variable continuum; spectral shape of objects down to 1/20 of 3C 273 up to 100-200 keV; spectra of very distant objects up to 10 keV; soft and hard X-ray excesses; Fe fluorescence and photoelectric absorption.
- Supernova remnants: spatially resolved spectra of galactic very extended ($\gg 1$ arcmin) remnants; spectra of remnants in the Magellanic Clouds.
- Clusters of galaxies: spatially resolved spectra of nearby ($z < 0.1$) clusters; studies of gradients in temperature and iron abundance.
- Normal galaxies: spectra of the extended emission discovered with the Einstein Observatory.
- Stars: spectra of stellar coronae.

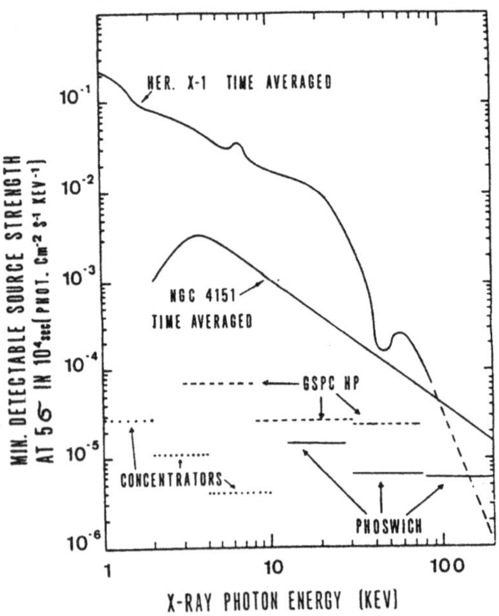

Fig.1 SAX NFI broad band sensitivity for an exposure time of 10^4 s.

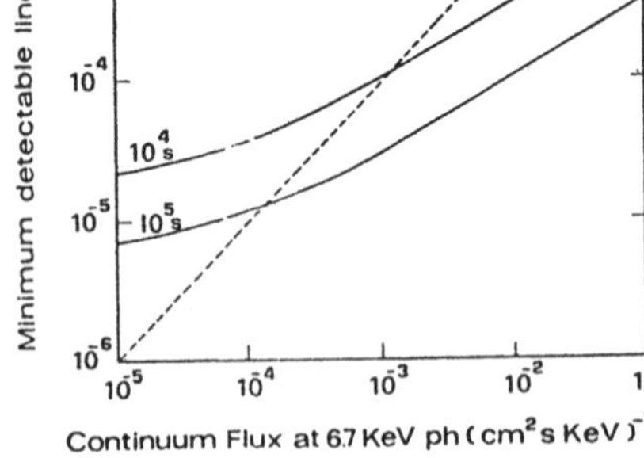

Fig.2 Detection limit (5σ) for the iron K line with the C/S, as a function of the continuum intensity. The dashed curve represents the intensity of a line with EW=100 eV.

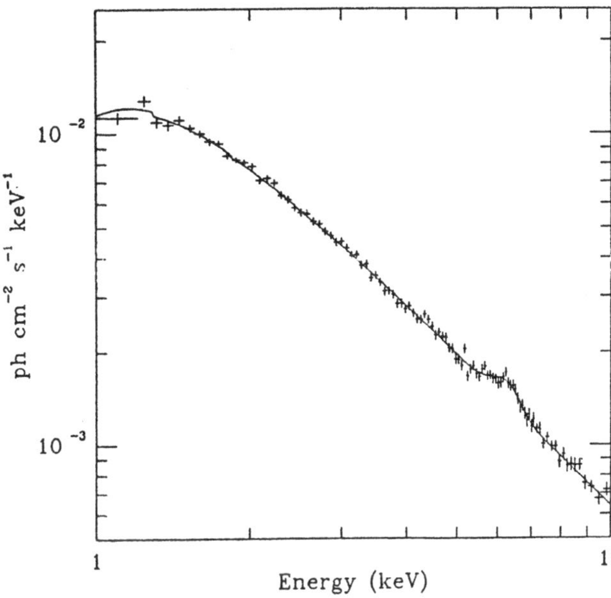

Fig.3 Results of a simulated 30.000 s exposure of the 3 ME units of the C/S to a source of 5 mCrab, with photon index -1.7 and a broad iron fluorescence line with EW=150 eV.

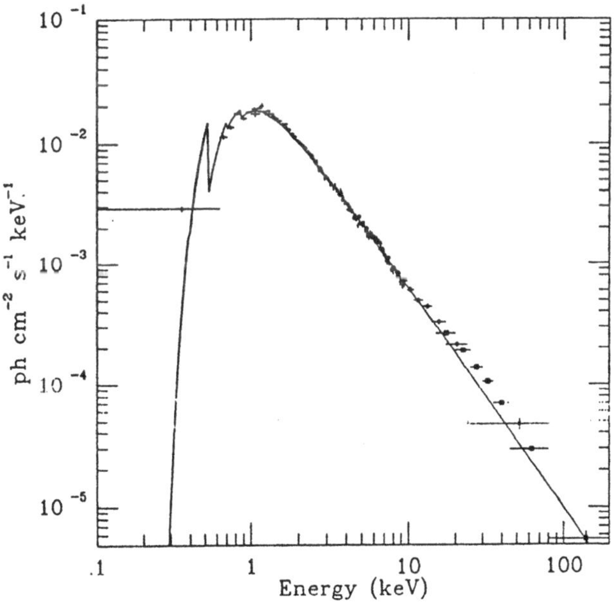

Fig.4 Results of a simulated 30.000 s point exposure of the C/S, the HPGSPC and the PDS to a 5 mCrab source with the spectral parameters of the Seyfert galaxy IC4329A. Note the excess at high energies above the fit with a single power law represented by the solid line.

Going into some detail on topics covered during this workshop, the SAX capabilities can be illustrated by selecting the iron K line and the full range continuum, the latter for its relevance to studies of wide band features imprinted in the spectra by matter in the neighborhood of compact sources. The iron K line can originate either from collisional ionization in hot thin plasmas, or from fluorescence. The first mechanism is operative mainly in sources such as SN remnants, clusters of galaxies and elliptical galaxies, stellar coronae, star formation regions - for iron with cosmic abundance the equivalent width (EW) is typically in the range 500 to 1000 eV. The second mechanism is operative mainly in compact sources driven by accretion, both galactic and extragalactic - the EW is then in the range up to 400 eV, depending on the amount and ionization degree of the fluorescent material and on the geometry. In the latter case the fluorescence should be accompanied by a certain amount of continuum reprocessed by scattering, which has been recently detected by the Ginga satellite as a hard tail, particularly in Seyfert galaxies.

The study of the iron line can be performed with two of the NFI, the C/S and the HPGSPC. Fig. 2 illustrates the detection limit (at 5σ) with the C/S: a line with EW = 100 eV can be detected down to a source strength of 5 mCrab in 10^4 sec, and down to 0.6 mCrab in 10^5 sec. The accuracy with which the (fluorescence) line parameters can be measured is illustrated in Fig. 3, representing a 30.000 sec simulated exposure of the Medium Energy C/S to a 5 mCrab source with photon index -1.7, line EW = 150 eV and profile as predicted for a flat disk rotating around a black hole seen face-on (Matt et al., 1990): at the 68% confidence level, a fit with a gaussian profile yields the line energy with an accuracy of 1%, the EW with an accuracy of 15% and the line width with an accuracy of 25%. Fig. 4 illustrates the ability of the three NFI used jointly to detect and measure the high energy excess: the simulation is for a 30.000 sec exposure to a source like the Seyfert galaxy IC 4329A (5 mCrab, spectral parameters as determined with Ginga by Piro et al., 1990). From this figure it is evident that a fit with a single power law is unacceptable, and, in particular, that at energies above 10 keV, the excess can be accurately determined. I would like to conclude by pointing out that the ability to study simultaneously both the fluorescence line and the reprocessed continuum in compact sources is rather unique to SAX among the various missions of the coming decade, a feature that partly compensates for the lower energy resolution of the GSPC as compared to the new generation of non-dispersive spectroscopic devices onboard of other satellites described in this workshop.

Acknowledgements. Thanks are due to L. Piro for help in producing the simulations.

REFERENCES

Butler, R.C., Scarsi, L.: *"The SAX mission"*, to appear in the Proceedings of the 1990 SPIE Conference

Matt, G., Perola, G.C., Piro, L.: 1990, *Astron. Astrophys.*, in press

Piro, L., Yamauchi, M., Matsuoka, M.: 1990, *Astrophys. J. Letters* **360**, L35

JET-X
A JOINT EUROPEAN X-RAY TELESCOPE FOR SPECTRUM-X

G.C. Stewart

Department of Physics and Astronomy, Leicester University, Leicester, U.K.

Abstract

The Joint European X-Ray Telescope, JET-X, is one of the core instruments of the scientific payload of the USSR SPECTRUM-X astrophysics mission due for launch in 1993. The JET-X instrument is described and its scientific performance and capability discussed particularly in the context of spectroscopy of the iron-line complex at 7keV,

1 INTRODUCTION

The Joint European X-ray Telescope, JET-X, forms part of the core payload of the USSR's SPECTRUM-X Project whose combined response extends over the range 20eV-100keV. JET-X is designed to study the X-ray emission of sources in the band 0.15-10 keV.

JET-X consists of two identical, coaligned X-ray imaging telescopes, each with a spatial resolution of 30 arcsecond or better. Focal plane imaging is provided by a cooled X-ray sensitive CCD detector which will combine high spatial resolution with good spectral resolution, with particular emphasis on high sensitivity and spectral resolution ($\frac{E}{\Delta E} \geq 50$) around the 7 keV Fe-line complex. An optical monitor is co-mounted with the X-ray telescopes to permit simultaneous observation and identification of the optical counterparts of X-ray target sources. A more complete description of the instrument design and specification is given in Wells *et al.*, 1990. Here I will concentrate on the scientific performance of the instrument, with particular regard to the capability of JET-X for spectroscopy of the 7keV iron line complex. I should note, however, that both the key hardware programmes on which this scientific capability so strongly depends, i.e. the mirrors and the CCD detectors, are progressing well and early results from both programmes show that targets for both spatial and spectral performance are being well met.

2 SCIENTIFIC CAPABILITY

2.1 Instrument Performance

2.1.1 Point Source Sensitivity

In Figure 1 the time taken to achieve a 5σ detection of the source count rate integrated over the instrument passband is shown as a function of the 0.5-10 keV flux for a source with a power-law spectrum of photon index 1.7, with low energy absorption corresponding to intervening cold matter with a neutral hydrogen column density of 10^{20}cm^{-2}.

Also shown in the figure is the confusion limit (calculated using the 1 source per 30 beamwidth criterion) for a detector with 20 arcsec beamwidth. The limit has been extrapolated from the observed source number - flux relation measured for extra-galactic X-ray sources with the Einstein Observatory by assuming a continuing Euclidean population of sources. The correction for the differing passbands of the instruments has been made assuming a power-law spectrum with photon index of 1.7 as is measured for the majority of local AGN. Recent results on fluctuations in the X-ray background from the GINGA satellite (Warwick and Stewart, 1990) show a normalisation of the source counts which is approximately 3 times higher than the EMSS, suggesting the possible existence of a population of cut-off X-ray sources which could not be detected by the IPC.

The time taken to reach this confusion limit, roughly 10^5s, is close to the expected orbital period of the SPECTRUM-X satellite, which therefore might be a typical exposure time, showing that the instrument throughput is well matched to its spatial resolution.

A by-product of the high sensitivity is that flux variability of approximately 10-20% on timescales of a few tens of seconds will be detectable in sources at flux levels of $\sim 1\mu$Jy. The simultaneous background measurement available with an imaging detector means that systematic errors will not substantially affect this type of measurement.

2.1.2 Spectral Capability

The energy resolution afforded by a CCD is well matched to the spectral properties of thermal plasmas with temperatures in the range $10^6 - 10^8$K . In addition to the strong iron lines at around 7 keV, other strong lines from various ionisation stages of iron (at $\sim$ 1keV), silicon, sulphur, oxygen, calcium and argon fall within the passband of JET-X. JET-X is capable of resolving the resonance lines from the He- and H- like ionisation stages of these elements.

The time taken to measure lines at 7 kev with equivalent widths of 2 keV, 1keV and 100eV, with respect to a power law continuum with a photon index of 1.7, is shown in Figure 2. A line with an equivalent width of 1keV can be detected at the 5 sigma level in 3×10^5s for sources ten times brighter than the confusion limit (e.g. high redshift clusters). A value of 750eV is typical for cosmic plasmas with temperatures in the 2-8 keV range. (At this observation length the line detection is still essentially photon limited and the limiting line intensity is $\sim 4 \times 10^{-6}$ ph cm^{-2}s^{-1}). For sources down to $\sim 100\mu$Crab strength lines of 100eV equivalent width can be detected.

2.1.3 Spatial Resolution

Objects larger than 500 Kpc, such as clusters of galaxies, will be resolved whatever their distance. Cooling flows, with typical extents of 100-200 Kpc will also be spatially resolved at all distances, while individual galaxies would be resolved to redshifts of 0.5. This is a significant advantage over, for example, a 2 arcminute resolution telescope.

2.2 Astrophysical Impact

The sensitivity and spectroscopic capability of JET-X are such that studies of astrophysical importance for almost all classes of astronomical object can be made. Sufficiently large samples for each class of object will be observable that a major impact in all areas of astronomy, from planetary studies to cosmology, in addition to the studies of 'classical' X-ray sources will be possible. JET-X, however, has been designed primarily for the study of faint, distant X-ray sources of cosmological importance. Here a small sample of the key problems which can be addressed with JET-X are outlined.

2.2.1 Faint Source Studies and the X-ray Background

The high sensitivity of JET-X to point sources (for example, luminous AGNs will be detected to redshifts ~ 4) will enable source counts to be studied directly at flux levels ~ 10 times fainter than achieved with the EINSTEIN observatory, allowing a cross-comparison with the ROSAT X-ray telescope deep survey results in a complementary energy region. This will be particularly important if a class of heavily obscured sources does indeed exist.

If the $\log N - \log S$ is Euclidean to this flux level then JET-X will have imaged at least 50% of the diffuse X-ray background. On the other hand fluctuations analyses (Fabian 1981, Hamilton and Helfand 1987) of EINSTEIN data suggest that the source counts flatten at flux levels just below the EINSTEIN limits. This is at the flux level directly addressed by JET-X. Flattening of the source counts also implies that a new class of faint X-ray sources (e.g. young galaxies) may be required to explain the background. If these sources have a luminosity function similar to that suggested one should be visible in each 10^5s JET-X exposure.

JET-X can also measure the spectrum of the residual X-ray background over the 0.1-10keV band on scales down to less than 1 arc-minute. Comparison of this residual background with the spectra measured for individual populations of X-ray emitters will further constrain possible models for the origin of the background.

2.2.2 Clusters

The high spatial resolution, coupled with the continuum spectroscopic capability of JET-X, will enable detailed studies of the temperature and density profiles of the gas in clusters of galaxies to be determined. The gravitational potential to which the gas is responding may thus be mapped, allowing the determination of the distribution of both the visible and dark matter in clusters. Measurements of cluster gas parameters when coupled with measurements of the microwave decrement through the

Sunyaev-Zeldovitch effect will give model independent estimates of the cosmological parameters H_0 and q_0. Measurements of the angular diameters of cooling flows and clusters as a function of redshift will also lead to estimates of these parameters.

JET-X can test models of the evolution of clusters and the enrichment of the intra-cluster gas by measuring the abundance of heavy elements as a function of radius in clusters and as a function of redshift.

JET-X will strongly constrain the cooling-flow rates in nearby clusters, the distribution of flow rate with radius and will also determine the magnitude of flows in clusters at high redshift.

2.2.3 Galaxies

The study of X-ray emission from normal and star-burst galaxies has to-date been very limited. The EINSTEIN observatory results suggested that most elliptical galaxies contain large amounts of hot gas. JET-X will be able to detect the most luminous ellipticals out to redshifts of order 0.25 and obtain spectra for typical galaxies out to 100 Mpc (e.g. the Coma cluster) in exposures of $< 10^5$s. If, as is suggested, this gas is in hydrostatic equilibrium, measurements of the radial temperature and density distribution of the gas will allow the mass distribution of these galaxies to be determined.

Studies of cooling flows in nearby elliptical galaxies will also be possible. The 'canonical' wisdom that elliptical galaxies contained little or no gas and that consequently no star-formation was taking place is now untenable. X-ray observations of the cooling gas in these galaxies, particularly for a range of galaxy environments, will be essential in understanding the true picture.

The dominant emission from spiral galaxies, which in general have lower luminosities than ellipticals, is not from hot gas but from discrete sources, in particular X-ray binaries. Spectroscopy of these discrete sources will be possible throughout the local group. Comparative studies of the source populations will provide new data for evolutionary calculations through studies of their luminosity functions, distribution of orbital periods, location in the galaxies. The relationship of the X-ray source properties to their metallicities can also be investigated.

The spectral difference between the discrete sources and the diffuse ISM (if our galaxy is representative of spirals) will mean that studies of the ISM in galaxies at distances of up to $\sim$ 100Mpc. will be possible. If these galaxies have gravitationally bound haloes with gas temperatures of $\sim 10^6$K, such as our galaxy, then it will be possible to derive the mass distribution within spirals as well as elliptical galaxies.

Star-burst galaxies have a much higher ratio of $L_x : L_{opt}$ than average and thus will be detectable out to distances of order 500 Mpc. The correlation between X-ray properties and the star-formation rate will mean that for an unbiased selection of star-burst galaxies the influences of effects such as inter-galaxy interaction can be studied. The relationship of X-ray emission to the star-burst phenomenon in nearby galaxies will be important in understanding the potential contribution of young galaxies to the X-ray background.

2.2.4 Active Galaxies

JET-X will, of course, be capable of measuring the 0.25-10keV continuum spectrum of at least several thousand (faint) AGN and QSOs which will be observed both as targets of pointed observations and will be detected serendipitously.

Of particular importance, however, may be the recent discovery (e.g. Nandra *et al.*, 1989, Pounds *et al.*, 1989) that AGN X-ray spectra are not simply featureless power-laws but that a substantial fraction show evidence for emission and absorption by cold iron imprinted on their spectra. This is another scientific area where the capabilities of JET-X are well-matched to the astrophysical problem. As has been extensively discussed at this meeting, if the emission features are produced by fluorescence in an accretion disc around the black hole then it is in principle possible to weigh that black hole using the technique of reverberation mapping. As is shown in figure 3 where simulations of the line profile expected from a $\sim 10^4$s observation of a source such as NGC5548 the spectral sensitivity and resolution of JET-X are easily sufficient to distinguish between a line which is only thermally broadened and one which has also been broadened by the relativistic and Doppler effects near the black hole.

The good spectral resolution of the instrument below 1keV when coupled with the ability to precisely determine the intrinsic continuum will mean that detailed information about the physical condition and location of the absorbing material intrinsic to the AGN will be obtainable for a significant sample of active galaxies. Some AGN show evidence for absorption by partially ionised gas. As the intrinsic luminosity of the source changes so does the ionisation of this gas. The sensitivity of JET-X to the absorption edges, and hence ionisation states, of intermediate Z elements will be important in determining the location, density and abundance of this material.

A substantial fraction of AGN show excess emission at energies below 1 keV. These soft excesses are believed to be caused by emission from the accretion disc surrounding the central source. The low energy response of JET-X is sufficient to detect and investigate the temporal and spectroscopic behaviours of these soft excesses.

3 Mission Profile

In a lifetime of 5 years JET-X will be capable of making $\sim 10^3 - 10^4$ observations of durations between $10^4 - 10^5$s. In addition to the detailed information obtained on the few thousand prime targets JET-X will serendipitously obtain information on $\sim 10^5$ X-ray sources. For at least 10% of these sources good quality spectra (sufficient, for example, to determine the power-law index of a serendipitous AGN to ~ 0.1) will be obtained. Multi-colour broad-band photometry will be possible for most of the remainder. For comparison, previous imaging X-ray observatory instruments, provided little or no spectral information for most of the few thousand sources observed serendipitously and were not sensitive in the spectral band of most interest.

The ROSAT X-Ray Telescope has a sensitivity similar to JET-X but only at energies < 2 keV and with limited spectral resolution. The ROSAT all-sky survey will provide invaluable information for optimising the scientific return from JET-X observations.

Scheduled for late 1993, well before either of the world class missions, AXAF and XMM, Spectrum-X will be one of the first of several international missions to exploit the rich astrophysical potential of X-ray spectroscopy. In comparison with contemporary instruments, JET-X will have a lower throughput, but a better limiting sensitivity, spectral and imaging resolution than SODART (also included in the core payload of SPECTRUM-X). It will have comparable throughput but much better spatial resolution than the Japanese/US ASTRO-D and larger throughput and higher resolution than the imaging spectrometer on the Italian SAX missions likely to be in orbit on the same time scale.

4 Acknowledgements

The JET-X instrument will be the product of the hard work and effort in the consortium institutes all over Europe. I gratefully acknowledge their contribution without which this paper would not be possible. In particular I thank Teresa Mineo whose work on the science simulator for JET-X formed the basis of much of the scientific content of this paper.

5 References

Fabian, A.C. 1981. *Ann. N.Y. Acad. Sci.* **375**, 235.
Hamilton, T.T. & Helfand, D.J. 1987. *Ap. J.* **318**, 93.
Nandra, K., Pounds, K.A., Stewart, G.C., Fabian, A.C. & Rees, M.J. 1989. *MNRAS* 236,39P.
Pounds, K.A., Nandra, K., Stewart, G.C. & Leighly, K. 1989. *MNRAS Submitted.*
Warwick, R.S, & Stewart, G.C. 1990, Proc. of 23rd Eslab Symposium, 727.
Wells *et al.*, 1990, in IAU Colloqium 115.

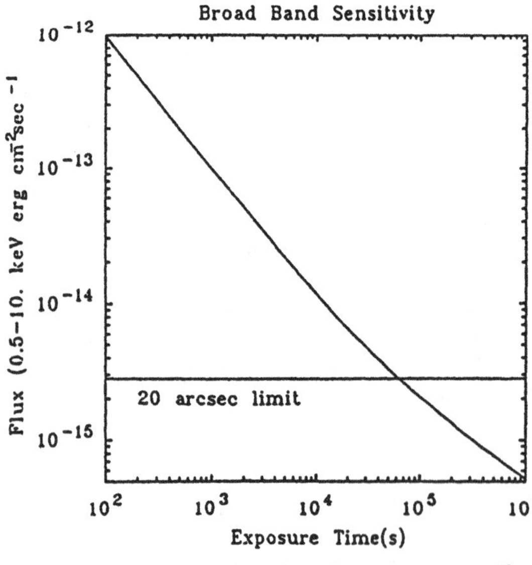

Figure 1: Time required for a 5σ detection of a point source. Also shown is the confusion limit for a 20 arc-second beamwidth assuming a Euclidean extrapolation of the Einstein log N - log S.

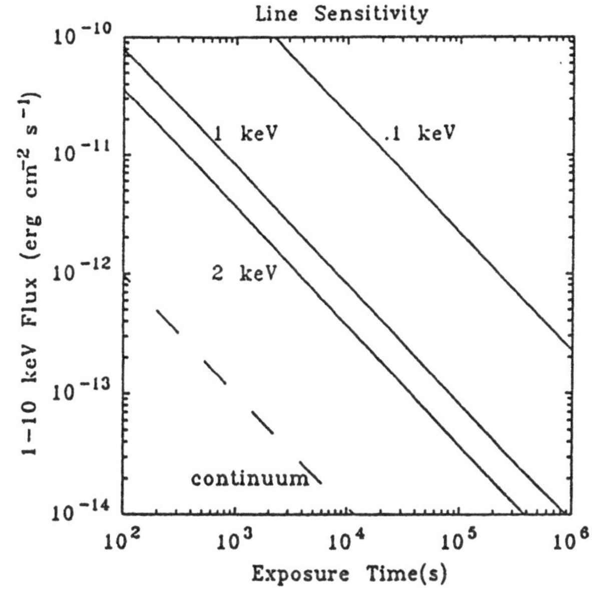

Figure 2: Time required by JET-X for a 5σ detection of narrow lines of varying equivalent widths as a function of the continuum intensity.

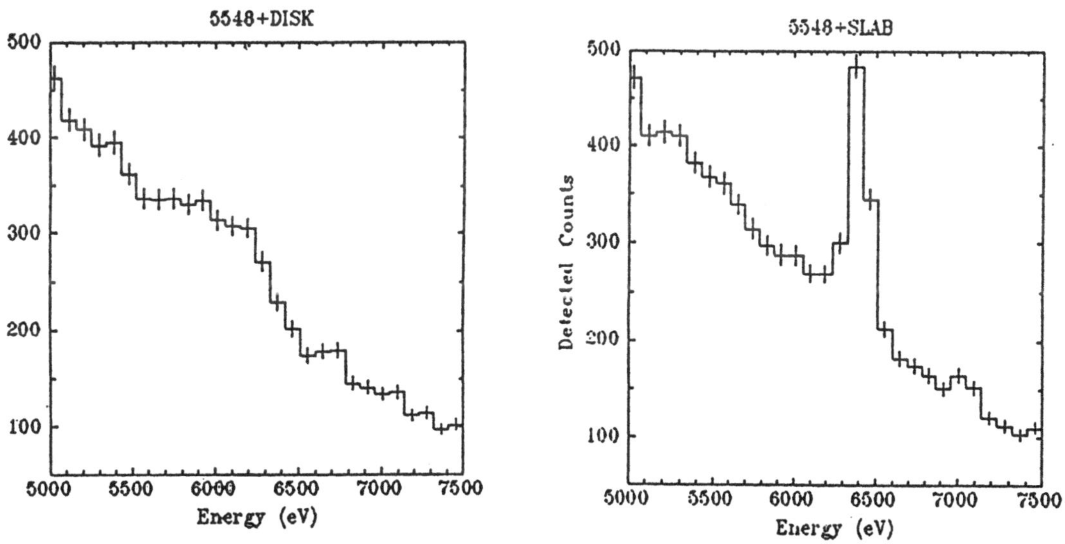

Figure 3: The sensitivity of JET-X to line profile variations. The response of JET-X to a 6.4 keV line with an equivalent width similar to that detected for NGC5548 with (a) only thermal broadening and (b) with the doppler and relativistic broadening expected for fluorescent reflection from an accretion disk around a massive black hole.

X-Ray Spectroscopy with the SODART/XSPECT Telescope

Herbert W. Schnopper
Danish Space Research Institute
Gl. Lundtoftevej 7, DK-2800, Lyngby

Abstract.

The SOviet-DAnish Röntgen Telescope (SODART) on board the Spectrum Röntgen Gamma (SRG) satellite will be equipped with three different instruments devoted to X-ray spectroscopy. Each of the two thin foil telescopes has an 8 m focal length, a 60 cm diameter, a 1 deg field-of-view (FOV), a <2 arcmin half-power width and ca. 1700 and 1200 cm^2 collecting area at 2 and 8 keV, respectively. Broad band (0.25 - 25 keV) imaging spectroscopy over the full FOV will be obtained from low- and high-energy imaging proportional counters with energy resolution as good as 28 and 14 percent (FWHM) at 1.5 and 6 keV, respectively, and with an efficiency >70 percent (2 - 25 keV). The central 18 arcmin of the FOV of one telescope will be viewed by an array of 19 cooled silicon detectors with energy resolution of 140 and 160 eV at 2 and 6 keV, respectively, and efficiency >85 percent (2 - 20 keV). Each detector element will subtend a 3.4 arcmin FOV. A full FOV objective crystal spectrometer in front of the other telescope will obtain high resolution imaged spectra of point and extended sources with an energy resolution of 2, 1 and 2 eV at 0.7, 2.5 and 6.7 keV, respectively.

Introduction.

SRG is to be the first of a series of new astronomical observatories to be launched by the Soviet Union under the sponsorship of the Academy of Sciences of the USSR. The expected launch date is mid-1994 and the observatory is expected to be operated for a period of 5 years. The satellite will be launched into a deep, highly eccentric orbit with a period of about four days. Long duration observations can be made for continuous periods of up to 80 hours. With all the operational constraints taken into account, there is pointing access to about 80 per cent of the celestial sphere at any time. The major facility on board is the SODART X-ray telescope and its associated focal plane instruments. Instrumentation in addition to SODART will cover the energy range from EUV to gamma rays. A complete description of the SRG system (Figure 1) will be found in Sunyaev (1990).

SPECTRUM RÖNTGEN-GAMMA

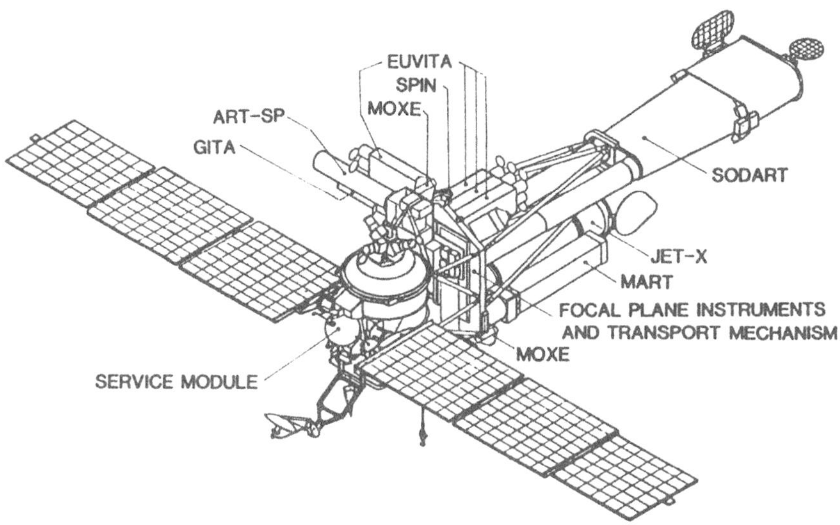

Figure 1. SPECTRUM RÖNTGEN-GAMMA (R.S. Kremnev)

SODART.

The system consists of X-ray optics, thermal and structural elements, focal plane instruments, a focal plane transport assembly and an objective crystal spectrometer (Figure 2). The elements of SODART provided by DSRI are called XSPECT. Brief descriptions of these elements follow.

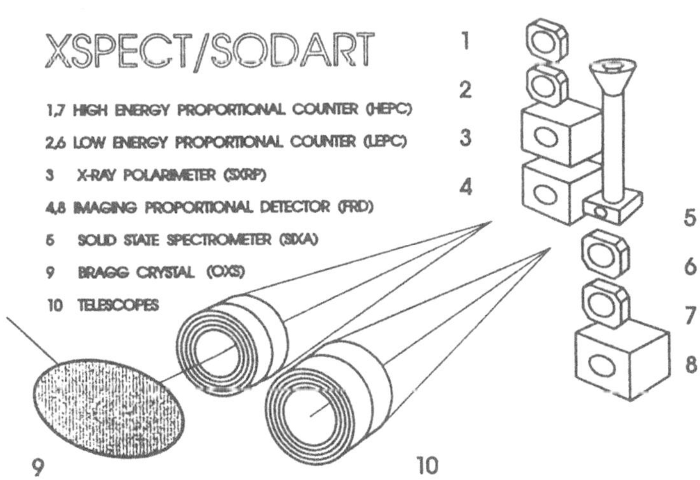

Figure 2. The telescopes and instruments of SODART

X-ray optics.

Serlemitsos, Petri, Glasser, and Birsa (1984) pointed out the advantages of using thin foil, conical approximations to Wolter 1 optics. A large fraction of the available aperture can be filled by densely nesting many shells of thin foils yielding a large collecting area and an extended energy range. The design goal for the XSPECT telescope is a Half Power Width (HPW) of <2 arcmin of which <20 arcsec comes from the conical geometry.
The parameters chosen for the XSPECT telescope are listed in Table 1 (Westergaard et al. 1990). Two such telescopes are to be provided for SODART. Figure 3 shows the collecting area which can be expected for a single SODART telescope. Curves for other telescopes are shown for comparison.

Table 1. XSPECT conical approximation telescopes.

Field of view	60	arcmin
Half-power width	<2	arcmin
Focal length	8	m
Diameter	60	cm
Reflecting surface	500	Å gold
Effective area	1700	cm^2 @ 2keV
	1200	cm^2 @ 8keV
	60	cm^2 @ 20keV
Shell material	0.3	mm aluminium
Outer shell diameter	60	cm
Inner shell diameter	16	cm
Shell separation (min)	0.5	mm
Shell length	20	cm
Number of shells	154	
Mass per telescope	90	kg
Total mirrored surface	75	m^2

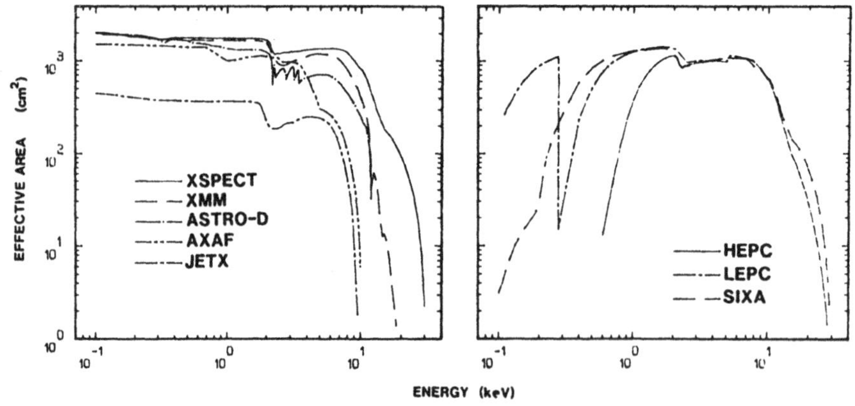

Figure 3. Left: On-axis collecting area for various telescopes.
Right: XSPECT collecting area with various detectors.(N.J. Westergaard)

Each XSPECT telescope consists of two sections, 1 alpha and 3 alpha, each of which is divided into quadrants (Figure 4). After assembly, each quadrant is mounted to a reference hub whose axis is made accurately parallel to the axis of a well collimated optical beam 25 cm in diameter. The reflected beam can be recorded by a CCD camera which can be placed at any distance along the converging beam. These optical test are used to control the adjustments of the ribs which guide the foils into the optimum position. Diffraction makes it impossible to measure the HPW directly by this method. When, however, a slit is used to scan the quadrant it is possible to measure the centroid of the image focused by the small portion of the illuminated foils. The distribution of these centroids is a good indication of the contribution to the HPW which comes from manufacturing tolerances and misalignment. Recent measurements on an assembled and aligned quadrant are well within specifications.

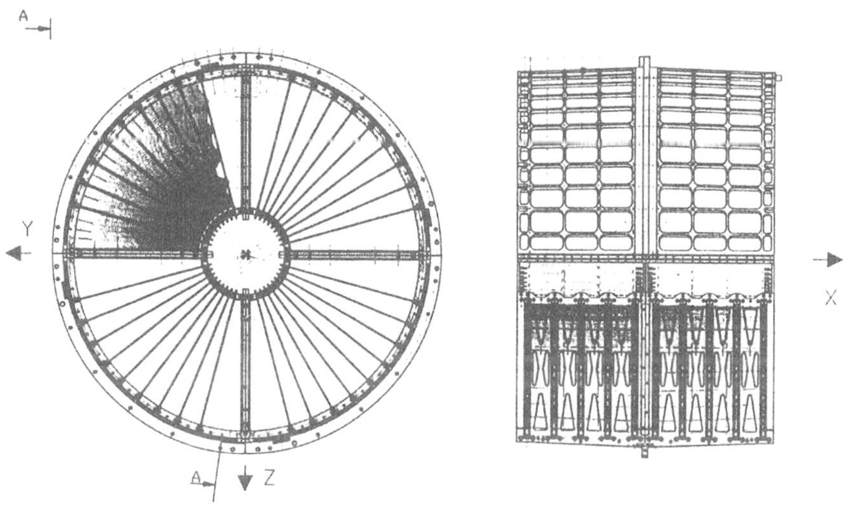

Figure 4. Structure of the XSPECT telescope. (J.Polny)

Focal plane detectors.

XSPECT MicroStrip Proportional Counters (MSPC).

In conventional proportional counters the electron avalanche takes place in the high field region defined by anode and cathode grids made from finely spaced wires. These designs usually lead to gain and position non-linearity, count rate restrictions and high operating voltages. The MSPC is a novel approach in which the wire grids are replaced by narrowly spaced, conducting microstrips (Oed, 1988; Budtz-Jørgensen et al., 1989) which are accurately deposited (+/-2 μm) on an insulating substrate. The XSPECT MSPC detector based on these principles is shown in Figure 5. A high- and a low-energy (HEPC and LEPC) detector will be provided for each telescope. The parameters chosen for them are listed in Table 2.

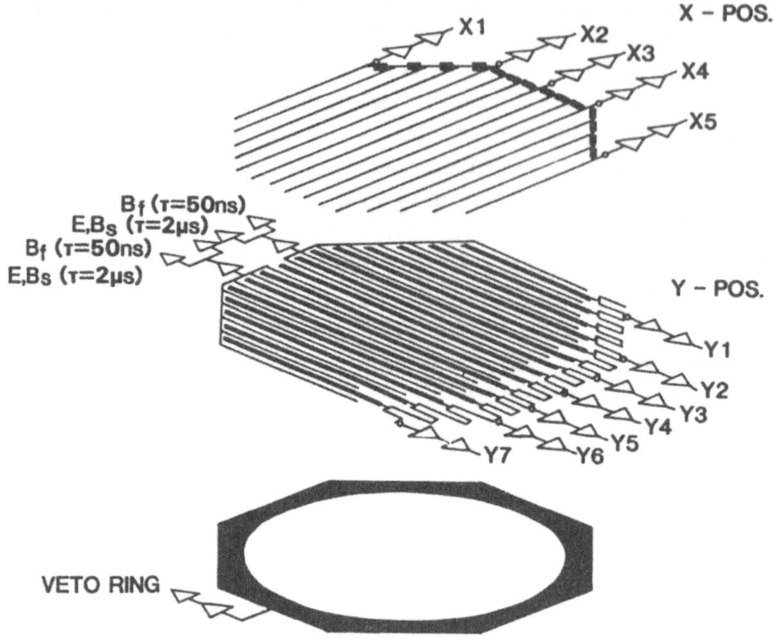

Figure 5. The XSPECT MSPC. (C. Budtz-Jørgensen)

Table 2. Proportional Counter Parameters.

Field of view arcmin	60 arcmin	
Diameter of active area	160 mm	
Diameter of window	140 mm	
Position resolution	<1 mm	
Time resolution	<5 μs	
Background rejection	>99 per cent	
Total memory	128 Mbytes	
Gas	Xe 90%, Propylene 10%	
Energy resolution	$0.32E^{1/2}$ keV	
	HEPC	LEPC
Energy range	2 - 25	0.5 - 8 keV
Dynamic range	625	800
Efficiency	>0.7, 2<E<15	>0.7, 1<E<8 keV
Gas thickness	4	2 cm
Gas pressure	1	0.5 atm
Window	13 μm Be	0.5 μm plastic 0.2 μm diamond 0.02 μm Al

X-direction positions are obtained from the charge induced on the wire grid which is suspended above the microstrip. The wires are connected with resistors and the charge division is read out by suitable spaced preamplifiers that insure the appropriate position sensitivity. Y-direction positions are obtained from the cathode strips on the microstrip which are connected by a resistive strip on the substrate. Energy, event risetime, and anti-coincidence discrimination are used to reject background. Results from energy- and position-resolution, and background rejection measurements are shown in Figure 6.

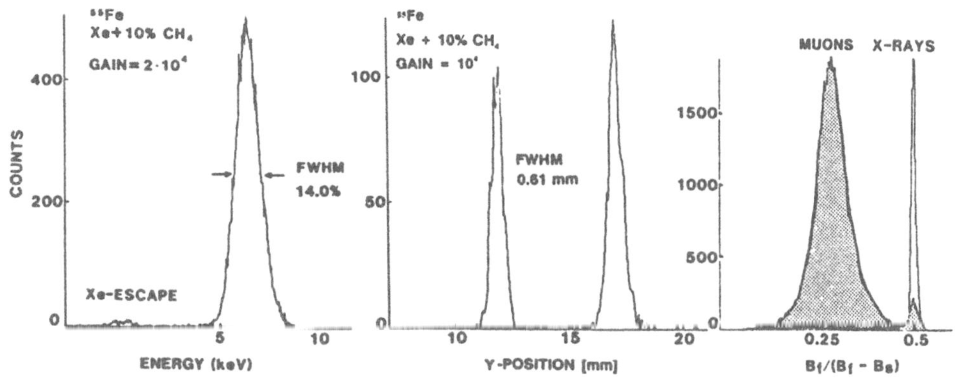

Figure 6. Energy, position and pulse-shape background resolution with a breadboard MSPC. (C. Budtz-Jørgensen)

SIlicon X-ray Array (SIXA).

A single pixel, lithium-drifted silicon [Si(Li)] solid-state detector was flown successfully on Einstein. It offers improved energy resolution over a gas filled detector since less energy is required to liberate an electron. There is, however, no equivalent to gas amplification and, to overcome room temperature electronic noise, it is necessary to cool the detector to ca. 100 K. On Einstein cooling was obtained from a block of solid argon which dissipated with time and limited the lifetime of the detector. A two-stage passive radiator connected to the detector with a heat pipe will, instead, be used for SIXA.

The SIXA array (Vilhu et al., 1990) covers an 18 arcmin FOV and consists of 19 hexagonal pixels each of which subtends 3.4 arcmin (Figure 7 and Table 3). The number of pixels is limited by the heat dissipating properties of the heat pipe and passive radiator. The detector is skulpted from a single piece of ion-implanted, rather that lithium-drifted, silicon.

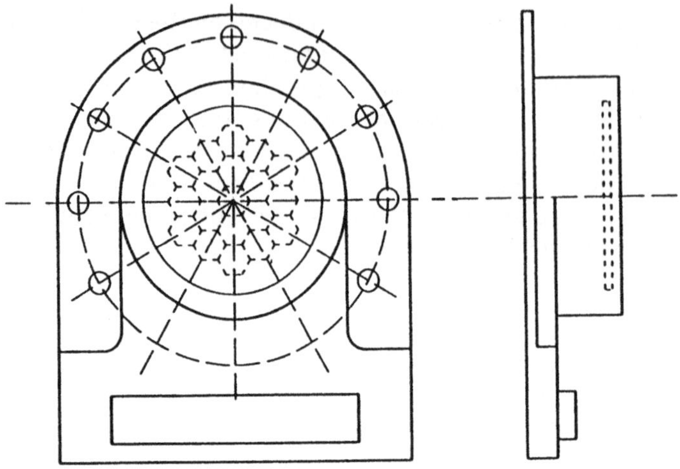

Figure 7. The SIXA array mounted on its cold finger. (H. Sipilä)

Table 3. SIXA Parameters

Field of view	18 arcmin
Diameter of active area	40 mm
Pixel diameter	8 mm
Number of energy channels	1000
Energy channel width	20 eV
Energy range	0.5-30 keV
Efficiency	>0.85; 0.5-30 KeV
Depletion depth	3 mm
Energy resolution	140 eV @ 2 keV
	160 eV @ 6 keV
	250 eV @ 20 keV
Operating temperature	110 K
Power consumption/Pixel	13 mW

Other focal plane detectors.

Two additional imaging proportional counters (FRD) (Sunyaev, 1990) and a Stellar X-Ray Polarimeter (SXRP) (Kaaret et al., 1989) complete the complement of focal plane instruments (Figure 7). They are not discussed in this paper.

XSPECT Objective Crystal Spectrometer (OXS).

The OXS concept separates the processes of energy dispersion and imaging (Schnopper and Byrnak, 1987). The large panel in front of the telescope consists of a mosaic of carefully aligned flat crystals on each side (Figure 2). It acts as a narrow pass filter and a mirror. The FOV of the telescope is reflected by the crystal panel which adds only a small contribution to the HPW. Each pixel in the FOV satisfies a specific Bragg angle on the crystal and, therefore, each pixel in the detector can be identified with a particular energy and position (Figure 8). The solid line in Figure 8a represents the locos of points in the FOV which satisfy the central Bragg angle defined in Figure 8b by the crystal reflectivity. The small angular range, $\Delta\theta$, over which the crystal has a significant reflectivity is on the order of a few arcmin and is shown by the band in Figure 8a.

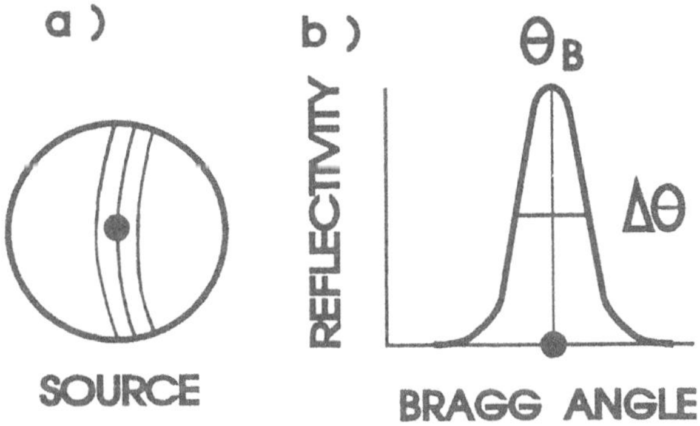

Figure 8. a: The locus of points in the FOV Which satisfy the Bragg condition.
b: The reflectivity of a crystal near the Bragg angle.

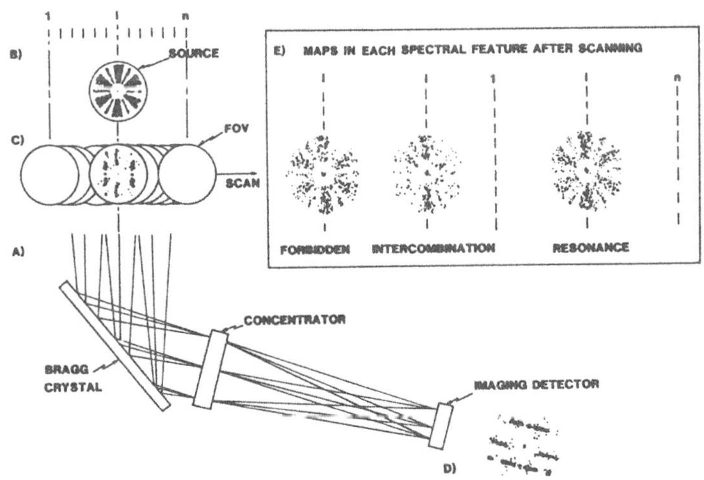

Figure 9. The OXS concept and the simulated response to an extended source containing a point source at its center. Images in the principal He-and H-like O lines are shown. (see text)

If an extended source (Figure 10b) emits several lines with closely spaced energies and the corresponding Bragg angle range is contained within the FOV, then the crystal will select those portions of the source which satisfy the reflection conditions for each of the lines (Figure 9c) and the telescope will image those portions on the detector (Figure 9d). A new Bragg angle footprint on the FOV is selected for every angular setting of the crystal with respect to the source (Figure 9c). During such a scan it will, in most cases, be necessary to adjust the pointing direction of the telescope to keep the reflected rays close to its axis and, thereby, avoid vignetting. Typical scans, when suitably corrected for the small drift of the spacecraft, will yield either the spectrum of a point source or energy resolved images of an extended source (Figure 9e).

It is not necessary that a high resolution spectrometer be able to be used over the entire energy range reflected by the telescope. By selecting a limited number of cosmically abundant ions and concentrating on the energy range which includes the H-and He-like emission, K absorption and neutral K emission of Fe, Ge, O and C (and, for Fe, the L emission), it will be possible to study a wide range of plasma parameters for both hot and cold matter.

Parameters chosen for OXS are given in Table 4. It is intended that MultiLayer (ML) structures be deposited on the surfaces of the LiF and Ge crystals to allow simultaneous measurements of the Fe and O (and perhaps C) energy ranges. Absorption in the detector window will limit the study of C to energies below the K edge. XSPECT has considerable sensitivity for energies below the C K edge (see Figure 4) but the MSPC and SIXA broad band detectors have almost no energy resolution at these energies. The ML crystals will, however, prove useful here.

Table 4. Crystals for the XSPECT OXS.

Crystal	Plane	2d (Å)	Range (keV)	R_c(rad)	E/ΔE (35<Θ<60 deg)
<u>Side 1</u>					
LiF	(220)	2.848	5.00-7.40	1.7×10^{-4}	1375-2300
ML Ni/C on LiF(220)	(001)	95	0.15-0.23	4.0×10^{-3}	60
<u>Side 2</u>					
Ge	(111)	6.532	2.19-3.23	2.9×10^{-4}	2575
RAP	(001)	26.121	0.55-0.81	5.0×10^{-5}	770
ML Ni/C on Ge(111)	(001)	80	0.18-0.27	4.3×10^{-3}	80

Sensitivities.

This topic is fully covered in the paper by Westergaard et al. (1991) at this Workshop.

Acknowedgements.

The staff at DSRI have made significant contributions to this work and I thank them warmly for their assistance in preparing the manuscript.

References.

Budtz-Jørgensen, C., Madsen, M.M., Jonasson, P., Schnopper, H.W., and Oed, A. 1989, SPIE Proc., **982**,

Kaaret, P., Novick,R., Martin, C., Hamilton, T., Sunyaev, R., Lapshov, I., Silver,

E., Weisskopf, M., Elsner, R., Chanan, G., Manzo, G., Costa, E., Fraser, G., and Perola, G.C. 1989, SPIE Proc., **1160**, 587-597.

Oed, A. 1988, Nucl. Instr. Meth., **A263**, 351-359.

Schnopper, H.W., and Brynak, B.P. 1987, Appl. Opt., **26**, 2871-76.

Serlemitsos, P.J., Petre, R., Glasser, C., and Birsa, F. 1984, IEEE Trans. Nucl. Sci., NS-31, 786-790.

Sunyaev, R. 1990, IAU Colloquium **123**, in press.

Vilhu, O. 1990, IAU Colloquium **123**, in press.

Westergaard, N.J., Byrnak, B.P., Christensen, F.E., Grundsøe, P., Hornstrup, A., Henrichsen, S., Henriksen U., Jespersen, E., Nørgaard-Nielsen, H.U., Polny, J., Schnopper, H.W., and Ørup, P. 1990, Opt. Eng., in press.

Capabilities of SODART/XSPECT for Iron Line Detection

N.J. Westergaard, H.U. Nørgaard-Nielsen, F.E. Christensen,
A. Hornstrup, C. Budtz-Jørgensen, B. Byrnak, H.W. Schnopper

Danish Space Research Institute
Gl. Lundtoftevej 7, DK 2800 Lyngby, Denmark

Abstract.

The Soviet mission Spectrum-X-Gamma is planned to be launched in 1993. The prime instrument is the SODART (Soviet Danish Röntgen Telescope) consisting of two telescope units, each with a thin foil nested mirror assembly and a number of focal plane instruments. These include gas detectors, a solid state detector and a polarimeter. One of the telescopes is equipped with an objective crystal spectrometer for high resolution spectroscopy of strong sources. The expected energy resolution and sensitivities for iron line detection are presented and compared with some current research goals.

Introduction.

The main advantage of the design for SODART is its throughput at energies around 8 keV which makes it well suited for the study of the Fe K line in X-ray sources to a flux limit of $S_{2\text{-}10keV}$ = 4×10^{-13} erg cm^{-2} s^{-1} (canonical AGN spectrum with insignificant absorption at 2 keV and equivalent width of ~100 eV). The telescope design and development have already been reported in Schnopper (1990) and Westergaard et al. (1990) but a short description is included here. More details can be found in Schnopper (these proceedings).

Crystal spectrometers have been included on earlier X-ray missions primarily for solar observations. The FPCS on Einstein had a very small effective area but still it produced some very significant results, e.g. for SNRs. The objective crystal spectrometer (OXS) (Schnopper and Byrnak, 1987) included in SODART is a new type of instrument for X-ray missions. The effective area per resolution element is quite large which makes the detection of narrow line features possible for a relatively large number of sources.

The Silicon X-ray Array (SIXA) detector (Vilhu et al. 1990) will have an energy resolution ~160 eV in the 8 keV region and the sensitivity means that a 10σ detection of the continuum in the 10 - 12 keV band is possible in 10^5 s or less for any source detected in the ROSAT All Sky Survey (assuming that the overall shape can be described by a power law) provided that $N_H < 10^{21}$ cm^{-2} for sources with photon index > 1.8.

Telescope description.

The optical system is an approximation to a Wolter I design in which the paraboloid and hyperboloid have been substituted with conical surfaces. The focal length is 8 m which together with the rather large number of mirror shells (154) give an effective area of 1190 cm^2 at 7 keV taking into account the obscuration of the telescope frame and spokes. The optical constants have been derived from measurements of the reflection coefficients for sample surfaces prepared in exactly the same way as the actual telescope mirrors (Hornstrup et al., 1990). The largest contribution to the focusing errors comes from large scale figure deviations in the foil material. The design goal of a half energy width (HEW) of 2′ is within reach, since the optical measurements show that the such deviations fall below 2′ and the tapering angle of the foils is well controlled. X-ray scattering measurements are underway (Christensen et al., 1991).

Focal plane instrumentation.

Each telescope unit has a slide with four instruments in the focal plane. The telescope with the Bragg panel (the OXS) has a high and low energy position sensitive microstrip gas counter (HEPC and LEPC) (Budtz-Jørgensen et al., 1990), a solid state pixel detector, SIXA (Vilhu et al., 1990 and Kamarainen et al., 1990), and a multi wire gas detector (FRD) (Sunyaev, 1990). The other telescope unit has two similar microstrip detectors, a polarimeter (Kaaret et al., 1989), and another FRD.

The Objective Crystal Spectrometer.

The high spectral resolution required for more detailed observations of the phenomena reported on in these proceedings (Tanaka, Koyama, Hughes, Kowalski, Ponman etc.) can be achieved with the SODART OXS. Its design is described in more detail by Christensen et al., 1990, and for a short introduction see Schnopper (these proceedings). For completeness in Table 1 we give the crystals and area (fraction of available surface) chosen for each of the two sides of the crystal panel. The crystals have been selected with regard to availability and coverage of interesting spectral features of Fe, S, Si, and O.

Table 1: Crystals on the OXS panel

Side 1		Side 2	
Crystal	Area	Crystal	Area
LiF(220)	100%	Ge(111)	33%
ML*	100%	RAP(001)	67%

*) A multilayer, an artificial crystal evaporated on the polished LiF crystals to make a "two band" spectrometer.

A simple estimate of the line detection capability for a point source is the product of the peak reflectivity and the effective telescope area available at the appropriate energy and Bragg angle. Fig. 1 summarizes the result. Here it is assumed that the optimal observation conditions are fulfilled i.e. that the rays reflected off the crystal panel propagate along the telescope axis for greatest efficiency. To maintain this situation it is required that the attitude of the spacecraft is changed every time the Bragg angle is shifted by more than 5′. A complete map of the iron line region from 6 keV to 8 keV would require about 200 pointings so a careful strategy for this kind of observation must be chosen.

The OXS instrument is best suited for narrow line features since, for a point source, only one resolution element is recorded at a time. The sensitivity drops for broad lines e.g. for a line width of 25 eV FWHM about 25% of the line flux falls in the resolution element and for 120 eV only 5%. The continuum detection (5σ) is possible in 10^4 s only for point sources stronger than $S_{2\text{-}10keV}$ = 2×10^{-10} erg cm^{-2} s^{-1}.

SIXA has an intermediate energy resolution and high efficiency which means that an observation with this instrument will give a clear indication of where to find the interesting features in the spec-

trum from a given source. It will also be important for a determination of the continuum leve-
laround the iron line due to property of OXS mentioned above.

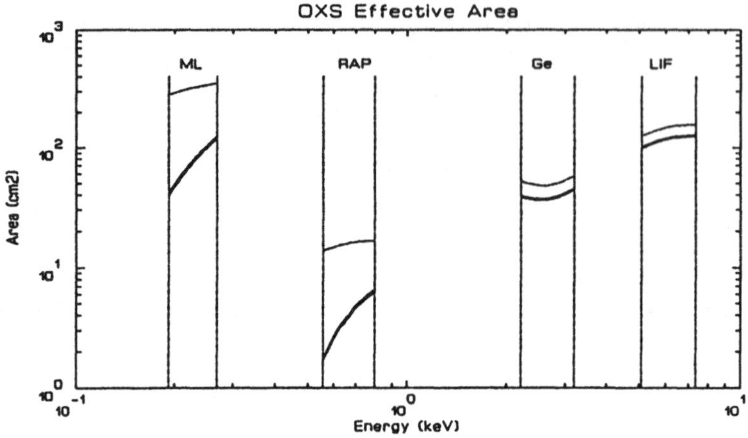

Fig. 1: The effective area of OXS for the four crystals. The dashed lines indicate the result for a 100% efficient detector. The full line shows the result when the LEPC is in the focus. The extent of the bands is given by the geometry of the spectrometer. For the band determined by the RAP crystal we are investigating the possibility of using a thinner window to improve the effective area by a factor of 2 - 3.

Line detection sensitivities.

The line detection sensitivities around 7 keV have been calculated using the numbers in Table 2 and including the extragalactic diffuse X-ray background. The particle background is estimated to 10^{-3} cts $cm^{-2}s^{-1}keV^{-1}$ and the design goal for the angular resolution is 2′ HEW. This number determines the size of the search cell for HEPC. An array element in SIXA has a size of 0.5 cm^2 and for OXS the small misalignments of the crystals will add to the width of the focal spot. The results are presented in Fig. 2. The circles represent the AGN from the EXOSAT data (Turner and Pounds, 1989). Although no line has been observed in most cases, iron lines have been assumed with a distribution of equivalent widths around 150 eV. For all the instruments the entire line flux is assumed to fall in a single resolution element. This is probably not true for OXS but there exists no secure measurement of the line width yet.

Table 2: Parameters used for the calculation of line detection sensitivities at 6.7 keV

Instrument	HEPC	SIXA	OXS
Effective area (cm^2)	1190	1190	125
Detector quantum efficiency (keV)	0.8	0.8	0.8
Detector energy resolution (keV)	0.9	0.15	0.9
Spectrometer resolution (keV)	0.9	0.15	0.006
Particle background (cts $cm^{-2}s^{-1}keV^{-1}$)	10^{-3}	10^{-3}	10^{-3}
Size of search cell (cm^2)	0.22	0.50	0.6
Fraction of flux in cell	0.5	0.8	0.6

Examples of candidates for observation.

Clusters of galaxies.

Iron line mapping can be obtained for nearby clusters with sizes larger than 5′ . If the surface brightness is sufficient, then an OXS observation can also give the details of the most important lines in the He-like spectrum as function of position. This is, however, possible only for very few clusters. The temperature distribution of the intracluster gas is very important particularly as a tracer for a cooling flow and the question of a homogeneous or clumpy flow can be investigated.

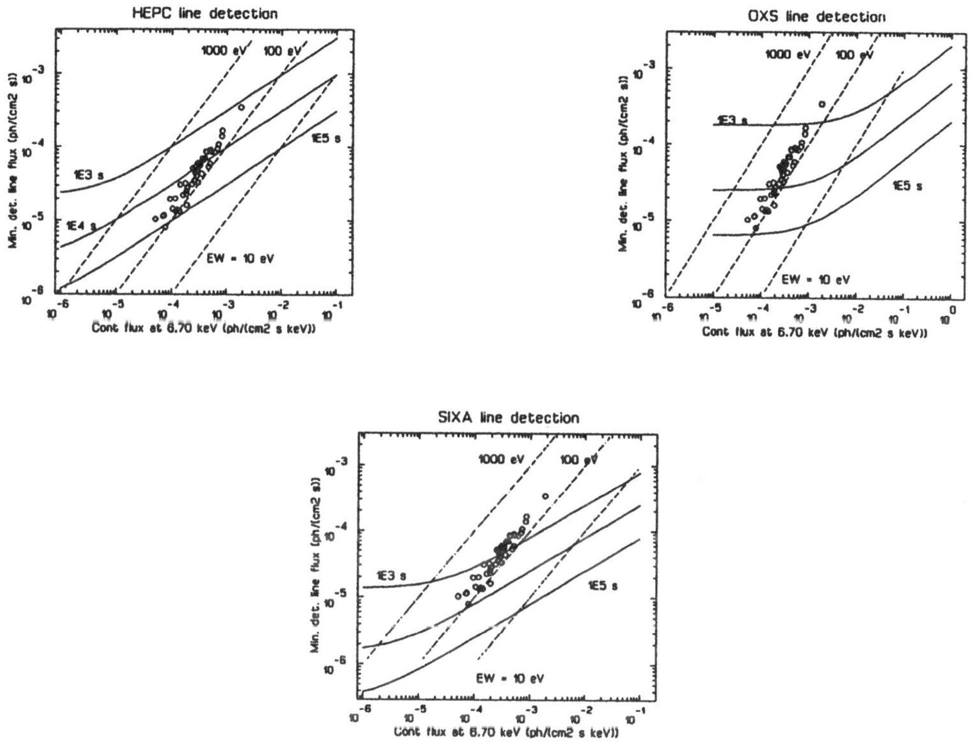

Fig. 2: The 5σ line detection sensitivities for observation times of 10^3, 10^4, and 10^5 s as a function of the continuum flux level. The lines for constant equivalent width are indicated. See text for circles. For OXS the time indicated applies if all the line flux falls in a single resolution element. Additional time must be allocated for broader lines.

The strength of the FeK_β line which has been observed with 'Tenma' in some clusters can be measured with much greater precision also by SIXA.

For a typical rich cluster the luminosity is in the order of 3×10^{45} erg s^{-1} in the 2 - 10 keV band and if the EW of the iron line is 500 eV it can be detected for $z < 1.8$ ($H_0 = 50$ km $s^{-1}Mpc^{-1}$) with SIXA. This is important for observing the evolution of the iron abundance in the clusters.

Active Galactic Nuclei.

If the line is narrow it will be possible to get the shape of the FeK line in great detail for the strong sources and for a quite large number of AGNs. In the light of the discussions in these proceedings it might be possible to observe the relativistic effects if the line comes from fluorescence in a rotating accretion disc and the possibility of an asymmetric line shape can be investigated.

Super Nova Remnants.

The iron line emission from the brightest young SNR (e.g. MSH11-54, Kepler, Tycho, and Cas A) is sufficiently strong, from 1.3×10^{-4} to 4.6×10^{-3} ph cm^{-2} s^{-1} (Singh and Hughes, 1990), that a mapping with OXS is possible with a cell size of $3'\times3'$. The iron lines probably originate from several ionization states and the energy resolution is good enough to separate these. In addition, the Doppler broadening can be measured or limits can be set.

Conclusion.

Many of the interesting questions in connection with the exact energy and strength of the iron line can be investigated with SODART which is particularly suited for this kind of investigations where angular resolution is of minor importance but high throughput is mandatory.

References.

C. Budtz-Jørgensen, A. Bahnsen, C. Olesen, M.M. Madsen, P. Jonassen, and H.W. Schnopper, SPIE Conf. 1344 "EUV, X- and gamma-ray Instrumentation for Astronomy", (1990) (in press).

F.E. Christensen, Z.S. Hua, A. Hornstrup, N.J. Westergaard, H.W. Schnopper, and L. Jalota, to appear in SPIE proceedings, San Diego (1991).

F.E. Christensen, B.P. Byrnak, A. Hornstrup, Z.S. Hua, and H.W. Schnopper, SPIE Conf. 1344, "EUV, X- and gamma-ray Instrumentation for Astronomy", (1990) (in press).

A. Hornstrup, F.E. Christensen, J. Garnæs, E. Jespersen, Z. S. Hua, and H.W. Schnopper, SPIE proceedings, vol 1333 (1990).

P. Kaaret, R. Novick, C. Martin, T. Hamilton, R. Sunyaev, I. Lapshov, E. Silver, M. Weisskopf, R. Elsner, G. Chanan, G. Manzo, E. Costa, G. Fraser, and G.C. Perola, SPIE Proceedings 1160, 587 (1989).

V. Kamarainen, H. Sipila, O. Vilhu, E. Lægsgaard, and I. Taylor, SPIE conference 1344 "EUV, X- and gamma-ray Instrumentation for Astronomy", (1990) (in press).

H.W. Schnopper, IAU Colloquium 123 "Observatories en Earth Orbit and Beyond", GSFC (1990) (in press).

H.W. Schnopper and B.P. Byrnak, Applied Optics, 26, 2871 (1987).

K.P. Singh and J. Hughes, private communication, 1990.

R. Sunyaev, IAU Colloquium 123 "Observatories en Earth Orbit and Beyond", GSFC (1990) (in press).

T.J. Turner and K.A. Pounds, MNRAS, 240, 833 (1989).

O. Vilhu, H. Sipila, V. Kamarainen, G. Leppelmeier, E. Lægsgaard, I. Taylor, and H. Schnopper, IAU Coll. 123 "Observatories en Earth Orbit and Beyond", GSFC (1990) (in press).

N.J. Westergaard, B.P. Byrnak, F.E. Christensen, P. Grundsøe, A. Hornstrup, S. Henrichsen, U. Henriksen, E. Jespersen, H.U. Nørgaard-Nielsen, J. Polny, H.W. Schnopper, and P. Ørup, Opt. Eng. 29, 658 (1990).

Iron Line Spectroscopy with the AXAF Calorimeter

Stephen S. Holt
Space Sciences Directorate, NASA/GSFC
Greenbelt, MD 20771, USA

Introduction

The AXAF calorimeter is a non-dispersive X-ray spectrometer that will be part of the AXAF focal plane instrumentation. In order to describe the revolutionary nature of such a device for astronomical X-ray spectroscopy in general, and for Fe lines in particular, it is necessary to define some terms.

Resolving power is defined R(E)=E/FWHM(E), so that R(E) can be interpreted as the photon energy E measured in units of its FWHM(E) resolution at E. R(E) is generally ascribed to be the prime attribute of a spectrometer, but it may not be the best parameter for describing its true utility. Dispersive spectrometers have the best possible resolving power, for example, but their typically low sensitivity has restricted their utility to only a handful of lines in a handful of strong sources.

In a recent review (Holt, 1990), I defined a parameter Φ to be an effective efficiency for potential detectors placed at the focus of a specific telescope (in fact, the AXAF telescope). In contrast to the quantum efficiency $\varepsilon(E)$ at a particular energy E, Φ was defined as the ratio of the photons actually detected to the those incident at the focus of the AXAF telescope over its entire bandpass, using a standard (i. e. typical Seyfert galaxy) spectrum. The combination of the two parameters R(E) and Φ is then a fairer measure of the utility of any spectrometer for use with a specific telescope than is the resolving power alone.

The generalization of Φ to an absolute parameter Ψ (rather than one related to a specific telescope) might be accomplished with a multiplicative factor that reflects the bandpass over which the instrument can be sensitive, so that instruments that take a bigger "bite" out of the spectrum get an appropriately higher factor. Here there is no dependence on the peculiarities of any particular telescope, as there is with Φ. I have arbitrarily chosen this normalizing factor to be the logarithm of its dynamic range, i. e. a detector that could measure all incident photons with perfect efficiency over exactly one order of magnitude would have $\Psi=1$. Because the sensitivity to a narrow line feature (i. e. narrower than FWHM(E)) will be proportional to the square root of most observational parameters (the quantum efficiency ε(E), the exposure, and R(E), for example) I should include an effective background parameter upon which the line sensitivity similarly depends, for completeness. Such a parameter that appropriately reduces to Ψ for zero background is $\Psi/(1+B(E)/S(E))$, where B(E) are the extra counts (whatever the cause) measured with the line counts S(E) within FWHM(E).

The current state-of-the-art in non-dispersive spectrometers for X-ray astronomy is the AXAF calorimeter (e. g. Holt, 1987). Such detectors have Ψ well in excess of unity because they can operate over a dynamic range of more than 100 with absolute quantum efficiency ε(E) approaching unity. Since non-dispersive detectors tend to have noise-dominated (approximately constant with energy) resolution, their resolving powers get better with increasing energy. The currently verified resolution of about 10 eV (FWHM) then corresponds to resolving powers of about 100 for Fe L lines near 1 keV, and about 700 for Fe K lines.

X-Ray Calorimetry

An X-ray calorimeter works by converting the energy of a single X-ray photon entirely into heat, and then allowing the precise measurement of the attendant temperature rise. For a calorimeter to function as a spectrometer for individual X-rays, the effective noise in the temperature measurement must be small compared to that temperature rise.

As displayed in Figure 1, the functional elements are an absorber and a thermometer mounted on a substrate that has small heat capacity C_0 at the heat sink temperature T_0, connected by a weak thermal link of conductance G_0 to the heat sink. When an X-ray of energy E is absorbed, the temperature of the substrate is increased by an amount $\Delta T = E/C_0$ in a time short compared to the time constant $\tau = C_0/G_0$ associated with the recovery of the substrate back to T_0. The random transfer of energy between the substrate and heat sink produces fluctuations in the energy content of the substrate. An elementary statistical calculation suggests that the mean square magnitude of these fluctuations should be $\langle\Delta E^2\rangle = kT_0^2C_0$, where k is Boltzmann's constant. Note that this is inde-pendent of G_0 and any details of the thermal link. These fluctuations can be thought of as $\sqrt{N}$ variations in the number of phonons contained in the substrate, and represent the background noise level against which the increase in energy due to an absorbed X-ray event must be measured. The advantages of low temperature operation are obvious, particularly when consideration is given to the fact that C_0 scales as T_0^3 with the proper choice of materials. As an example, a piece of silicon 0.5 mm on a side and 25 μm thick with heat capacity 4×10^{-15} J/K at 100 mK would have a limiting fluctuation noise of about $\Delta E_{rms} = 0.2$ eV.

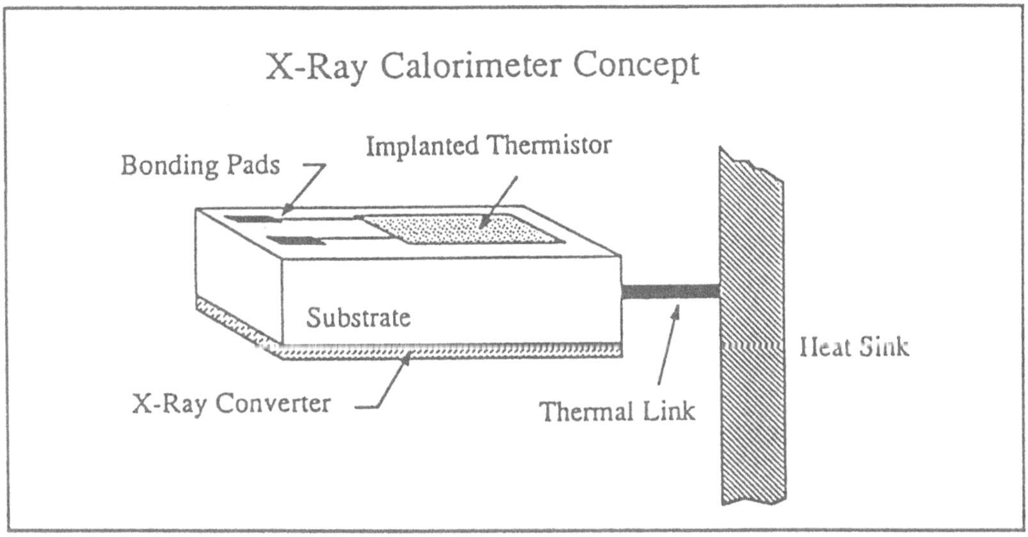

Figure 1: Schematic of the X-ray calorimeter concept.

The most important additional noise sources are those which might arise from the conversion of X-ray energy to heat (e. g. if some energy is trapped in states that are long-lived compared to the rise time of the heat pulse), and from the measurement of the temperature. Conversion noise may (in principle) be made arbitrarily small, but the Johnson noise in ion-implanted thermistors provides an unavoidable increase in the minimum system noise achievable for this thermometry technique, as demonstrated in the seminal work of Moseley, Mather and McCammon (1984) on this subject, based on analyses of noise sources in infra-red bolometers. These authors demonstrate that in the case where only Johnson noise is an important addition, the expression for achievable noise can be approximated with a multiplicative factor χ (with a value of about 2 for well-chosen thermistor parameters), so that $\Delta E_{rms}(FWHM) = 2.35\chi[kT_0^2C_0]^{1/2}$.

The baseline system for AXAF has an array of 32 such individual calorimeters in the telescope focal plane, covering a total field-of-view of approximately 1 arc minute. It is necessary to utilize an array in order to simultaneously cover this large a field-of-view while keeping the heat capacity in each element low, and it offers the advantage of introducing some imaging into the calorimeter system.

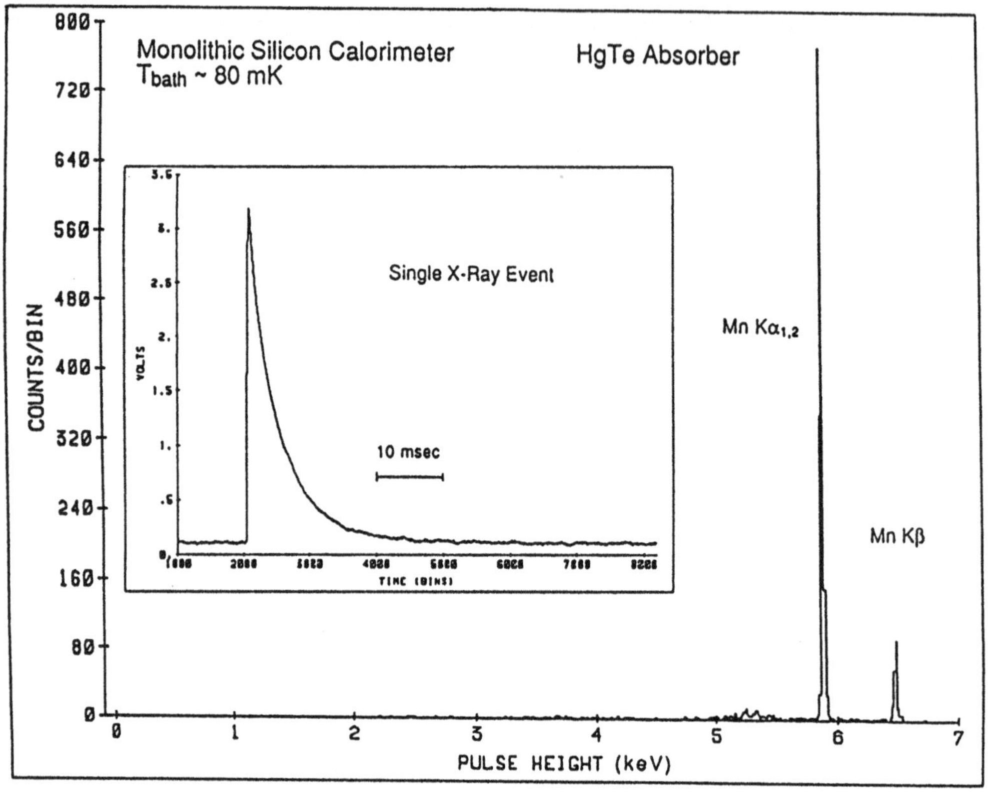

Figure 2: Response of a prototype AXAF X-ray calorimeter to X-rays from Fe^{55} (Mn K). The inset demonstrates the temporal response to a single X-ray. The pulse height spectrum demonstrates the clear separation of the Kα and Kβ peaks, and can be fit with a system resolution of 11.8 eV FWHM for these lines. The small broad peak near 5.2 keV is due to X-rays that were absorbed in the silicon substrate (which was not completely covered by the HgTe absorber for this test). In more recent tests, we have achieved resolution as good as 7.3 eV for Fe^{55}.

The detailed solid-state nature of silicon prevents it from satisfying the criterion of allowing the complete degradation of all the deposited energy in a time small compared with τ, so that it is not an ideal material for the absorption and thermalization of X-rays. It does have a low heat capacity and it is well-suited for accommodating implanted thermometers, however, so that a good compromise for AXAF might be an X-ray absorber mounted on a silicon substrate. HgTe has been baselined as this absorber, and some superconductors could also be usable. A recent discussion of such design issues for the AXAF calorimeter is given in McCammon et al. (1990). Figure 2 demonstrates the achievement of approximately 10 eV (FWHM) for Fe55 with a system close to the baseline AXAF geometry.

Required Resolving Power

A good starting place for determining of a sensible requirement for the AXAF resolving power derives from an examination of coronal equilibrium spectra for temperatures $>10^6$ K. In general, the identification and interpretation of features superposed on the continuum requires increasingly better spectral resolution as the features become weaker, or as they become confused with other features from which they cannot be resolved.

Figure 3 contains some representative coronal equilibrium spectra as measured by detectors with different resolving power. At temperatures $>10^8$ K, an equilibrium thermal spectrum is dominated by bremsstrahlung continuum, and exhibits strong isolated Fe K-emission at energies > 6.6 keV, as well as Fe L lines near 1.2 keV. Note, however, that the "completely ionized" lower-Z elements exhibit strong recombination lines: H-like Si XIV (at 2.0 keV) and S XVI (at 2.6 keV), each with ~30 eV equivalent continuum widths, are barely discernible with ~100 eV FWHM, but ~10 eV FWHM can also uncover O VIII at 0.65 keV and even separate Ne X and Mg XII from the Fe L-band emission. At temperatures near 10^7 K, where the power emitted in the lines and the continuum are comparable, ~100 eV FWHM can separate the K-emission complexes of the abundant elements (here dominated by He-like rather than H-like analogs for Z > 10), while 10 eV uncovers most of the detail that 1 eV can fully resolve. At 10^6 K, the equilibrium spectra are line-dominated; here O and Ne may be recognized with 100 eV, but ~10 eV FWHM is necessary to extract quantitative information.

Such atomic lines provide diagnostics for electron temperature, ionization temperature, density, mass motion and elemental abundances. The model-dependent interpretation of these diagnostics can also be used to distinguish among thermalization, shock heating or photoionization, as well as the extent to which the plasma is in the ionizing or recombining phase, so that even its history may be gleaned from its current status. The required minimum sensitivities depend, of course, on the details of the scientific problems that the observations are intended to address. Examples of the utilization of these diagnostics can be found in many references (e. g. Bahcall and Sarazin 1978; Pradhan and Shull 1981) and in many papers in these Proceedings.

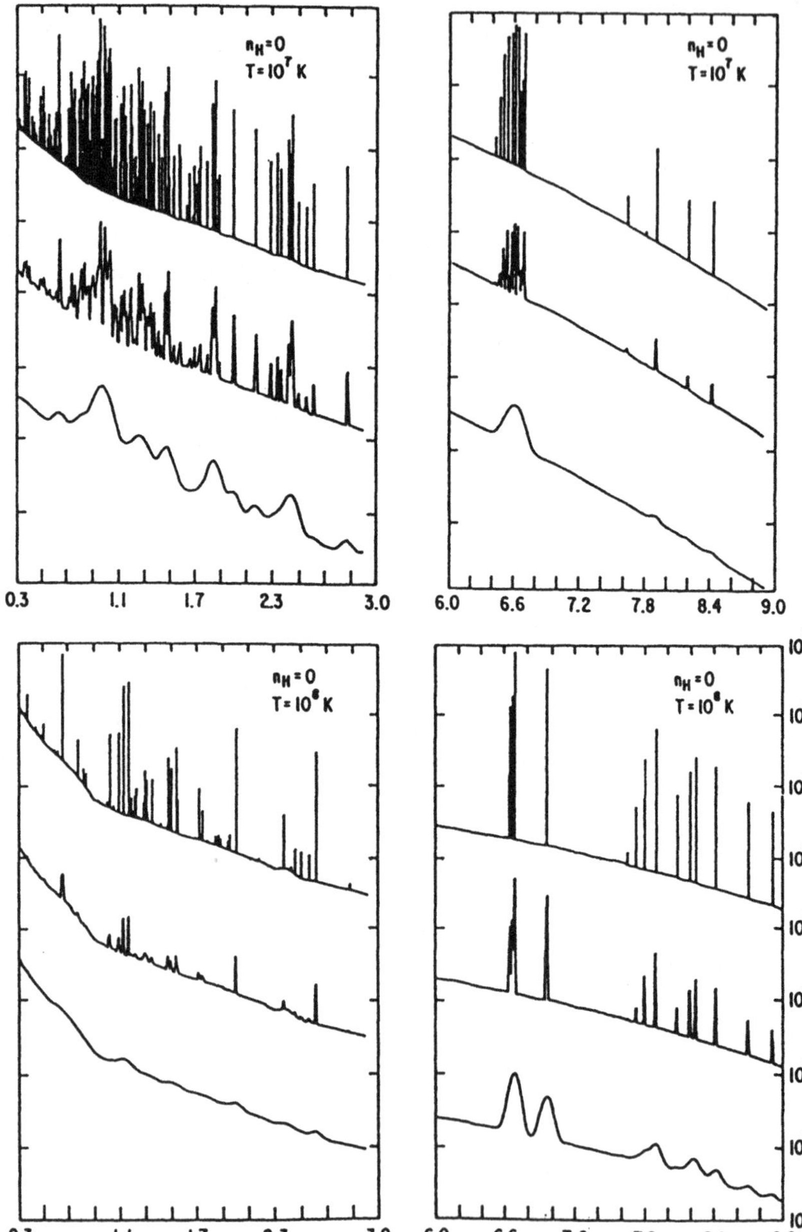

Figure 3. Coronal equilibrium spectra (arbitrary logarithmic intensity) of plasmas with solar abundances, in the energy bands (keV) containing Fe L- and Fe K-emission, as viewed with detectors having FWHM resolutions of 1, 10 and 100 eV (upper, middle and lower trace of each panel).

For L-shell to K-shell transitions, the most important lines are those from:

- the fluorescence of cold (neutral) material,
- the analog of Lyα from hydrogen-like material, and
- the resonance (r), forbidden (f) and intercombination (i) lines of He-like material.

Fluorescence and Lyα-analog recombination are clearly the low and high temperature limiting cases, and the He-analog lines are important because there is a wide range of plasma conditions and temperatures for which there is substantial population of this ionization state. Here the ratio (i+f)/r can be a useful diagnostic for electron temperature, for example, while f/i and f/(i+r) are more sensitive to density.

The energy of the Lyα analog is, of course, Z^2 times the 10.2 eV for hydrogen, so that demanding equivalent performance of a single spectrometer for all Z would require its resolution FWHM(E) to scale the same way. For a spectrometer with constant FWHM(E), the required value would be that for the lowest Z that is expected to be generally useful. For the most important Z>2 elements (i. e. 7<Z<27), the energy separation between the above lines has a very simple analytic approximation (Holt, 1987): those between Lyα and resonance, and resonance and neutral, are each about 10Z, while those between resonance and intercombination, and intercombination and forbidden, are each about Z. This means that for Z>7, ~10 eV is sufficient for separating (if not completely resolving) these lines. All of the above would suggest, that ~10 keV represents an important threshold for an X-ray spectrometer.

To obtain a general capability for the measurement of thermal or natural broadening, however, the resolution would have to be improved substantially to reach the next threshold for qualitatively new spectroscopic capability. In the following, all the subscripts are of the standard logarithmic variety, i. e.. T_8 is short-hand for $(T/[10^8\ K])$, Z_1 is atomic number Z divided by 10, and v_3 is (v/[1000 km/s]). The natural widths of K-resonance transitions are $\sim 10^{-2}(Z_1)^4$ eV, so only those of Fe (among the abundant elements) will be as broad as ~1 eV (forbidden lines will be even narrower, of course). For plasmas in equilibrium, the thermal FWHM broadening of the K-emission lines is $\delta E \sim 2(T_8[Z_1]^3)^{1/2}$ eV, so that ~1 eV is necessary to measure widths for Z<26 and $T<10^8$ K. The speeds of the same ions are only $\sim 300(T_8/Z_1)^{1/2}$ km/s, but there can be bulk motions at least an order of magnitude larger in supernova remnants, active galactic nuclei and stellar winds, where the FWHM broadening is $\delta E \sim 4v_3(Z_1)^2$ eV. Here the AXAF calorimeter, with FWHM resolution of ~10 eV, will suffice for both emission and absorption features of Fe, as well as for those of lower-Z constituents for $(v_3)>>1$.

The arguments for the resolution required for absorption features are almost identical to those for emission features. Elemental abundance measurements via absorption edges can be performed with ~100 eV FWHM for Z>10, but ~10 eV is required for the abundances of low Z elements, and ~1 eV for the populations in arbitrary ionization states. The special difficulty in the detection of absorption lines is that their natural widths (the same as their emission counterparts) are so narrow that a large optical depth results in only a small equivalent width. The best opportunity for their detections (i. e. the largest equivalent width for the same column density) is for situations where the bulk motion is such as to produce ~unit optical depth over the widest possible energy band that is small compared with the detector bandpass; the O VIII trough measured by the Einstein OGS in the BL Lac source PKS 2155-306 is just such a feature (Canizares and Kruper, 1984), but with optical depth >1, i. e. too large to measure abundances.

Required Sensitivity

General requirements for the spectroscopy of a large number of sources include the ability to investigate a reasonable portion of the telescope bandpass with a respectable level of sensitivity. Simulations can be used to demonstrate that a minimum of 10^{3-4} detected photons are usually required for detailed spectral analysis. The "logN-logS" relation for extragalactic sources with slope -1.5 (e. g. Maccacaro et al. 1982) crosses the flatter trace (with slope -0.7) for sources in the galaxy, including binary accretors and supernova remnants, at about 10 photons reaching the focal plane of the AXAF telescope per sec. Since the total number of sources which will provide >1 photon per second at the AXAF focal plane approaches 10^4, the Φ>.6 of the AXAF calorimeter will allow sufficient numbers of photons to be collected for detailed spectral analysis from thousands of targets in typical observing times of 10^4 s or less.

A totally objective (albeit incomplete) measure of the capability of a spectrometer is its sensitivity to the equivalent continuum width (EW) of a single narrow spectral feature. Here, since we are discussing a feature at a specific energy, it is appropriate to use the properties of the detector at the energy E rather than parameters that are averaged over all energies to which the detector is sensitive. If the feature is narrower than FWHM(E), the detectable EW will scale as the inverse square root of the following parameters for any spectrometer:

- the source strength,
- the exposure time, and
- the detector parameters $\varepsilon(E)$ and R(E).

For a scanning spectrometer (such as one that utilizes a rocking Bragg crystal), the same formalism may be used if $\varepsilon(E)$ is the mean efficiency to the feature over the entire energy range that must be scanned during the exposure time. Additionally, the effect of background (either non-X-ray or X-rays multiplexed from other energies) on the detectable EW can be approximated with a multiplicative factor containing the ratio of the background counts to source counts, as indicated in the first section of this paper.

Figure 4 demonstrates the 5-σ detection criterion applied to isolated features as a function of observation time, for two particularly interesting energies. Because the scaling laws go like the square roots of most parameters, the formal sensitivities are not that much better for slightly better $\varepsilon(E)$ or $R(E)$; it is only in comparison with dispersive spectrometers that the sensitivity differences are truly dramatic.

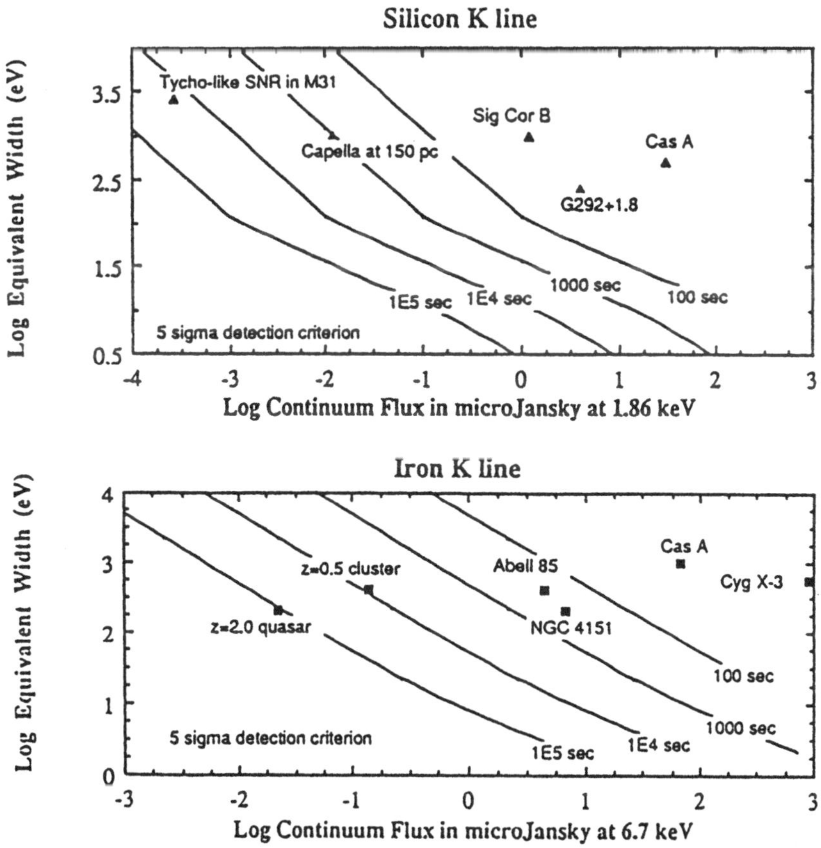

Figure 4. AXAF X-ray calorimeter sensitivities. Equivalent widths detectable at 5-σ are plotted as a function of continuum flux for two important emission lines, for accumulation times between 10^2 and 10^5 s.

Summary

It is relatively straightforward to provide simulations of calorimeter data based upon conventional wisdom or popular models. The coronal equilibrium spectra exhibited in Figure 3 provide a few such examples. A more speculative one, for the pre-discovery of a hot ISM in NGC4151 (which also includes suitably randomized counting statistics) is given in Holt (1987); here the "discovery" of the hot ISM is accomplished by separating its He-like Fe K-emission from the Fe K-fluorescence that arises in cold X-ray-absorbing clouds.

As impressive as the sensitivities discussed in the last section appear to be, they still cannot convey the new qualitative spectroscopic capability of the calorimeter. The generality of an instrument that simultaneously samples the entire telescope bandpass with high spectral resolution and high sensitivity cannot be overemphasized. The following are important astrophysical problems that can be addressed uniquely by the AXAF calorimeter:

- The detection of X-ray line emission from the hot intercloud medium of AGN -- a direct X-ray redshift measure
- The determination of absolute elemental and ionization state column densities via absorption edges in the broad line clouds of AGN and the ISM
- The measurement of resonant absorption lines (if present) in systems such as AGN and OB stellar winds
- The tracing of plasma diagnostics and elemental abundances in low surface brightness areas of SNR
- The determination of the sites at which Fe line emission originates in all compact binary X-ray sources (i. e. in each one separately)
- The detailed history of cooling flows in clusters of galaxies and in giant elliptical galaxies
- Sensitivity high enough to survey large samples of each source category

Acknowledgements

The development of the calorimeter has been accomplished through the imagination and hard work or many individuals at the Goddard Space Flight Center and at the University of Wisconsin. The key scientists currently working to develop the AXAF calorimeter are Mike Juda, Rich Kelley, Greg Madejski, Dan McCammon, Harvey Moseley, Paul Plucinsky, Andy Szymkowiak, and Jihan Zhang.

References

Bahcall, J. N., and Sarazin, C. L. 1978, *Ap. J.* **219**, 781

Canizares, C. R., and Kruper, J. 1984, *Ap. J. (Lett.)* **278**. L99

Holt, S. S. 1987, *Ap. Lett. and Comm.* **26**, 61

Holt, S. S. 1990, in *Proc. IAU Coll. 115, High Resolution X-Ray Spectroscopy of Cosmic Plasmas* (P. Gorenstein and M. Zombeck, eds), Cambridge Univ. Press, p.346

Maccacaro, T. et al. 1982, *Ap. J.* **253**, 504

McCammon, D. et al. 1990, in *Low Temperature Detectors for Neutrinos and Dark Matter III* (Edition Frontieres: Gif Sur Yvette), in press

Moseley, S. H., Mather, J. C., and McCammon, D. 1984, *J. Appl. Phys.* **56**, 1257

Pradhan, A. K., and Shull, J. M. 1981, *Ap. J.* **263**, 477

OPERATIONAL CAPABILITIES OF AXAF FOR IRON LINE DIAGNOSTICS

D. A. SCHWARTZ
Harvard–Smithsonian Center for Astrophysics

1. INTRODUCTION

The Advanced X–ray Astrophysics Facility (AXAF) will offer us all great opportunities for iron line studies. With multiple gratings and focal plane instruments, and with uniquely high spatial and spectral resolution, it is not easy to give a simple summary of the capabilities. It will provide high resolution spectroscopy, $E/\Delta E \geq 1000$ for the K, L, and M line regions. The CCD imager provides simultaneous broad field of view and high spatial resolution (arcsec) observations of the K and L lines, with modest energy resolution. The microcalorimeter allows high spectral resolving power in 5×20 arcsec pixels, with high throughput. The throughput of both the CCD and calorimeter (net effective areas of 100 to 1000 cm^2) will be useful for detailed studies of thousands of sources, although other observatories will have comparable or greater effective areas for moderate resolution spectroscopy of K–shell iron lines.

To limit the scope of this talk, I will discuss only the parameters of AXAF itself rather than spectra of hot plasmas and astrophysical objects, which is done most capably by all the other talks we have been hearing. More general reviews of the AXAF mission have been given by Tananbaum (1986), and Weisskopf (1988). It should also be kept in mind that the numerical values I present are design parameters. They are reasonably mature since AXAF is an approved program actually under development. However, they should not be regarded as highly precise, since some modest changes during development and calibration of the observatory are to be expected. The different scientific instruments have a varied status of the NASA programmatic commitment to include them on the initial AXAF launch, but I will ignore such distinctions in the remainder of this talk.

In using AXAF for Fe line observations, there are also several operational factors you may wish to consider (although for many purposes these may be ignored and standard defaults used). These include selection and focussing of one of the scientific instruments, image reconstruction and location on the celestial sphere, calibration, and the criteria for event selection and for background rejection and estimation.

2. OBSERVATORY DESCRIPTION

The high resolution mirror assembly (HRMA), consists of six nested pairs of paraboloid–hyperboloid shells, aligned to a common on–axis focus 10 meters from their intersection plane. The three outer pairs are to be coated with Ni, while it is planned that the three inner pairs will be coated with Au. Ir is also being studied as a possible alternative to Au, since its greater density offers improved reflection efficiency at higher energy.

The most powerful and unique features of AXAF are related to the optical quality of its mirrors. Figure 1 shows the minimum expected encircled energy fraction inside a diameter of 1 arcsec (squares) or of 10 arcsec (plus signs).

Figure 2 gives the product of this imaging performance with the expected effective area. This definition of effective area is the quantity needed to plan an experiment, and is of course smaller than the usual quotation of the total number of photons arriving anywhere in the focal plane (c.f., other papers in this session). The squares and plus signs show the minimum required effective area for which photons are imaged in 1 or 10 arcsec diameter, respectively, as in figure 1. However, considerable incentive bonuses will be given to the glass contractor for reducing mid–frequency (i.e., spatial frequencies of order 1 mm^{-1}) and micro–roughness errors. These would result in the performance goals, shown as points without connecting lines, for an encircled image of 1 arcsec (diamonds) or 10 arcsec (triangles) diameter.

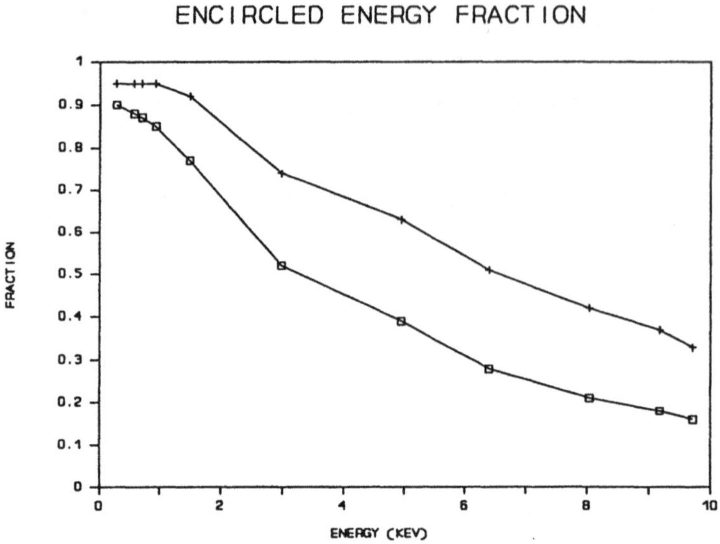

Figure 1 The minimum specified encircled energy fraction, vs. energy in keV, within a 1 arcsec (squares) or 10 arcsec (plus signs) diameter. The points show the specific energies at which the specification applies.

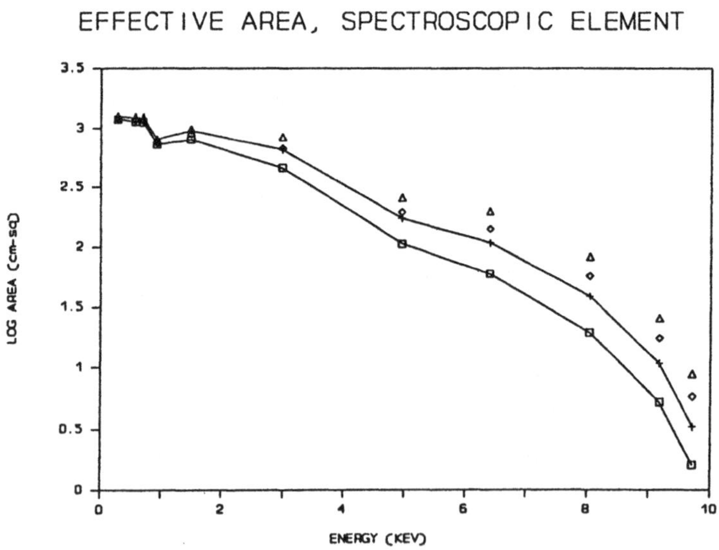

Figure 2 The minimum required effective area for focussing within an encircled diameter of 1 arcsec (squares) or 10 arcsec (plus signs). The goal for enhanced performance is shown by the diamonds (1 arcsec) and triangles (10 arcsec), without connecting lines.

The HRMA is mounted on an optical bench via moveable actuators. These provide six degrees of freedom, so that the X-ray beam can be pointed to any spot of any focal plane instrument, focussed along the optical axis, and tilted in two directions (e.g., to be perpendicular to the surface of the detector). The rotation of the HRMA about the optical axis is degenerate in terms of optical properties (by design, to be verified by calibration), and the redundancy of the HRMA actuators is such that this rotation is the first capability which is lost.

The aspect system allows reconstruction of the X-ray image to a fidelity which would be 0.5 arcsec rms diameter, and to locate it on the sky to within a 1.0 arcsec rms radius. (The actual image reconstruction and location is also degraded by the imperfections of the mirror and the science instruments.) Registration to the sky is accomplished by a CCD star camera. Five stars and three fiducial lights will be simultaneously and continuously measured by the camera. Two 2-axis gyros will be read out into telemetry every 0.1 sec, giving the equation of motion to very high relative precision, for the reconstruction of the X-ray image.

At any given time, AXAF can point anywhere within the 85% of the sky which excludes a 45° cone about the sun. A sunshade on the facility ensures that no sunlight ever illuminates the front end of the telescope assembly, while also avoiding the reflection of any light within 30° of the star tracker axis. The solar panels extend as wings, and can be rotated about their long axis. It is planned to control the roll angle of the spacecraft about the target direction in order to keep the solar panels perpendicular to the line to the sun. The orientation of the grating dispersion relative to the sky might be of interest if spectra are to be taken of several objects in the same field of view. In such a case the observer would want to specify the spacecraft roll angle. This can always be accommodated by observing at a particular time of the year, or by using orbit eclipse only. The star camera will have a properly designed light shade to work within 15° of the sunlit earth. Such a shade does not allow any point inside the baffle assembly to view both the edge of the earth and any optical element.

AXAF will be in a low earth orbit, nominally circular at 600 km. Earth occultations will limit uninterrupted views of a target to typically 3000 to 4000 seconds. This may be a minimum time to do significant spectroscopic observations of the brightest few thousand sources, while ten times as much exposure may be used for weaker sources or more detailed studies; e.g., with the Bragg crystal spectrometer. Two on–board tape recorders can each store 12 hours of data, which will provide complete recovery of the telemetry stream with 2 or 3 TDRSS contacts per day. The capability for storing commands on–board AXAF is sized at 3 days. The relatively low demands on TDRSS for telemetry and commands should simplify AXAF operations, and allow flexibility for scientific innovation.

AXAF will contain four different instrument packages in the focal plane, along with two transmission grating assemblies. Either grating assembly, (or neither), can be rotated into the optical path. Any of the four instrument assemblies can be chosen, and the optical axis of the HRMA positioned at any point in their entrance apertures, and brought to any focus relative to the detector surfaces. The quantization of these motions is actually about 0.025 mm, corresponding to 0.5 arcsec transverse to the focal point and comparing to a nominal 0.1 mm depth of focus. The specific spectroscopic capabilities of the instruments will be discussed later.

3. HRMA CALIBRATION

The various investigations planned for AXAF will be limited in many cases by systematic effects, rather than the statistical uncertainties of the photon counting. A unique feature of the AXAF mission is its planned ground calibration program. The X–ray test facility at Marshall Space Flight Center (MSFC) is being upgraded to include a 1700 foot long X–ray pipe, and an enlarged vacuum chamber to hold the HRMA and science instruments on an internal optical bench structure. At this distance the X–ray source will appear to be about 0.2 arcsec in size, and will strike the mirrors within a cone of half angle 2 to 4 arcmin, instead of exactly on–axis.

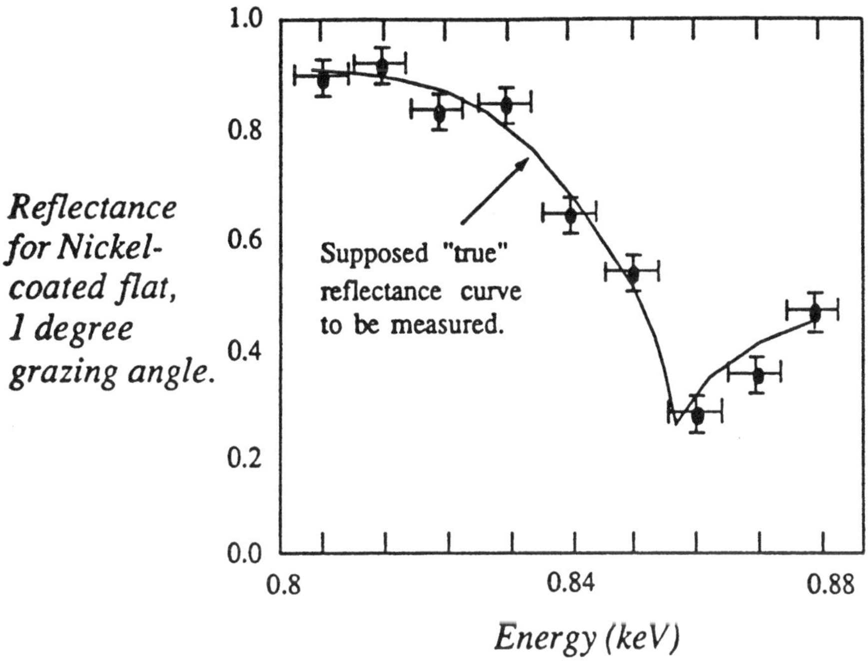

Figure 3 Simulation of synchrotron measurement of reflection efficiency, in the vicinity of the Ni L–edge (from Graessle 1990).

The strategy of calibration is that on the ground we will attempt to measure the effective areas and spectrometer throughputs as precisely as possible as a function of energy. In the sky, we do not have either any flux standards nor any monochromatic sources, although AXAF may help to establish such standards. Conversely, the final calibrations of imaging performance and spectrometer resolution must be done on–orbit, due to the difficulties of reproducing on the ground the zero–g, parallel incident radiation conditions encountered on–orbit.

The key feature of our calibration plans is to make a very accurate measurement of the shape of the effective area curve using a synchrotron facility. These measurements will be made on polished flats, coated at the same time as each of the 12 pieces of glass forming the AXAF mirrors. Figure 3 (taken from Graessle 1990) gives simulated data, illustrating the reflectivity measurements in the vicinity of the Ni L–edge. Several Fe XVIII lines are directly in this region.

The reflectivity will be measured at the six mean on–axis grazing angles of the nested array, in order to over–determine the optical constants. From the reflectivities and the geometrical dimensions of the mirrors, the predicted shape of the effective area is calculated. It is then normalized by means of the MSFC facility measurement, which is carried out using the Ni L–lines excited by an electron beam. This measurement is actually a broad beam average over several lines, and including a small bremsstrahlung component, all of which must be modelled since the MSFC facility detectors will have only modest energy resolution.

4. SPECTROMETER DESCRIPTIONS

AXAF is planned to carry six different spectrometers: three transmission gratings, two non–dispersive focal plane instruments, and a Bragg crystal spectrometer in the focal plane. The SRON/MPE low energy transmission grating spectrometer (LETGS; Brinkman et al. 1987) is a 1000 l/mm grating in a toroidal configuration covering all six mirror pairs. The grating is made from free standing Au, and covers the 2 to 140 Å region. The primary readout used with this grating is expected to be the high resolution camera (HRC), provided by SAO. The HRC contains two 100 mm microchannel plates arranged corner–to–corner with the LETGS dispersion along their common diagonal. The MIT high energy transmission grating spectrometer (HETGS; Canizares et al. 1987) consists of two different kinds of gratings, on a common holder. The three outer mirror elements are covered by the medium energy transmission grating (METG). This is a 1600 l/mm grating with Ag lines, covering the 0.4 to 5 keV region. The three inner mirror shells are covered by the high energy transmission grating (HETG). This is a 5000 l/mm grating, made of Au lines, and covering the 1 to 8 keV range. The HETG and METG are tilted at a small angle with respect to each other, so that their dispersed spectra do not interfere when read out with a 2–dimensional imaging detector. Either the LETGS or HETGS can be rotated behind the exit aperture of the HRMA, with the grating facets closely approximating a toroid normal to the exiting X–ray trajectories.

The dispersion direction of the LETGS is the same as the mid–line of the two HETGS gratings, and both imaging focal plane instruments have their long axis positioned so that either grating can be read out by either imager. Since the focal plane instruments cannot follow the Rowland circle, the observer should consider at what wavelength the resolution is to be optimized, and choose to offset the zero order spot from the Gaussian focal plane accordingly, in order to obtain the narrowest dispersed image at the desired wavelength.

Goddard Space Flight Center is providing an X–ray spectroscopy (XRS; Holt 1987) experiment consisting of an array of micro–calorimeters in the focal plane. Each of 32 elements covers 5 x 20 arcsec solid angle, arranged in three rows of 12 units to form a 1 x 1 arcmin square, but with the four corners omitted. The calorimeters respond from 0.3 to 10 keV with considerable throughput, and with 10 eV energy resolution. The most precise calibration and use of this spectrometer will be with the X–ray image defocussed on the order of ~2 arcsec, to average over (possible) irregularities in the quantum efficiency.

The AXAF CCD imaging spectrometer (ACIS; Nousek et al 1987), being prepared by Penn State/MIT, provides an imaging array covering roughly 15 arcmin with 0.5 arcsec resolution, and simultaneously providing non–dispersive energy resolution ranging from 70 to 120 eV over its 0.4 to 10 keV bandpass. In general, an X–ray CCD splits charge among multiple pixels, for at least a small fraction of events. The raw telemetry data typically consists of the amplitudes measured in a small array of pixels. A given data set can be multiply processed with different algorithms, to maximize either the detective quantum efficiency or the energy resolution. The latter algorithm would select an event only when occurring in a single pixel, as indicated by a "zero" readout from the surrounding pixels. ACIS includes a separate imaging array of six CCD chips along a line. This will be the readout of choice for most HETGS observations, since the quantum efficiency of the CCD is somewhat better than that of the HRC over most of the HETGS region, and since its energy resolution provides discrimination against higher order spectra.

MIT is also preparing a Bragg crystal spectrometer (BCS) experiment, consisting of three crystals covering the energy range 0.5 to 3 keV. A rotating drum allows selection of a TAP, RAP, or PET crystal to cover the desired energy range and to rock about the selected energy. Position sensitive proportional counters read out X-rays which satisfy a Bragg selection, and the position information can be used for spectroscopy of extended sources as well as of point sources. For point sources, the proportional counters can be moved off the Rowland circle, and the positional information used to trace the X-ray reflection geometry, enhancing the energy resolution.

Another potential spectrometer must be mentioned, because it is one of the key features of AXAF compared to all the other missions discussed in this conference: A replacement instrument spectrometer. This instrument is currently undefined; however, by definition it will exceed any of the existing spectrometers very significantly in some capability.

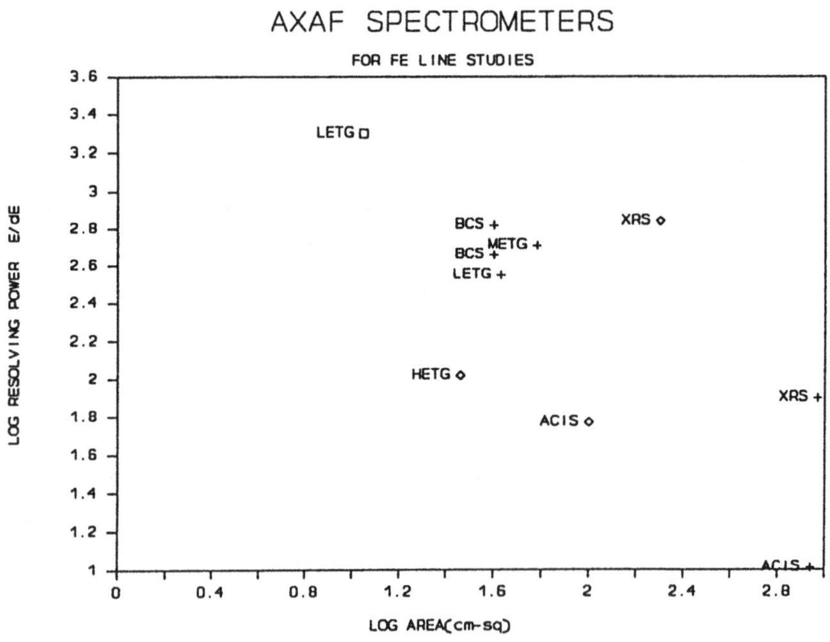

Figure 4 Capabilities of the AXAF spectrometers for the regions of the 2s–2p lines, taken at 100 Å (square), the L–lines, taken at 0.8 keV (plus signs), and the K–lines, taken at 7.0 keV (diamonds).

Figure 4 plots the available resolving power vs. effective area for the AXAF spectrometers at representative energies of the three X–ray Fe–line regions. A single representative energy is taken for each group: for the L lines in particular the situation is more complex than shown because of the rich spectrum of lines spanning 0.7 to 1.5 keV, and because the HRMA, Bragg crystals, and gratings all have spectral features and considerable efficiency and resolution changes over the span of this region. The area considers the net throughput of the mirrors, spectrometers, and detectors. For the dispersive instruments and the CCD the imaging element is taken to be 1 arcsec diameter, while for the XRS a 10 arcsec diameter is used.

Each of the six spectrometers offers some superior capability for iron line observations. For the 2s–2p region around 100 eV, the LETGS provides the only coverage. It gives very high resolution, ~1000, with low effective area, ~10 cm^2, and is suitable only for point sources. For the L–line complex, the BCS gives the highest energy resolution, and gives the only capability for extended sources. The METG and LETGS will also naturally be used because of their broad energy bandpass, and because they do have sufficient energy resolution for many plasma diagnostics. The XRS energy resolution of only ~100 cannot directly give diagnostics in this energy region, but because of its high throughput it may naturally be used to fit parameters of models; e.g., in cases where we think we know what line energies to expect. Due to their high throughput, both XRS and ACIS will play a role in exploring spectra of unknown objects or categories, to see which have indications of lines that should be followed up by the higher resolution instruments. In the K–line region, XRS is the premier instrument in both resolving power and effective area. The CCD imager will be called on for studies of clusters of galaxies, supernova remnants, and in order to do phase–resolved spectroscopy of the bright galactic binary sources.

At this meeting we have heard much about what can be learned from the observations of the X–ray iron lines, and about their ubiquity in astrophysical settings. The theme of this talk has been that AXAF can be used in a wide variety of observations to extract such information. It is appropriate that we all begin detailed consideration now of how we will use this international facility.

ACKNOWLEDGEMENTS

I thank other members of the NASA/MSFC, TRW prime contractor, SAO, and instrument PI teams, who are working so hard for so long to make this mission possible. I thank C. Canizares and M. Weisskopf for permission to use several figures for my oral presentation.

REFERENCES

Brinkman, A. C., van Rooijen, J. J., Bleeker, J. A. M., Dijkstra, J. H., Heise, J., De Korte, P. A. J., Mewe, R., and Paerels, F. 1987, Ap. Letters & Comm., 26, 73.

Canizares, C. R., et al. 1987, Ap. Letters & Comm., 26, 73.

Graessle, D. E. 1990, unpublished report, "The use of synchrotron radiation in the calibration of AXAF: feasibility study," SAO–AXAF–90–032.

Holt, S. S. 1987, Ap. Letters & Comm., 26, 73.

Nousek, J. A., Garmire, G. P., Ricker, G. R., Collins, S. A., and Reigler, G. R. 1987, Ap. Letters & Comm., 26, 35.

Tananbaum, H. 1986, Ann. N. Y. Acad. Sci., 470, 338.

Weisskopf, M. 1988, Space Sci. Rev., 47, 47.

THE PROPOSED ISRAELI X-RAY SATELLITE

G. Shaviv ,T. Weller, J. Felsteiner, B. Rosner, M. Livio, N. Lupu,
I. Gertner, H. Abramovitz, N. Markevich, and N. Niv

The Asher Space Research Institute,
Technion-Israel Institute of Technology, 32000 Haifa, Israel

1. INTRODUCTION

About two years ago the Israel Space Agency issued a call for proposals for an experiment on board an Israeli satellite. The constraints were to use the Ofek 2 platform without any change in its design and specifications. The constraints were: an experimental volume of 0.5 m^2, total weight of payload 20 kg, total power consumption 20 W, about 0.1° pointing accuracy and about 0.1° drift in direction stability, survival in acceleration of 35g during take off, and mission duration of two years.

We propose to use the platform for a hard X-ray telescope. The technological and scientific objectives and our guiding policy are:

a) The development of a new technology for X-ray astronomy based on light satellites, new type of detectors and arrays of detectors as well as telescope structure and electronics. We show that light satellites are the optimal platform for high energy experiments since due to their low mass they create a particularly low background.

b) A scientific mission designed to perform special astrophysical observations of current interest in astrophysics in general and X-ray astronomy in particular, for which a low background is essential and complementary to known existing or planned experiments.

The specifications of the instrument are :
Energy range: 3-30 keV.
Total area: 426 cm^2 at 10 keV.
Background count rate: 3.5×10^{-5} counts keV^{-1} cm^{-2} s^{-1}.
Energy resolution: $\sim$ 1 keV FWHM over the entire energy range at $-20°C$.
Time resolution: few milliseconds.

2. THE SCIENTIFIC MISSION

The temporal variability on many timescales of X-ray sources is very omnipresent, implying that the most powerful sources in the universe are not at rest or in a steady state. The dramatic fact is that variability extends over all timescales, from milliseconds (in some compact X-ray sources) to months (for active galactic nuclei and quasars). Time variability provides a multitude of information about the objects.

The proposed X-ray telescope, which we call the Israeli X-Ray Telescope (or in short IXT) covers a range in energy (3-30 keV) not covered by HEAO-II and which will be covered neither by ROSAT nor by AXAF (see Fig. 1 where a comparison is shown). A part of our energy range (below 20 keV) will be covered by the proposed Monitoring X-Ray Experiment (MOXE, Priedhorsky *et al.* 1990), and this can provide an interesting collaboration between the two experiments. The Italian SAX astronomy satellite will also cover the energy range of the proposed telescope. The particular energy range is so chosen (among other things) because of the Fe line at 6.7 keV and the cyclotron lines at about 16-24 keV.

The telescope will be able to perform the following observations:

(1) Imaging with a moderate angular resolution (about 1 arcminute) and broadband spectroscopy over the energy range 3-30 keV.

(2) Continuum and high resolution line spectroscopy in the range 3-30 keV.

(3) Studies of time variability of the spectra of bright sources, over timescales ranging from milliseconds to months.

(4) Studies of long term variabilities of various X-ray sources, by means of periodic, relatively long observations.

(5) Detection of sources fainter by a factor 4-5 than the detection limit of HEAO-II, due to a much lower background. The expected background count rate is 3.5×10^{-5} counts keV^{-1} cm^{-2} s^{-1} over the entire range of 3-30 keV.

(6) The prominent hard X-ray source classes that can be observed with the proposed telescope include: bright low mass X-ray binaries, X-ray transients, Crab-like SNR, γ-ray bursters, AGN, several clusters of galaxies and iron line diagnostics.

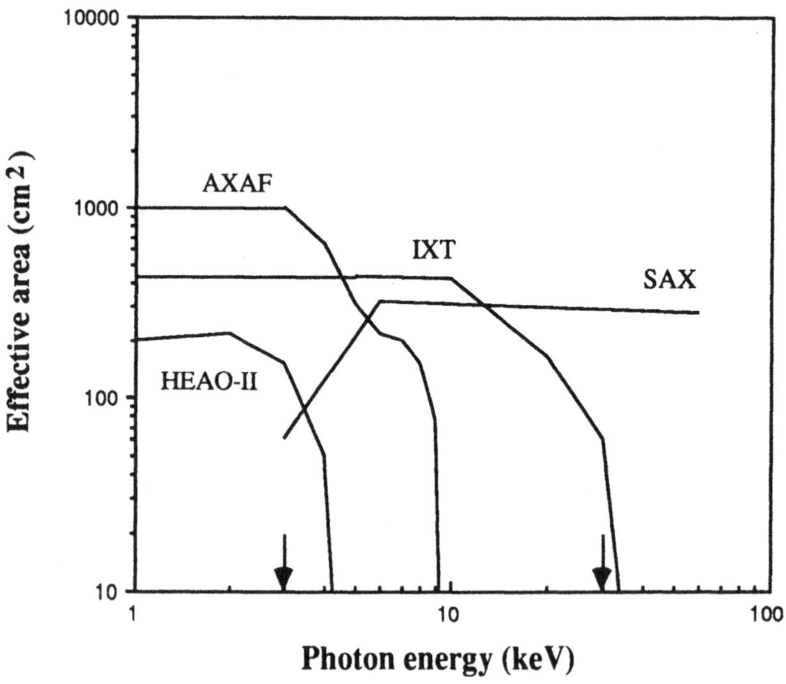

Figure1. The effective area of HEAO-II,AXAF, SAX and the present telescope IXT, as a function of energy. The arrows mark the relevant energy range of the present detector. Note however, that AXAF and HEAO-II use mirrors to concentrate the X-ray radiation.

3. PRINCIPLES OF OPERATION

The collimator and the structure of the telescope are the major factors in the weight (and in the background) while the detectors and the electronics behind it are the major factors in the power consumption. A schematic view of the telescope is given in Fig. 2, and the method of operation of the screens is given in Fig. 3.

The basic unit detector has the dimensions of $1.0 \times 1.0 \times 0.05$ cm^3, exposing the large surface to the source. An array of 440 detectors is formed by placing the detectors at equal distances from one another. The coverage obtained in this way is about 25%. The collimator differs from standard ones in the following sense. In the standard case there is a big detector and the collimator covers parts of it. In the present case the detectors cover only a part of the area and consequently the collimator may cover the rest.

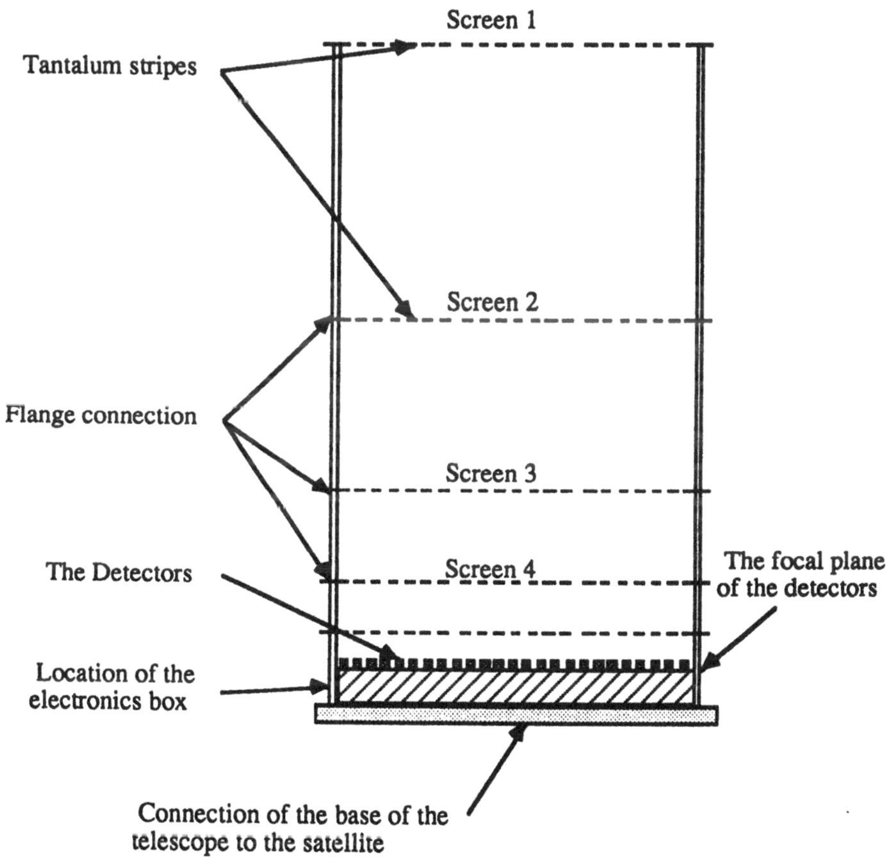

Figure 2. A schematic view of the telescope.

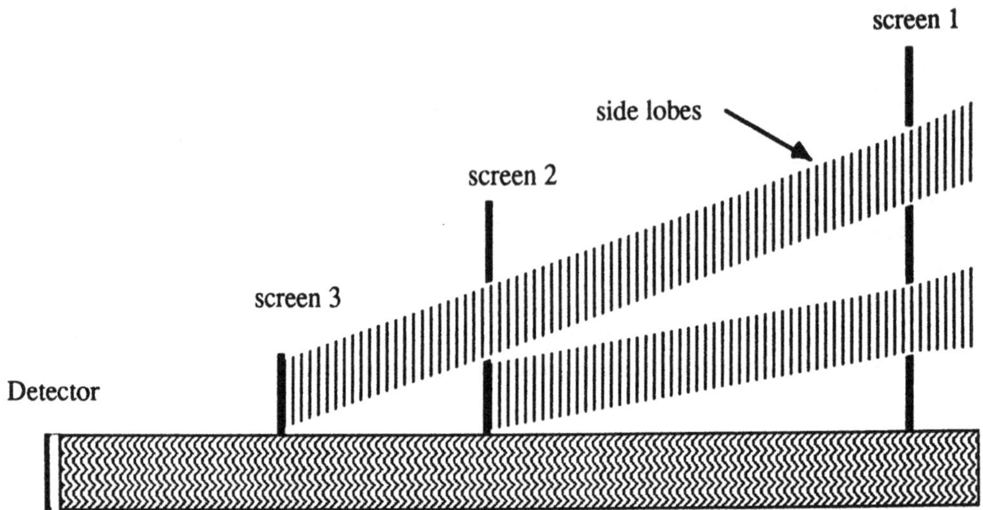

Figure 3. The operation of the screens. The screens are put in such distances from one another as to shade the detector from the photons that may enter through the 'wrong' opening.

4. THE COLLIMATOR

The most massive part of any X-ray telescope is the collimator. The total mass of the collimator depends on the following factors: The metal out of which it is made, the required spatial resolution and consequently the design and the energy range in which it is planned to operate. Simple estimates of the required weight of a typical collimator, like for example a honeycomb, are well above the weight constraint. Consequently, we resort to the following concept in the design of collimators.

The constraints on the length of the payload is about 1 m and its diameter 50 cm. The size of a single detector is 1 cm^2. Assuming a total length of 100 cm for the collimator, the theoretical limit on the spatial resolution is 0.57°. This figure suits the pointing accuracy of the present satellite without any need for additional control systems.

The material out of which the collimator is made is chosen on the basis of the following factors: absorption power of X-ray radiation, weight and structural properties. The best metal in terms of absorption per gram is tin. However, tin has poor structural properties, and anticipating problems of strength we reject this possibility . The next best metal is cadmium . This metal is very difficult to work with and is also a poison. Also, cadmium is not a good absorber below its K-edge which happens to fall inside the energy range of interest here. Our choice therefore is tantalum. We give in Table 1 the required data for estimating the weight and thickness of the collimator. The required weight is calculated for an attenuation factor of 10^5. It may be preferred to prepare each screen from a sandwich of two metals: the first one is aluminum, and it serves to support the second one, which is tantalum. In view of the present total weight, which is less than the allowed one, we even investigated the possibility to add another metal with an intermediate Z-value between the tantalum and the aluminum and to do so only in the last screen. In this way the screening from the self-emitted X-ray lines is improved. The numbers given in the table are approximate and the actual numbers will be somewhat different, since it is clear that the metallic screens will be obtained in the closest available widths found in the market.

Table 1

Basic Atomic Data of X-Ray Absorption by Different Metals

Metal	Density (gm/cm^3)	Atomic Absorption of X-ray μ (30keV)(cm^2/gm)	Required width (mm)	Total Screens Weight (kg)
Sn	5.75	41.1	0.49	5.500
Cd	8.65	38.1	0.35	5.933
Bi	9.8	31.45	0.37	7.188
Pb	11.34	29.8	0.34	7.586
W	19.3	22.98	0.26	9.837
Ta	16.6	21.76	0.32	10.388
Fe	7.86	8.09	1.81	27.942
Al	2.7	1.107	38.50	204.201

5. THE MAIN SUPPORTING STRUCTURE OF THE TELESCOPE

The screens are held by a supporting cylindrical structure. Although the basic units are rectangular, the demand on the stiffness of the entire structure favors a cylindrical shape. This is therefore the chosen form. Besides supporting the screens, the cylinder shields the detectors from the side. However, a too massive structure will generate a large background and hence is destructive to the physics. The material used for the main structure is aluminum. The full length of the telescope is divided into several small cylinders which are connected to one another with flanges at the location of the screens. The thin-walled cylindrical parts obtain their stiffness from the flanges and from integral stringers. Laboratory tests have verified the structural properties of this shape and assure us that the telescope will be able to withstand the launch loads very easily despite of its very light weight.

In summary, the weight of the telescope is made up of three parts: (a) the aluminum structure (cylinder, screens and flanges), (b) the tantalum structure (cylinder and screens), and (c) the electronics. The present estimate of the weights of the most massive parts is:

Cylinder:	Aluminum	1.06 kg
	Tantalum	0.24 kg
Screens:	Aluminum	0.48 kg
	Tantalum	10.39 kg
Total Weight:		12.17 kg

The above weight estimate does not take into account nuts, bolts, flanges, thermal blankets, passive cooler and electronics. However, we do not expect these items to have crucial weight constraints on them.

6. THE DETECTORS

The detection system will be constructed from an array of X-ray detectors operating at −20°C. Each element of the array will be composed of a Hamamatsu 1723-09 PIN silicon photodiode with an active area of 1 cm^2 and a thickness of 0.5 mm, a preamplifier and an amplifier built in a modular way and in the size of the element itself.

In view of our measured efficiencies at various energies we calculated the effective area to be 426 cm^2 at 10 keV, 166 cm^2 at 20 keV and 61 cm^2 at 30 keV. A thin aluminized mylar will protect the surface area of the detector from light. The number of 440 detectors is a result of the constraint of a total diameter of 50 cm.

The preamplifiers will be of Amptec-225 units (or their equivalent) which are space and military specified and are known to be reliable under extreme temperature and pressure conditions. They require a very low power for normal operation and hence suit best our purpose.

Two other ideas that are presently being considered are: (*i*) to put additional detectors not in front of any aperture so as to measure the background only, and (*ii*) to use in the array several detectors with different sizes and hence different spectral response, to improve handling of the background even more.

7. ADVANTAGES OF THE PRESENT DESIGN

Most X-ray satellites use NaI scintillators coupled to photomultipliers, or cryogenically cooled Ge diodes for high energy resolution, or gas proportional counters. The detector system proposed here is significantly more compact than any of the above mentioned systems having the same total collecting area. In addition, the requirements of voltages and power are also modest. (A conventional system would be beyond the allowed limits of weight and power for the Israeli experiment). Furthermore, the energy resolution of our system is superior to that of a NaI system. Finally, the proposed system is made of many small (relatively cheap) units. The ability to get information from every detector increases the flexibility in the use of the telescope. The failure of a few units will not affect significantly the performance of the array unlike the breakdown of a photomultiplier or a proportional counter. Actually, due to the relatively low power required and the internal structure of the single unit, the space survivability of each detector is good and we expect a long operation time free of failures. In other words, the redundancy and reliability of the detection system is very high.

During test calculations, we discovered that the background counts are very low, so low in fact that we reach the source confusion limit. Consequently, we changed the design of the screens so as to create the capability of a lobster-eye imaging. This capability is one of the advantages of the present design.

8. EVALUATION OF THE BACKGROUND COUNTS

Due to the intense radiation in space and the weak source strengths of astronomical objects at the energies under consideration the observations are necessarily background dominated. This has been already well realized by several investigators (*e.g.* Gehrels 1985; Dean *et al.* 1986, 1989). The signal to background ratio is therefore one of the most crucial design parameters of the telescope. Grossly speaking, three main factors are contributing to the low background of the proposed telescope. The first is related to the particular detectors we use. The thickness of the Si PIN diodes (0.5 mm) matches the photon energy range under consideration. Namely, the efficiency of the detectors is high in the 3-30 keV energy range and rapidly decreases for higher photon energies. This thickness also ensures that the cosmic ray protons or other heavier ions impinging on the detectors will produce energy signals above the 30 keV limit. Low energy signals and electronic noise will be rejected by a suitable discriminator, and signals above the high energy limit will be vetoed out by means of software (and not by means of anticoincidence). Note that it is the combined effect of quantum efficiency and thickness of the detector which does the job. Second, the major source of background is X-rays generated in the strucure by cosmic rays. Finally, an important factor is the total mass of the satellite and the telescope. The total mass of the HEAO-II satellite, for example, was about 3000 kg while the mass of the Israeli satellite is only 200 kg. This fact alone contributes a factor of about 15 to the relative background counts since the rate of conversion of cosmic radiation (protons and electrons) into γ-rays which scatter in the satellite is proportional to the total number of atoms. The fact that the collimator and the tube which holds it are very light is therefore an advantage.

The background counts were evaluated by means of a Monte Carlo computer program. The satellite and the telescope are represented in the program by means of a cylinder which consists of three regions made of different materials. The outer part is made of 200 kg of aluminum and represents the satellite material surrounding the telescope. The plane of the detectors is simulated by 440 square detectors, each of 1 cm^2 area and 0.5 mm thick, placed in the plane and representing the detector array. Each silicon detector is assumed to have the natural density of silicon and its response to a photon of a given energy is the one found in the laboratory. The regions above and below the silicon disk are filled with 15 kg of tantalum, representing the X-ray shielding. Photons that reach the detectors after going through the screens and the structure of the telescope are considered as background. The number of 15 kg of tantalum used in the calculation is an upper limit to the mass of the screens. The aluminum and tantalum are assumed to occupy the actual dimensions of the satellite. Their effective density is calculated in such a way as to preserve the total number of atoms.

The background has the following sources: cosmic X-rays and γ-rays, cosmic charged particles, and the albedo of the Earth. The external radiation hits the structure of the satellite or the telescope, produces a γ-ray or an X-ray, which propagates subsequently in the satellite and can reach the detectors with an energy in the range of interest. Photons which produce energy signals within the 3-30 keV range in the detectors are considered as background counts. The direct contribution of protons and electrons is negligible because they are much too energetic to create signals in the detectors with energies in our range. The contribution of the direct diffuse X-ray background depends on the opening of the telescope, is simple to evaluate and is not included in this calculation.

The treatment of the background due to charged particles is different from the one applied to X-rays and γ-rays. The high energy particles give rise to a mass emissivity function $\varepsilon(E_\gamma)$ photons s^{-1} keV^{-1} g^{-1} (Gehrels 1985) which is assumed to be isotropic. In the first step the mass emissivity is found as a function of energy and location for monoenergetic particles. The source function was then fed into a second Monte Carlo code to find out the detector counts due to this emissivity of γ-ray photons. The results were then plotted and a fit to the background counts as a function of the proton energy $N(E)$ was obtained. We found that the following expression

$$N(E) = 6.8 \times 10^{-4}(E/\mathrm{MeV})^{-0.887}$$

describes very well the background as a function of proton energy (the least squares fit has $R = 0.98$). Of course, the numerical values of the parameters are determined by the particulars of the telescope and the satellite and are found numerically by means of the Monte Carlo program. In the second step the integral

$$\int_{E_0}^{\infty} N(E)I(E)dE$$

was evaluated. $I(E)$ is the flux of high energy protons and E_0 is the threshold energy. Again, photons which produce energy signals within the 3-30 keV range in the detectors are considered as background counts.

The total background count rate originated from cosmic X-rays and γ-rays and cosmic charged particles was found to be 3.5×10^{-5} counts keV^{-1} cm^{-2} s^{-1}. The conclusion is that the thin detectors as well as the light satellite and telescope structure are indeed responsible for the low background.

9. THE PERFORMANCE OF THE DETECTORS AND PREAMPLIFIERS

The performance of the detectors, namely, their energy resolution as a function of the temperature and energy as well as the stability, gain etc., are all very important. Along with the behavior of a single detector we have to evaluate the performance of the array of detectors.

Radiation measurements at several ambient temperatures with several types of silicon PIN photodiodes using several preamplifiers were carried out. In all the experiments the detector and its preamplifier's FET were cooled. For a detector of 1 cm^2 area and 0.5 mm thickness the capacitance is about 30 pF. The results for the energy resolution as a function of temperature, for measurements made at 13.4 keV (Rb K_α), are shown in Fig. 4. The results reflect the performance of the detectors and preamplifiers. As can be seen, a resolution of about 1 keV can be reached between about $-30°$C and $-20°$C. The statistics of the charge carriers in the detector contribute to the energy resolution a term which depends on the square root of the photon energy. Since this contribution is smaller than the electronics noise in our system throughout our whole energy range, the resolution is almost independent of the photon energy.

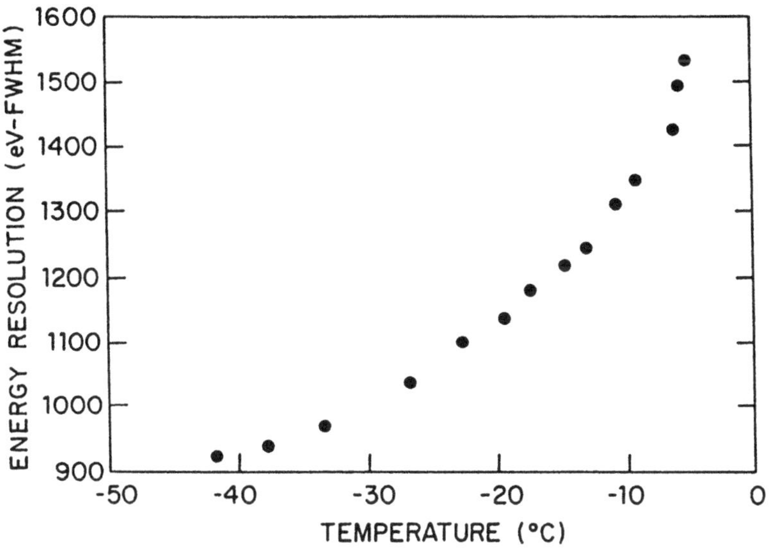

Figure 4. Energy resolution using a silicon PIN photodiode (of 1 cm^2 area and 0.5 mm thickness) as a function of temperature, measured at 13.4 keV.

When one uses an array of detectors for photon counts and spectral analysis, a major issue is the relative calibration between the detectors. This can be done in several ways and we have not yet finalized our design. The traditional way used for calibration in astronomy is to look at a standard source and measure it. Thus, observing a given known source will allow us to compare all the detectors and find the gain of each one of them in each channel. Another possibility is to use a radioactive source which emits the proper X-rays at known energies slightly outside our energy range. Such a source must have the proper lifetime or else it may decay before the satellite ends its life. We have found several such sources and we are presently examining their use in the telescope for continuous calibration.

REFERENCES

Dean, A.J. *et al.*, 1986. *Nucl. Instr. Meth.*, A**242**, 342.

Dean, A.J. *et al.*, 1989. *Astr. Astrophys.*, **219**, 358.

Gehrels, N., 1985. *Nucl. Instr. Meth.*, A**239**, 324.

Priedhorsky, W., Fenimore, E.E., Moss, C.E., Kelley, R.L. & Holt, S.S., 1990. *Los Alamos preprint No.* 1159-20.

Index of Authors

Lecture Notes in Physics

Vol. 350: G. Tenorio-Tagle, M. Moles, J. Melnick (Eds.), Structure and Dynamics of the Interstellar Medium. Proceedings, 1989. XXI, 537 pages. 1989.

Vol. 351: A. Dervieux, B. Larrouturou (Eds.), Numerical Combustion. Proceedings, 1989. VII, 481 pages. 1989.

Vol. 352: K.B. Wolf (Ed.), Lie Methods in Optics II. Proceedings, 1988. XII, 197 pages. 1989.

Vol. 353: M. Barthes, J. Léon (Eds.), Nonlinear Coherent Structures. Proceedings, 1989. X, 280 pages. 1990.

Vol. 354: J. De Coninck, F. Dunlop (Eds.), Wetting Phenomena. Proceedings, 1988. VII, 112 pages. 1990.

Vol. 355: R. Lima, L. Streit, R. Vilela Mendes (Eds.), Dynamics and Stochastic Processes. Proceedings, 1988. V, 298 pages. 1990.

Vol. 356: R. Wehrse (Ed.), Accuracy of Element Abundances from Stellar Atmospheres. Proceedings, 1988. V, 109 pages. 1990.

Vol. 357: J.L. Lumley (Ed.), Whither Turbulence? Turbulence at the Crossroads. Proceedings, 1989. V, 535 pages. 1990.

Vol. 358: J. Kovalevsky, Astrométrie moderne. VI, 294 pages. 1990.

Vol. 359: K. Kirchgässner (Ed.), Problems Involving Change of Type. Proceedings, 1988. XII, 207 pages. 1990.

Vol. 360: Y. Mellier, B. Fort, G. Soucail (Eds.), Gravitational Lensing. Proceedings 1989. XV, 315 pages. 1990.

Vol. 361: P. Gaigg, W. Kummer, M. Schweda (Eds.), Physical and Nonstandard Gauges. Proceedings, 1989. IX, 310 pages. 1990.

Vol. 362: N.E. Kassim, K.W. Weiler (Eds.), Low Frequency Astrophysics from Space. Proceedings. XII, 280 pages. 1990.

Vol. 363: V. Ruždjak, E. Tandberg-Hanssen (Eds.), Dynamics of Quiescent Prominences. Proceedings, 1989. XI, 304 pages. 1990.

Vol. 364: T.T.S. Kuo, E. Osnes, Folded-Diagram Theory of the Effective Interaction in Nuclei, Atoms and Molecules. V, 175 pages. 1990.

Vol. 365: M. Schumacher, G. Tamas (Eds.), Perspectives of Photon Interactions with Hadrons and Nuclei. Proceedings. IX, 251 pages. 1990.

Vol. 366: M.-J. Goupil, J.-P. Zahn (Eds.), Rotation and Mixing in Stellar Interiors. Proceedings, 1989. XIII, 183 pages. 1990.

Vol. 367: Y. Osaki, H. Shibahashi (Eds.), Progress of Seismology of the Sun and Stars. Proceedings, 1989. XIII, 467 pages. 1990.

Vol. 368: L. Garrido (Ed.), Statistical Mechanics of Neural Networks. Proceedings. VI, 477 pages. 1990.

Vol. 369: A. Cassatella, R. Viotti (Eds.), Physics of Classical Novae. Proceedings, 1989. XII, 462 pages. 1990.

Vol. 370: H.-D. Doebner, J.-D. Hennig (Eds.), Quantum Groups. Proceedings, 1989. X, 434 pages. 1990.

Vol. 371: K.W. Morton (Ed.), Twelfth International Conference on Numerical Methods in Fluid Dynamics. Proceedings, 1990. XIV, 562 pages. 1990.

Vol. 372: F. Dobran, Theory of Structured Multiphase Mixtures. IX, 223 pages. 1991.

Vol. 373: C.B. de Loore (Ed.), Late Stages of Stellar Evolution. Computational Methods in Astrophysical Hydrodynamics. Proceedings, 1989. VIII, 390 pages. 1991.

Vol. 374: L. Ting, R. Klein, Viscous Vortical Flows. V, 222 pages. 1991.

Vol. 375: C. Bartocci, U. Bruzzo, R. Cianci (Eds.), Differential Geometric Methods in Theoretical Physics. Proceedings, 1990. XIX, 401 pages. 1991.

Vol. 376: D. Berényi, G. Hock (Eds.), High-Energy Ion-Atom Collisions. Proceedings, 1990. IX, 364 pages. 1991.

Vol. 377: W.J. Duschl, S.J. Wagner, M. Camenzind (Eds.), Variability of Active Galaxies. Proceedings, 1990. XII, 312 pages. 1991.

Vol. 378: C. Bendjaballah, O. Hirota, S. Reynaud (Eds.), Quantum Aspects of Optical Communications. Proceedings 1990. VII, 389 pages. 1991.

Vol. 379: J.D. Hennig, W. Lücke, J. Tolar (Eds.), Differential Geometry, Group Representations, and Quantization. XI, 280 pages. 1991.

Vol. 380: I. Tuominen, D. Moss, G. Rüdiger (Eds.), The Sun and Cool Stars: activity, magnetism, dynamos. Proceedings, 1990. X, 530 pages. 1991.

Vol. 381: J. Casas-Vázquez, D. Jou (Eds.), Rheological Modelling: Thermodynamical and Statistical Approaches. Proceedings, 1990. VII, 378 pages. 1991.

Vol. 382: V.V. Dodonov, V.I. Man'ko (Eds.), Group Theoretical Methods in Physics. Proceedings, 1990. XVII, 601 pages. 1991.

Vol. 384: M.D. Smooke (Ed.), Reduced Kinetic Mechanisms and Asymptotic Approximations for Methane-Air Flames. V, 245 pages. 1991.

Vol. 385: A. Treves, G.C. Perola, L. Stella (Eds.), Iron Line Diagnostics in X-ray Sources. Proceedings, 1990. IX, 321 pages. 1991.

Batch number: 10640288

Printed by Printforce, the Netherlands